The Ethics of Protocells

Basic Bioethics
Glenn McGee and Arthur Caplan, editors

A complete list of the books in the Basic Bioethics series appears at the back of this book.

The Ethics of Protocells

Moral and Social Implications of Creating Life in the Laboratory

edited by Mark A. Bedau and Emily C. Parke

The MIT Press
Cambridge, Massachusetts
London, England

© 2009 Massachusetts Institute of Technology

All rights reserved. No part of this book may be reproduced in any form by any electronic or mechanical means (including photocopying, recording, or information storage and retrieval) without permission in writing from the publisher.

MIT Press books may be purchased at special quantity discounts for business or sales promotional use. For information, please email special_sales@mitpress.mit.edu or write to Special Sales Department, The MIT Press, 55 Hayward Street, Cambridge, MA 02142.

This book was set in Sabon by SNP Best-set Typesetter Ltd., Hong Kong and was printed and bound in the United States of America.

Library of Congress Cataloging-in-Publication Data

The ethics of protocells : moral and social implications of creating life in the laboratory / edited by Mark A. Bedau and Emily C. Parke.
   p. ; cm.—(Basic bioethics)
Includes bibliographical references and index.
ISBN 978-0-262-01262-1 (hardcover : alk. paper)—ISBN 978-0-262-51269-5 (pbk. : alk. paper) 1. Artificial cells–Research–Moral and ethical aspects. I. Bedau, Mark. II. Parke, Emily C. III. Series.
[DNLM: 1. Biotechnology–ethics. 2. Cell Proliferation–ethics. 3. Bioethical Issues. QU 375 E84 2009]
TP248.23.E862   2009
174'.957–dc22

                                                        2008041190

10 9 8 7 6 5 4 3 2 1

Contents

Series Foreword

We are pleased to present the twenty-fifth book in the series Basic Bioethics. The series presents innovative works in bioethics to a broad audience and introduces seminal scholarly manuscripts, state-of-the-art reference works, and textbooks. Such broad areas as the philosophy of medicine, advancing genetics and biotechnology, end-of-life care, health and social policy, and the empirical study of biomedical life are engaged.

Glenn McGee
Arthur Caplan

Basic Bioethics Series Editorial Board
Tod S. Chambers
Susan Dorr Goold
Mark Kuczewski
Herman Saatkamp

Preface

Protocells are tiny, self-organizing, evolving entities that spontaneously assemble and continuously regenerate themselves from simple organic and inorganic substrates in their environment. A number of scientific teams around the world are racing to create protocells, and success is expected within a few years.

Protocells will raise a number of social and ethical issues, involving benefits to individuals and to society, risks to human health and the environment, and transgressions of cultural and moral prohibitions. This volume contains the thoughts of a diverse group of experts who explore the prospect of protocells from a variety of perspectives. These perspectives include applied ethics in analytical philosophy, continental philosophy, and anthropology as well as political and social commentary. The book raises broad questions for a broad audience, without necessarily drawing final conclusions.

We produced this book because we believe the social and ethical issues raised by the prospect of protocells are complex, and involve many interesting open questions. We hope that this volume will contribute to finding responsible solutions to these issues. Our aim is to engage and inform all stakeholders and to prepare them for navigating the uncharted waters ahead.

Our intended audience is multidisciplinary and includes a wide variety of people. One primary target audience is the protocell research community, including those interested in the commercial potential of protocells, and scientists working in related areas of biotechnology and nanotechnology. It also includes policy makers, political scientists, applied ethicists, activists concerned with bioethics, science, and technology studies or ethical, legal, and social issues (ELSI) concerning biotechnology, and those concerned with the convergence of nanotechnology, biotechnology, information technology, and cognitive science. In addition, we aim to engage those in the general public who are concerned about social and ethical implications of biotechnology.

Our own interest in this project arose from our participation in the EU-funded FP6 project on Programmable Artificial Cell Evolution (PACE). As part of the PACE project, we organized a series of workshops on the social and ethical implications of protocell research and development; half of the chapters in this volume originated as presentations at those workshops. We are grateful to PACE for supporting our work on these issues. We are also grateful to the European Center for Living Technology and to Los Alamos National Laboratory, for their hospitality during those workshops.

The chapters in this book do not provide all the answers, but they do raise many important questions about the socially and ethically complex project of creating living technology. We will judge this book a success if it helps our society to flourish in a future with many new forms of living technology, including protocells.

Mark A. Bedau
Emily C. Parke

The Ethics of Protocells

1

Introduction to the Ethics of Protocells

Mark A. Bedau and Emily C. Parke

Protocells are microscopic, self-organizing, evolving entities that spontaneously assemble from simple organic and inorganic materials. They are also known as artificial cells; however, that phrase is sometimes used to refer to things like artificial red blood cells, which are more inert than alive. By contrast, protocells are alive; they are similar to single-celled organisms like bacteria, in that they grow by harvesting raw materials and energy from their environment and converting it into forms they can use, they sense and respond to their environment and take steps to keep themselves intact and pursue their needs, and they reproduce and ultimately evolve. They are a new kind of technology that can, for all intents and purposes, be considered literally alive. Indeed, they are sometimes called "protocells" to emphasize their similarity to simple single-celled life forms. But protocells are simpler than any existing bacterium. And unlike bacteria, they are not natural but artificial, and exist only through human creation. Or at least this is what protocells will be like when they exist, for they do not exist now. However, that will change sooner than many people realize. Teams of scientists around the world are racing to create protocells, and incremental success is continually yielding more and more lifelike systems. The creation of fully autonomous protocells is only a matter of time.

Protocells are capturing growing public attention. New companies for creating artificial life forms are now being created in Europe and America,[1] and the commercial and scientific advances are attracting increasing media attention.[2] The increasing pace of breakthroughs in protocell science will increasingly heighten public interest in their broader implications.

The prospect of creating protocells raises some pressing social and ethical issues. Protocells will offer new benefits to individuals and to society and vast new economic opportunities, but they also have the potential to pose risks to human health and the environment, as well as to transgress cultural and moral norms. Because protocells are living matter created from nonliving matter, they will be unlike any previous technology humans have created, and their development will take society into uncharted waters. This book aims to inform interested parties about these new developments and to promote an open and responsible process of evaluating the prospect of protocells.

The essays in this volume were written by a variety of experts who explore different aspects of the social and ethical implications of protocells. The authors reflect many different professions and intellectual traditions, so the chapters provide a diversity of perspectives on the relevant issues. The aim is not primarily to settle all questions, for that aspiration is unrealistic today. The questions are too complex and unexplored, and the discussion too preliminary to produce unequivocal answers. By providing relevant detailed information that reflects the complexity of the issues, this book can foster constructive reflection and discussion that will help stakeholders become informed and involved. Since the creation of living technology will, in time, profoundly change the world we live in, we think the stakeholders include all concerned individuals.

The State of the Art in Protocell Research

Nobody has yet created a protocell, but research aimed at this goal is actively under way. A comprehensive survey about the state of the art in protocell science is well beyond the scope of this introduction; such a survey can be found in the book *Protocells: Bridging nonliving and living matter* (Rasmussen et al., 2008). Protocell research strategies can fall into one of two categories: the "top-down" and "bottom-up" approaches. The top-down approach involves creating new kinds of life forms by modifying existing ones. The bottom-up approach involves creating living systems from nonliving materials, or "from scratch."

In 2002, J. Craig Venter and Hamilton Smith publicized their intention to create a partly manmade artificial cell, with $3 million in support from the U.S. Department of Energy (Gillis, 2002; Zimmer, 2003); the Department of Energy subsequently increased its support by an order of magnitude (Smith, personal communication). Venter and Smith are using the top-down approach to simplify the genome of the simplest existing cell with the smallest genome: *Mycoplasma genitalium*. Sequencing showed that the 580 kb genome of *M. genitalium* contained only 480 protein-encoding genes and 37 genes for RNA species, for a total of 517 genes (Fraser et al., 1995). Random shotgun gene knockout experiments subsequently determined that approximately 300 of those genes were essential for *M. genitalium* to survive and reproduce in laboratory conditions (Hutchison et al., 1999).

Venter and Smith have used off-the-shelf DNA synthesis technology to construct an entirely artificial genome for *M. genitalium* (Gibson et al., 2008), having first proved the methodology on the significantly smaller genome of a virus (Smith et al., 2003). They have also transplanted the genetic material from one bacterium into another, and expressed the transplanted genes (Lartigue et al., 2007). The final step is to put those two pieces together, and transplant a fully synthetic whole genome into a bacterium, express the genome, pass on the synthetic genome to daughter cells, and initiate a lineage of artificial cells. When the bacterium cytoplasm expresses that synthetic DNA, it will grow and reproduce and thus start a lineage of bacteria that has never existed anywhere before (Gillis, 2002). Once perfected, the process can be repeated to add new genes that perform various useful functions, such as generating hydrogen for fuel or capturing excess carbon dioxide in the atmosphere. By the time this book appears in stores, we expect that Venter and his associates will have succeeded in expressing a wholly synthetic *Mycoplasma* genome inside a donor *Mycoplasma* cell.

Venter recently (in 2005) created a company, called Synthetic Genomics Inc., to commercialize artificial cells. The company aims at producing, among other things, bio-factories for environmentally friendly forms of energy. Synthetic Genomics has applied for a patent for life forms created with the minimal *Mycoplasma* genome that they have discovered (USPTO, 2007). At the same time, a group of new similar companies (e.g., Amyris Biotechnologies, Codon Devices, LS9 Inc.) are reengineering existing life forms for various purposes, such as bio-factories for pharmaceuticals and other useful but rare chemicals, calling their work "synthetic biology."

The synthetic biology approach to making protocells is a logical extension of our experience with genetic engineering over the past thirty years. It has the virtue of allowing one to take advantage of all the complexity and biological wisdom produced by millions of years of evolution and currently embodied in the biochemistry of living organisms, without needing to fully understand the underlying mechanisms that produced them. It has the corresponding disadvantage that its insights will be constrained by the same evolutionary contingencies. If the fundamental problems simple cells must face have essentially different solutions, these will not be discovered by building on the minimal genetic requirements of *Mycoplasma*.

The alternative, bottom-up approach to creating protocells sacrifices the head start existing life forms provide, and instead attempts to create protocells de novo, entirely from nonliving materials. The goal is to create an aggregation of molecules that is simple enough to form by self-assembly, but complex enough to grow, reproduce itself, and evolve without using any products of preexisting life forms such as ribosomes and enzymes. The advantage of this approach is that, freed from the contingent constraints within existing life forms, it can explore a broader canvas of biochemistries and thus eventually reach a more fundamental understanding of the molecular mechanisms required for life.

Several research groups are currently working toward building protocells from the bottom up (Rasmussen et al., 2008). Many groups are expressing more and more complicated genetic networks inside vesicles, typically using an enzyme found in living cells, and often using all of the cellular material found in cell-free extract, including ribosomes and hundreds of additional enzymes. More strict bottom-up groups avoid using materials derived only from living cells, although they do use DNA, RNA, and some other molecules distinctive of life on Earth.

Most bottom-up protocell research draws inspiration from existing cellular biology. The three primary elements of most protocells are the formation of enclosed membranes from amphiphilic lipids, the replication of information-carrying molecules such as DNA or RNA through a templating process, and the harvesting of chemical energy to construct cellular structures from small molecules that are transported across the membrane from the environment into the cell (Szostak, Bartel, & Luisi, 2001; Pohorille & Deamer, 2002). Nucleic acids like DNA and RNA will replicate under appropriate laboratory conditions, but a container is needed to maintain reactants at sufficient concentrations and keep molecular parasites out. A population of such containers encapsulating differently functioning molecular species would allow competing reaction systems to evolve by natural selection. So, the general consensus in the protocell community is that a protocell must have three functionally integrated chemical systems: an informational chemistry with some combinatorial molecule like DNA that programs vital cell functions, an energy-harvesting chemistry (metabolism) that drives all the chemical processes in the cell, and a self-assembling container (cell wall) that keeps the whole cell intact.

The spontaneous growth and replication of lipid bilayer vesicles has already been demonstrated in the laboratory (Walde et al., 1994), as has the synthesis of information-carrying molecules inside lipid vesicles (Oberholzer, Albrizio, & Luisi, 1995; Pohorille & Deamer, 2002). Furthermore, self-replicating RNA molecules can be encapsulated in self-replicating lipid vesicles. Monnard and Deamer (2002) encapsulated T7 polymerase and a 4,000-base-pair plasmid inside lipid vesicles, and added the ribonucleotides ATP, GTP, CTP, and UTP to the surrounding broth. By carefully cycling the temperature in a polymerase chain reaction (PCR) machine, they were able to achieve RNA synthesis inside the vesicles. This worked because the lower temperature permitted the RNA substrates to cross the vesicle membrane and the higher temperature activated the polymerase, catalyzing the production of more RNA.

Those using the bottom-up approach to protocells have no qualms about using external, manual methods to accomplish cellular functions that natural life forms carry out internally and autonomously, for such artifices can achieve important intermediate milestones and then can be removed at a later stage. Hanczyc, Fujikawa, and Szostak (2003) have used external, manual extrusion and serial transport methods to achieve a complex vesicle system that repeatedly grows and divides. This system uses the clay montmorillonite, which is known to catalyze the polymerization of RNA from activated ribonucleotides. In Hanczyc's system, the clay serves two functions: It catalyzes vesicle growth by converting fatty acid micelles into vesicles, and it promotes RNA encapsulation into the vesicles. The enlarged vesicles can then be manually extruded through tiny pores, which causes them to divide into smaller vesicles that retain their original content. Putting this all together produces a complex vesicle-based system consisting of lipids, genetic material, and mineral catalysts that can be made to undergo continual cycles of growth and division.

Much effort in bottom-up protocell research is directed at finding an appropriate RNA replicase, that is, an RNA molecule that functions both as a repository for genetic information and as an enzyme directing replication of RNA itself. A single molecule that performs both critical functions could vastly simplify the protocell's biochemistry, because an RNA gene for the replicase could be copied using that very same RNA molecule as a catalyst. A promising initial step toward this goal includes using in vitro selection to find ribozymes (RNA enzymes for breaking down RNA) that act as primitive polymerases (enzymes for building up RNA) (Ekland & Bartel, 1996; Bartel & Unrau, 1999). Key remaining challenges are to find an RNA replicase efficient enough to accurately replicate all of itself, and to get this function to work quickly inside a self-replicating vesicle.

A third challenge in the bottom-up approach to protocells is to combine the functioning of the genetic, metabolic, and container chemistries so that the entire system evolves as a unit. Rasmussen, Chen, Nilsson, and Abe (2003) proposed a protocell in which these three chemical systems are explicitly combined in a very simple fashion. The unnaturalness of this design is especially striking in several ways. First, because of various advantages it affords, the synthetic and completely artificial nucleotide PNA (Nielsen, Egholm, Berg, & Buchardt, 1991) replaces RNA and DNA. Second, the protocell has no proteins. The PNA programs cellular functionality, but not by producing structural proteins or catalysts. Instead, the sequence of nucleotides directly modulates the amphiphile synthesis that drives the growth and reproduction of the protocellular containers. Third, these containers might not be vesicles formed by bilayer membranes but micelles, which are many orders of magnitude smaller than vesicles. In contrast to the aqueous interior of vesicles that usually house protocellular components, Rasmussen's design encapsulates the protocell's metabolic and genetic chemistry in the oily micellular interior and in the micelle-water interface. Rasmussen's design illustrates how far bottom-up protocell designs have deviated from any form of life known today.[3]

The top-down and bottom-up approaches represent alternative tendencies in protocell research, but they are not necessarily diametrically opposed. Many lines of research combine elements of both; a good example is research using cell-free extracts. Cell-free extracts are complex mixtures containing the cytoplasmic and nuclear material of living cells. Different cell-free extracts can be made, containing different cytoplasmic or nuclear components. For example, a cell-free translation extract contains all the cellular components needed to synthesize proteins from messenger RNA (mRNA). Recent work with cell-free extracts has yielded some impressive results. Noireaux and Libchaber (2004) created a vesicle system in which two genes are transcribed and translated into proteins that function in tandem. They encapsulated a plasmid with two genes inside vesicles containing cell-free extract. One gene produced green fluorescent protein (GFP), and the other produced alpha hemolysin, a protein that forms pores in bilayer membranes. This pore-forming protein spontaneously embedded itself inside the vesicle membranes. The resulting pores permitted amino acids to continually resupply the GFP synthesis process controlled by the second gene. This sustained GFP production for an order of magnitude longer than occurred without the pores.

Natural living cells do not only contain many genes producing proteins with interlocking cellular functions; these genes often regulate one another, in complex genetic networks. Yomo and colleagues (Ishikawa et al., 2004) have demonstrated the first vesicle system with a two-stage functioning genetic network. They also encapsulated a plasmid with two genes in a vesicle containing cell-free extract. One gene produces T7 RNA polymerase, and this enzyme produces mRNA for GFP from the second gene. They demonstrated that these two genes acted in concert, with the polymerase significantly promoting the production of the GFP. A natural generalization of their achievement is to encapsulate genetic networks with three or more stages.

Risks and Benefits of Protocells

There are two main motivations behind protocell research. One is pure science. The ability to synthesize life in the laboratory will be a scientific milestone of immense proportions. Achieving it will mark a profound understanding of the biochemical systems that embody life, and it will also open a fast track to a series of further fundamental insights. Every step along the way toward building a protocell allows researchers to learn more and more about life and living systems. The ability to create such systems in the laboratory is an acid test of the extent to which one really understands how such systems work.

But there is also a practical motivation behind creating protocells. Natural cells are much more complicated than anything yet produced by man, and many people believe that the next watershed in intelligent machines depends on bridging the gap between nonliving and living matter (e.g., Brooks, 2001). So, making protocells that organize and sustain themselves and evolve in response to their environment will lead to the creation of technologies with the impressive capacities of living systems. The promise of harnessing those capacities for social and economic gain is quite attractive.

Many technological, economic, and social benefits can be expected to follow the creation of protocells, because protocells are a threshold technology that will open the door to qualitatively new kinds of applications. Pohorille and Deamer (2002) note numerous pharmacological and medical diagnostic functions that protocells could perform. One application is drug-delivery vehicles that activate a drug in response to an external signal produced by target tissues. Another function is micro-encapsulation of proteins, such as artificial red blood cells that contain enhanced hemoglobin or special enzymes. A third application is multifunction biosensors with activity that can be sustained over a long period of time. After reviewing these examples, Pohorille and Deamer (2002, p. 128) conclude that "Protocells designed for specific applications offer unprecedented opportunities for biotechnology because they allow us to combine the properties of biological systems such as nanoscale efficiency, self-organization and adaptability for therapeutic and diagnostic applications. . . . [I]t will become possible to construct communities of protocells that can self-organize to perform different tasks and even evolve in response to changes in the environment." It is easy to expand the list of hypothetical protocell applications with possibilities ranging from molecular chemical factories and the metabolism of environmental toxins to smart defenses against bioterrorism and a cure for heart disease (e.g., protocells that live in our bloodstream and keep it clean and healthy by ingesting atherosclerotic plaque).

Protocells are the tip of an iceberg of opportunities provided by living technology. Living systems have a remarkable range of distinctive useful properties, including autonomous activity, sensitivity to their environment and robustness in the face of environmental change, automatic adaptation and ongoing creativity. There is increasing need for technology that has these features; such technology could be said to be literally "alive." Conventional engineering is hitting a complexity barrier because it produces devices that are nonadaptive, brittle, and costly to redesign. The only physical entities that now exhibit self-repair, open-ended learning, and spontaneous adaptability to unpredictably changing environments are forms of life. So, the future of intelligent, autonomous, automatically adaptive systems will be living technology. And protocells will be the first concrete step down this path.

The prospect of protocells also raises significant social and ethical worries. Ethical issues concerning the creation of artificial forms of life have a long history, dating back at least to the artificial production of urea in 1828, the first manmade organic compound synthesized from inorganic materials. Concerns about nanostructures proliferating in natural environments were expressed in the nanotechnology community almost a generation ago (Merkle, 1992), and in the past decade Bill Joy's cautionary piece about the combination of nanotechnology with genetic engineering in *Wired* (Joy, 2000) sparked extensive commentary on the Web. Similar public concerns have surfaced over the minimal cells of Venter and Smith's research, which are sometimes dubbed "Frankencells." Mindful of this public outcry, Venter halted research and commissioned an independent panel of ethicists and religious leaders to review the ethics of synthesizing protocells (Cho et al., 1999). When this panel subsequently gave a qualified green light to this line of research, Venter and Smith resumed their project (Gillis, 2002). This move attracted some commentary on editorial pages (e.g., Mooney, 2002). Events like these are increasingly bringing the social and ethical implications of protocells to the attention of a wider and wider audience.

One of the most widespread worries about protocells is their potential threat to human health and the environment. Bill Joy's *Wired* article did not mention protocells specifically, but he worried about essentially the same thing: molecular machines with the ability to reproduce themselves and evolve uncontrollably. Referring to the dangers of genetic engineering and Eric Drexler's (1986) warnings about the dangers of self-reproducing nanotechnology, Joy concludes that "[t]his is the first moment in the history of our planet when any species, by its own voluntary actions, has become a danger to itself—as well as to vast numbers of others." He describes one key problem thus:

"Plants" with "leaves" no more efficient than today's solar cells could out-compete real plants, crowding the biosphere with an inedible foliage. Tough omnivorous "bacteria" could out-compete real bacteria: They could spread like blowing pollen, replicate swiftly, and reduce the biosphere to dust in a matter of days. Dangerous replicators could easily be too tough, small, and rapidly spreading to stop—at least if we make no preparation. We have trouble enough controlling viruses and fruit flies.

Joy later adds the threat of new and vastly more lethal forms of bioterrorism to the list of health and environmental risks of protocells.

These possible dangers stem from two key protocell properties. First, since they self-replicate, any danger that they pose has the potential to be magnified on a vast scale as they proliferate and spread in the environment. Second, because they evolve, their properties could change in ways that we never anticipated. For example, they could evolve new ways of competing with existing life forms and evading our control methods. This potential for open-ended evolution makes the long-term consequences of creating them extremely unpredictable. Much of the positive potential of protocells stems from their ability to self-replicate and evolve, and the very same power raises the specter of life run amok.

There are obvious possible strategies for coping with these risks. One is simply to contain protocells strictly in confined areas and not let them escape into the environment. This is a familiar way of addressing dangerous natural pathogens (for example, the Ebola virus). Another method is to take advantage of the fact that protocells are artificially created and build in mechanisms that cripple or control them. A third method is to make them dependent on a form of energy or raw material that can be blocked or that is normally unavailable in the environment, so that they would survive in the wild only if and when we allow them to. A fourth method is to engineer protocells to have a strictly limited life span, so that they die before they could do any harm. Or one could engineer them to remain alive only on receiving regular external signals, or to die when an externally triggered on/off switch is tripped. For example, there is evidence that magnetic fields can be used to turn genes on and off (Stikeman, 2002). A further form of crippling would be to block their ability to evolve. For example, one could hamper protocell evolution by preventing recombination of genetic material during reproduction. Merkle (1992) has also proposed encrypting artificial genomes in such a way that any mutation would render all the genetic information irretrievable. A further suggestion is to put a unique identifier (a genetic "bar code") inside each protocell, so that we can identify the source of any protocell that does damage and seek redress from the responsible parties.

Such measures would not placate concerns about protocells, though. All such safeguards are fallible and costly. No containment method is perfect, and more effective containment is more expensive.

Another indirect cost of stringent containment procedures is that they would significantly hamper protocell research. This, in turn, would impede the growth of our knowledge of how protocells work and what beneficial uses they might have. Many envisioned benefits require protocells to inhabit our environment or even our bodies. Such applications would be impossible if all protocells were isolated inside strict containment devices. Furthermore, methods for crippling or controlling protocells could well be ineffective. When humans have introduced species into foreign environments, it has often proved difficult to control their subsequent spread. More to the point, viruses and other pathogens are notorious for evolving ways to circumvent our methods of controlling or eradicating them. This implies that protocells could experience significant selection pressure to evade our efforts to cripple or control them. Another kind of social cost of crippling protocells is that this would sacrifice a key benefit of living technology—taking advantage of life's robustness and its flexible capacity to adapt to environmental contingencies.

Themes in the Chapters

The chapters in this book weave together various themes, in three broad sections. The first section treats risk, uncertainty, and precaution with protocells. The second section draws lessons from recent history and related technologies. The third section explores how society should approach ethical questions in a future with protocells. Many chapters discuss issues that fall into two or all three of these categories, so chapters are placed within individual sections somewhat loosely.

Protocell technology is too novel to have yet generated a substantial literature. One way to start thinking through their social and ethical implications is to identify basic methodological lessons about good and bad arguments for and against protocells, as Boniolo argues in chapter 17. Another strategy is to empirically compare related technologies that have received significant public attention. A number of new biotechnologies are already pressuring our values and preconceptions about life. Stem cell research, cloning, and genetic engineering, for example, are increasingly changing the nature of the life forms that exist in our world. This is causing more and more people to think about the importance and sanctity of life, and the very notion of what it is to be alive. A number of chapters draw lessons from our track record with related biotechnologies, including genomics and the chemical industry (chapter 4), modern agriculture (chapter 2), biomedical science (chapter 16), synthetic biology (chapters 9, 11, and 14), and nanotechnology (chapter 8). These related areas are mirrors in which we can glimpse our future with protocells.

Another way to triangulate our future with protocells is to examine their implications for intellectual property rights. Pottage, in chapter 10, identifies some distinctive aspects of patents for protocell innovations, and explores how protocell patents disturb the political and institutional foundations of our current intellectual property practices. Hoping to foster the generation and diffusion of protocell innovations, some take inspiration from the open-source movement in software, typified by the Linux operating system, created through the collaboration of a wide network of interested individuals and then distributed for free. Hessel argues in chapter 11 that open-source biology would allow science and commerce alike to benefit from innovations in living technology while minimizing the potential for harmful applications.

Each new technology has heightened public awareness and sensitivity about potential technological risks. Most would agree that decision makers should take account of the public's concerns about protocell technology. However, doing so is not easy. A number of chapters concentrate on the complexity of the relationship between public perception and the progress of new technologies. Citing examples from recent history, in chapter 7 Durodié discusses the simultaneous importance and difficulty of making sure that the public's actual voice is heard, rather than just the voices of interest groups that purport to represent the public but are actually driven by their own private agendas.

Several chapters examine the relationship between the public and those actively engaged in protocell research. The latter group involves not only scientists but also entrepreneurs. Some scientists and businesspeople proactively engage other stakeholders and the general public, elicit their views and concerns, and respond to them. Others try to lay low, out of the public eye, for as long as possible. Johnson, in chapter 2, argues for the wisdom of proactive engagement with stakeholders, drawing on examples from recent history. The public's role in decision making about new technologies is a contentious issue, for reasons examined by Hantsche in chapter 12. Hantsche emphasizes the difficulty of giving due weight to societal values when making decisions about our future with protocells, since there is no clear precedent on which to base these decisions.

One natural response to the complexity of the social and ethical questions about protocells is to engage the assistance of professional ethicists. Precedents for this include the earmarking of a small fraction of the human genome project budget for studies of the ethical, legal, and social implications of sequencing the human genome, and the existence of a number of professionally staffed institutes and centers for examining social and ethical implications of nano- and biotechnologies. But the advice of such studies and institutes is no quick fix that resolves ethical dilemmas and absolves scientists from making difficult ethically laden decisions. In chapter 15, Gjerris emphasizes that enlisting the help of professional ethicists will not eliminate the ethical problems raised by protocells but might, in fact, magnify them.

Chapter 14, by Rabinow and Bennett, examines the two main existing modes in which professional ethicists interface with the scientific, regulatory, and policy-making communities, and advocates adopting a new, third mode of interaction that is especially appropriate for synthetic biology. Khushf makes a complementary point in chapter 13, arguing that ethical reflection must be an integral part of the development of protocell science and technology. Scientists have often viewed their work as ethically neutral, appealing to a division of labor in which their job is to learn how nature works, and leaving it to policy makers to decide how this knowledge should be deployed. Khushf argues that this traditional separation of the scientific and ethical realms is no longer sensible, and that scientists need to get their hands dirty with the ethical complexities of their scientific investigations.

Protocells present an unusual challenge for society not only because their consequences are uncertain, but also because they force us to confront our views about life and its creation. In chapter 3, Bedau and Triant examine some related concerns, including the worries that protocells are inherently bad because they are unnatural, that they violate the sanctity of life, or that their creators are "playing God." Addressing the risks associated with the creation of life is an extremely value-laden undertaking. Hauskeller develops this point with details about comparable experience with biomedical science (chapter 16).

Addressing public concerns about protocells and making appropriately balanced decisions requires understanding the risks and benefits involved. With the potential for great benefit comes the potential for great harm and abuse. Because fully functional protocells have not yet been created, their actual risks are merely hypothetical today. Cranor, in chapter 4, discusses how society perceives risk, using examples from the chemical and genomics industries to draw conclusions about the perceived acceptability of protocell technology and the proper way to manage the accompanying risks.

The analysis and evaluation of the risks and benefits of protocells is so complex that deciding what to do is challenging. The stakes are high and any decision will have both good and bad consequences. The traditional tools policy makers use to weigh risks and benefits are decision theory and risk analysis, but Bedau and Triant point out that these methods offer scant help when we are in the dark about the ultimate consequences of our decisions.

A growing number of people have been arguing that society's traditional policy- and decision-making tools should be augmented with the precautionary principle, which states that precaution should be the overriding concern when considering new technologies. Several chapters deal with precaution as a decision-making tool, discussing the precautionary principle as a framework for evaluating critical policy issues involving new technologies. Parke and Bedau review various arguments for and against the precautionary principle in chapter 5, and in chapter 6 Sandin develops and defends a new virtue-based understanding of the precautionary principle. Durodié, in chapter 7, counters the contemporary popularity of precautionary thinking, arguing from recent history that too much precaution has led to bad decision making. And chapter 3 argues that following the precautionary principle is itself too risky, because society cannot afford to forgo all the potential benefits of new technologies like protocells.

* * *

The creation of protocells promises to alter our world forever. Protocells could bring many impressive benefits for human health, the environment, and defense, and dramatically accelerate basic science. But they could also create new risks to human health and the environment, and enable new forms of bioterrorism. So, their potential upside and downside are both quite large.

In addition, creating life from scratch will fundamentally shake public perceptions about life and its mechanistic foundations, undermining certain entrenched cultural institutions and belief systems. Society should weigh all of these significant consequences when deciding what to do about protocells.

This book should help. The chapters offer a diversity of perspectives on the relevant issues and discuss the key questions. This volume does not indicate final solutions, but rather initiates an open, informed, and responsible discussion of the prospect of protocells. Rather than being the last step in understanding the moral and social implications of creating protocells, this book is the first.

Notes

1. For example, Synthetic Genomics Inc., ProtoLife Srl., and Codon Devices.

2. In addition to the publication of Michael Crichton's popular novel, *Prey*, about artificial cells, numerous news accounts have appeared in the past few years in, for example, *Science, Nature, The New York Times, The Wall Street Journal, The Chicago Tribune, NOVA, The Scientist*, and *New Scientist*.

3. Rasmussen's fully integrated design has not yet been experimentally realized; only pieces of the picture have been achieved in the laboratory, at the time of this writing.

References

Bartel, D. P., & Unrau, P. J. (1999). Constructing an RNA world. *Trends in Cell Biology, 9*, M9–M13.

Brooks, R. (2001). The relationship between matter and life. *Nature, 409*, 409–411.

Cho, M. K., Magnus, D., Caplan, A. L., McGee, D., & the Ethics of Genomics Group. (1999). Ethical considerations in synthesizing a minimal genome. *Science, 286*, 2087.

Crichton, M. (2002). *Prey*. New York: HarperCollins.

Drexler, K. E. (1986). *Engines of creation: The coming era of nanotechnology*. New York: Doubleday.

Ekland, E. H., & Bartel, D. P. (1996). RNA catalyzed nucleotide synthesis. *Nature, 382,* 373–376.

Fraser, C. M., Gocayne, J. D., White, O., Adams, M. D., Clayton, R. A., Fleischmann, R. D., et al. (1995). The minimal gene component of Mycoplasma genitalium. *Science, 270,* 397–403.

Gibson, D. G., Benders, G. A., Andrews-Pfannkoch, C., Denisova, E. A., Baden-Tillson, H., Zaveri, J., et al. (2008). Complete chemical synthesis, assembly, and cloning of a Mycoplasma genitalium genome. *Science, 319,* 1215–1220.

Gillis, J. (2002). Scientists planning to make new form of life. *Washington Post,* November 21, A01.

Hanczyc, M. M., Fujikawa, S. M., & Szostak, J. W. (2003). Experimental models of primitive cellular components: Encapsulation, growth, and division. *Science, 320,* 618–622.

Hutchison, C. A., Peterson, S. N., Gill, S. R., Cline, R. T., White, O., Fraser, C. M., et al. (1999). Global transposon mutagenesis and a minimal Mycoplasma genome. *Science, 286,* 2165–2169.

Ishikawa, K., Sato, K., Shima, Y., Urabe, I., & Yomo, T. (2004). Expression of cascading genetic network within liposomes. *FEBS Letters, 578,* 387–390.

Joy, B. (2000). Why the future doesn't need us. *Wired, 8* (April). Available online at: http://www.wired.com/wired/archive/8.04/joy.html (accessed August 2008).

Lartigue, C., Glass, J. I., Alperovich, N., Pieper, R., Parmar, P. P., Hutchison III, C. A., et al. (2007). Genome transplantation in bacteria: Changing one species to another. *Science, 317,* 632–638.

Merkle, R. (1992). The risks of nanotechnology. In B. Crandall & J. Lewis (Eds.), *Nanotechnology research and perspectives* (pp. 287–294). Cambridge, MA: MIT Press.

Monnard, P. A., & Deamer, D. (2002). Membrane self-assembly processes: Steps toward the first cellular life. *The Anatomical Record, 268,* 196–207.

Mooney, C. (2002). Nothing wrong with a little Frankenstein. *Washington Post,* December 1, B01.

Nielsen, P., Egholm, M., Berg, R. H., & Buchardt, O. (1991). Sequence-selective recognition of DNA by strand displacement with a thymine-substituted polyamide. *Science, 254,* 1497.

Noireaux, V., & Libchaber, A. (2004). A vesicle bioreactor as a step toward an artificial cell assembly. *Proceedings of the National Academy of Sciences of the United States of America, 101,* 17669–17674.

Oberholzer, T., Albrizio, M., & Luisi, P. L. (1995). Polymerase chain reaction in liposomes. *Chemistry & Biology, 2,* 677–682.

Pohorille, A., & Deamer, D. (2002). Protocells: Prospects for biotechnology. *Trends in Biotechnology, 20,* 123–128.

Rasmussen, S., Bedau, M. A., Chen, L., Deamer, D., Krakauer, D. C., Packard, N. H., & Stadler, P. F. (Eds.). (2008). *Protocells: Bridging nonliving and living matter.* Cambridge, MA: MIT Press.

Rasmussen, S., Chen, L., Nilsson, M., & Abe, S. (2003). Bridging nonliving and living matter. *Artificial Life*, *9*, 269–316.

Smith, H. O., Hutchinson III, C. A., Pfannkoch, C., & Venter, J. C. (2003). Generating a synthetic genome by whole genome assembly: PhiX174 bacteriophage from synthetic oligonucleotides. *Proceedings of the National Academy of Sciences of the United States of America*, *100*, 15440–15445.

Stikeman, A. (2002). Nanobiotech makes the diagnosis. *Technology Review*, May, 60–66.

Szostak, J. W., Bartel, D. P., & Luisi, P. L. (2001). Synthesizing life. *Nature*, *409*, 387–390.

USPTO. (2007). Patent application 20070122826: Minimal bacterial genome. Published May 31, 2007.

Walde, P., Wick, R., Fresta, M., Mangone, A., & Luisi, P. L. (1994). Autopoietic self-reproduction of fatty acid vesicles. *Journal of the American Chemical Society*, *116*, 11649–116454.

Zimmer, C. (2003). Tinker, tailor: Can Venter stitch together a genome from scratch? *Science*, *299*, 1006–1007.

I

Risk, Uncertainty, and Precaution with Protocells

2

New Technologies, Public Perceptions, and Ethics

Brian Johnson

New technologies have been part of the development of human society since the dawn of our history. Discovery by experimentation, although not known as "science" until recent times, seems to be a distinctly human trait, probably derived from primate ancestors who experimented with tools. Until recent times, adoption of technologies has been very slow, with stone tools, joinery, and the wheel appearing in Neolithic times and the woodscrew and other iron fastenings and tools thousands of years later. With the development of formal and organized science after the Renaissance, discovery and innovation have accelerated to the bewildering levels that we see in the twenty-first century, with major breakthroughs occurring with ever-increasing frequency.

But discovery and innovation do not necessarily lead to acceptance and adoption of the products of a new technology. The waysides are littered with examples of technological daring that never appealed to markets, from nineteenth-century pneumatic railways to the Sinclair C5 personal transport, and more recently genetically modified (GM) foods in Europe. This chapter explores possible public attitudes toward new technologies, such as those that might flow from the work of the PACE project to construct protocells (PACE, 2004), and attempts to answer the following questions:

Can we identify why the public choose one product over another and in some cases either reject or accept a whole technological field; for example, stem cell technology in the United States versus Europe?

What are the key factors that drive public and political sentiment, and can we pick up signals from these communities at an early stage in technological development?

How does the public deal with the inevitable tradeoffs between risks, benefits, and the ethical and social implications associated with new technologies?

Public Perceptions of New Technologies and Products

From surveys of attitudes toward biotechnology, we now recognize that the public (or perhaps more accurately "publics"), acting as individuals or collectively, carry out a kind of risk/benefit analysis when confronted with a novel technology and its products (Hoban, 1998). This analysis is increasingly well informed, especially since the advent of the Internet, but a part of it is probably intuitive, based on snippets of information, however inaccurate, gleaned from the media and by word of mouth.

The risk part of the analysis is the classical "risk = hazard x exposure" equation familiar to engineers building, for example, vehicles or bridges. In the case of public risk/benefit analysis, the equation is often populated not by measured parameters but by perceptions of the hazard, and exposure to the hazard. This may seem an irrational and unscientific approach to risk assessment, but it is in fact similar to the risk assessment carried out by most regulatory authorities when faced with decisions that cannot be made using only measured parameters, either because the science has not yet been done or cannot be done. A topical example of this is in the regulation of GM organisms (GMOs), where regulators have little "hard" information about all the factors they take into account. They, like the public, use the depth and breadth of their experience to make informed judgments about the risks associated with GMOs (House et al., 2004). Perhaps surprisingly, they rarely get risk assessment wrong. Decisions taken on the human health risks and environmental safety of GMOs are periodically reviewed and monitored by regulators and academic researchers (see the European Union Joint Research Centre Web site, and Stewart & Wheaton, 2001). So far these reviewers have shown that regulatory decisions are sound, although in some cases they found unauthorized releases of GMOs.

Unlike regulatory authorities who focus mainly on risk per se, the public also take into account the real and perceived benefits to themselves. Acting as individuals, many people are prepared to take relatively high risks if benefits are attractive enough. People use mobile phones knowing that there is a risk that the microwaves generated may damage their brains, they drive cars knowing the high risks involved on crowded roads (especially in developing countries), and they continue to smoke tobacco, drink alcohol, and take recreational drugs knowing the potential risk of disease. Most individuals tend to reject technologies and products that are perceived to involve high risk and low benefit, a reaction that typifies attitudes toward GM foods in Europe. When acting collectively through formal political systems or nongovernmental organizations (NGOs), the public is perhaps more conservative and prepared to take even fewer and smaller risks, even when benefits are high. These collective decisions are usually expressed through the development of regulations governing product quality and safety, especially when considering food and machinery. So, for example, we now have such a large portfolio of risk-averse rules and regulations that the UK government has set up a specialist task force to try to reduce them (the so-called Better Regulation Taskforce).

Do Introductions of New Technologies Follow Patterns?

The ways in which civil society greets innovative and potentially risky technologies can be illustrated by some examples from the past. When electricity was harnessed in the mid-nineteenth century, it was first used to power public lighting. Though there were early attempts in the 1860s to replace gas lighting in domestic homes, even those who could afford the infrastructure initially rejected the introduction of electricity into their homes. This was in fact a rational decision, because the electricity companies in effect were offering to bring an invisible and potentially lethal force into homes, using primitive unguarded switches and terminals. This was hardly an attractive prospect to those who were familiar with the use of candles and gas to light their houses (despite the fact that these sources were responsible for most house fires!). It was not until the early twentieth century that domestic adoption of electrical power took off commercially, significantly in the homes of wealthy entrepreneurs, that is, people prepared to take high risks (for an example, see Wolverhampton Electricity Supply, 2005). With better switches and safer wires and terminals, and subsidies from the public sector, the public began to use electricity in their homes, with major growth in the sector between 1910 and 1930, some seventy years after the first reliable generators and appliances were developed.

Motor transport similarly developed very slowly, not just because engines were noisy, weak, and unreliable (which they were), but also because the first automobile brakes were adapted from those used on horse-drawn carts and were virtually useless above five miles per hour. In the nineteenth century, this led to a large number of accidents at the bottoms of hills, which were widely reported in the press (for an account, see Chapman, 2005). Because brakes were so poor, the UK parliament passed the Locomotive Act in 1865, setting speed limits at four miles per hour. This was not repealed until 1895, some thirty years later, effectively delaying the development of the automobile until better brakes were developed. There was also a perception, again hotly debated in the press and in scientific journals, that braking forces associated with high speeds would damage human bodies, a perception that lasted until the beginning of the twentieth century. When mass production of vehicles with reliable, efficient brakes and effective warning systems was developed, cars and trucks replaced horses within thirty years. This was around sixty years after the first motor transport was developed.

A more recent example of adoption of technologies occurred in the 1980s with the introduction of transgenic crops as an exciting breakthrough that would feed the poor and make food cheaper in the developed world. They were promoted in the United States as routine and safe, with little public discussion and widespread adoption by farmers. In Europe, the story was very different, with strong public opposition and vigorous campaigns by NGOs who adopted GM technology as a major "lightning rod" issue, encompassing issues of food and environmental safety, and control of food supplies by large multinational corporations. Frequent and sometimes dramatic mistakes by industry, such as the accidental release of unlicensed GMOs, were enthusiastically reported in the media, adding fuel to the fire. Perhaps uniquely, the issue settled around the technology rather than the products, with the public perceiving that they were being asked to take risks but could see no benefit for themselves, only for the farmers who grew the crops, and the companies that sold the seeds. Some twenty years after the first transgenic crops were made, they still have not been widely adopted in Europe, although they are grown in small areas in Spain and some Eastern European countries (see the European Union Joint Research Centre Web site for details).

Digital technologies have been adopted much more quickly than the preceding examples. From the discovery and early development of digital information technology in the 1960s, it has taken a mere four decades for information technology to be integrated into the lives of most people in both the developed and developing worlds. This technology is rapidly replacing mechanical devices for controlling aircraft and automobiles, and has vastly increased the capacity of individuals both to communicate with each other and to mine information on a global scale. Surveys of public views of digital technologies have consistently revealed that people perceive almost all aspects as highly beneficial and, not surprisingly, show real enthusiasm for any gadget that might either save time and money or provide entertainment, even though risks are associated with their use. These range from physical risks from radiation from digital devices or failure of digital control systems (risks that are common to other electronic devices but are magnified by the widespread use of digital devices), to mental health risks and other risks associated with the Internet. Significantly, there is widespread media support for these technologies, with whole sectors of the media being devoted solely to promoting digital technologies, such as the racks of computer and music magazines to be seen in any bookstore. Digital technologies are therefore not perceived as dangerous, despite risks associated with their use. This demonstrates that the simplistic perception (promulgated, for example, by some biotechnology companies) that the European public is intrinsically "risk-averse" is incorrect.

Patterns run through the introduction of new technologies to the public: They can be seen in the preceding examples and can also be found with the introduction of pesticides, therapeutic drugs, and radioactive substances.

First, there is usually a long lag between discovery and adoption, often caused by public skepticism and media scrutiny. Second, a new technology often produces a flush of new and largely unregulated products, because regulation often lags way behind innovation. Electrical appliances multiplied exponentially shortly after the widespread implementation of electricity in the home. A wide range of devices were even designed to administer electric shocks to people in the belief that these were beneficial to health. There were experimental forms of lighting and heating, many of which later proved to be dangerous.

When motor transport first became popular, many vehicles had very poor steering and braking characteristics, and several were also prone to catching on fire. The range of early transgenic crops included some than were allergenic. Many of the early pesticides, such as DDT and organophosphates, were either potentially poisonous or could produce unpleasant side effects, and some drugs like thalidomide had potentially debilitating side effects. So shortly after commercial products are developed from new technologies, problems often arise and the real risks start to become clearer, triggering close public scrutiny and debate. Media debates on controversial technologies and products are inevitable and have a long history, as seen in the preceding examples. Risks are almost always exaggerated and technologists and industry have little control over press coverage. The European debate on GM crops was dogged by unsubstantiated and exaggerated claims, some from scientists themselves, about both their potential benefits and their potential impacts on humans and the environment (see, e.g., Krebs, 2000; Trewavas & Leaver, 2001; Observer, 2003). The current debate on stem cell technology (Mieth, 2000; Coghlan, 2003) is another example of this phenomenon, in which premature and somewhat exaggerated claims on all sides of the debate are avidly reported.

In response to controversy and debate, there is often a phase of voluntary codes of practice developed by industry in consultation with civil society, politicians, and NGOs. Some, such as the Recombinant DNA Advisory Committee's codes of practice set up after the Asilomar Conference of 1975 on the use of recombinant DNA, have been successful (see Berg (2004) for an analysis of the Asilomar Conference and its consequences). Other codes may be effective at first but frequently fail to deliver public confidence, as loopholes are found and a few unscrupulous companies may break codes of practice. The pesticides codes of practice in the UK in the 1960s and 1970s went through just such a phase, when new and largely untested pesticides were released, damaging wildlife and persisting at potentially dangerous levels in wildlife and human food chains (Carson, 1962). The UK straw-burning code of practice is another example in which voluntary measures aimed at stopping farmers burning vast quantities of straw in the open air were flouted, destroying large numbers of wildlife-rich hedgerows and leading to serious air pollution that even affected western parts of continental Europe across the North Sea.

The Emergence of Regulation

Civil society can respond to the failure of voluntary codes of practice either by rejecting the technology outright or by setting up statutory regulatory systems. In the case of pesticides, the UK government introduced strict regulations within a statutory framework while at the same time setting up a regulatory committee (the Advisory Committee on Pesticides) to advise on individual substances and to give industry strategic guidance. In the case of straw burning, the UK government eventually banned the practice by law in 1993 (UK Statutory Instrument No. 1366, 1993). Regulations now govern many of the older technologies such as electricity supply and motor transport, covering an increasingly wide range of impacts that the technology may have on civil society. The safety and continuity of electricity supply is now tightly regulated, and road and vehicle safety is regulated in detail at the European level. Product safety is much improved, increasing consumer confidence in the technologies, and not surprisingly industry is content that regulation is applied. Making technologies safer also changes public perception of them from being "out of control" toward being in the control of users, when used according to the instructions and safety recommendations now found with almost every technological product. Nevertheless, there remains a healthy public skepticism about safety and other aspects of new technological developments, with vigorous debate about the issues.

New Technologies in Biology

Debates about whether electricity or motor transport should be widely used are now consigned to history, largely because these technologies are perceived to be "safe" even though motor transport is still the most dangerous way to move from one place to another, especially for young drivers (World Health Organization, 1999; Crashtest.com, 2006). Arguably, public skepticism and debate about new developments is currently at its strongest in the area of biological science and technology, especially developments stemming from the new genetic and genomic sciences. There are intense debates about the impacts of these new technologies on the environment and human safety, and about the morals and ethics of modifying genomes. The application of transgenic technologies to crop and animal breeding has stirred up a maelstrom of controversy, especially about the morals and ethics of transferring genes between distant genera and across barriers that exist in nature between the plant, animal, bacterial, and fungal kingdoms. The GM debate has also brought to a head public and political concerns about ownership of genetic techniques and the control of human and livestock food supplies. Whether or not the public and politicians should be concerned about these technologies is often a subject within the debate itself, with the promoters of biotechnology arguing that the techniques are safe and in public control, and opponents arguing that the risks are too great and that profit, rather than public good, is the real driver of development. Whatever the truth, the fact remains that there is not yet a consensus on safety and control of GM organisms. Public concern about transgenic organisms should concern those scientists working at the cutting edge of biology and chemistry, such as protocell researchers trying to synthesize self-replicating entities from chemical components. There are many parallels to the GM debate, for example, that protocells could pose environmental risks if released, that they may be capable of evolutionary adaptation, and that control of the technology could be patented and in the hands of industry and researchers. Added to this is the same situation that applied to GMOs in the early stages of development: To date there are few, if any, codes of practice that apply to protocell research, and little development of regulation.

The Morality, Ethics, and Regulation of Genetic Modification

During the debate on transgenic organisms in the United Kingdom, several prominent campaigners raised the issue of the morality of moving genes across natural reproductive barriers. In the past, society appears to have had few problems with the morality of genetically modifying natural organisms through selective breeding. Animal breeding in particular has not, until fairly recently, been considered to be an area of deep moral concern by the public. Most people have found the production of a wide range of genetic curiosities such as pet dogs and birds acceptable, although these organisms can have gross phenotypic distortions (such as dogs with miniature legs, and roller and tumbler varieties of pigeons) that make them unable to survive in the wild, and can cause considerable distress and discomfort. Moral and ethical concern has centered on the conditions in which pet, farm, and experimental animals are kept, rather than their genetic composition and phenotypic traits. The argument that introducing a single trait in an organism by transgenic methods is morally unacceptable seems rather weak against this historical background of selective animal breeding.

It has also been argued that putting "animal" or "fungal" genes into plants is "unnatural" and may be morally unacceptable to, for example, vegetarians or religious people who have a taboo on some kinds of meat like pork. Geneticists find these arguments rather shallow because we now know that genes do not have "animal" or "plant" characteristics, but are simply chemical codes that govern metabolism and structure. Furthermore, the discovery that plants, animals, and fungi have a significant proportion of their genes in common, and that even very specialized traits can be governed by the same genes in different genera, suggests that giving genes "animal," "fungal" and "plant" labels is far too simplistic. It is also worth noting that species barriers are naturally very "leaky" in a genetic sense, with, for example, chunks of bacterial genomes being incorporated into plants and animals, and rare hybridizations giving rise to new species in which genes from both distant ancestors suddenly find themselves sharing the same novel genome. Even the proteins that allow plant cells to replicate appear to have originated in viruses (Villarreal & DeFilippis, 2000), and the evolution of mitochondria, the powerhouses of plant and animal cells, involved wholesale transfer of bacterial genomes, allowing advanced life to exist in an atmosphere growing increasingly rich in oxygen from plants (Gray, Burger, & Lang, 2001).

As humans, we are part of the natural fauna of Earth, and we have a moral imperative to survive. We have in the past modified the genetics of a number of key ancestral crop and animal species toward this end, and we feel morally justified in having done so. In my view the argument that genetic modification is somehow unnatural and therefore morally wrong is flawed, partly because we are natural and what we do to other species may be considered a natural consequence of our being on Earth. However, as Heidegger (1977) pointed out, there is a moral dimension to the impacts of our actions on ecology and natural resources. Heidegger contended that humans are morally justified in exploiting nature, but only if we do not destroy the resources that continue to provide sustenance. He believed that we have a moral right to exploit natural resources only when we have a deep understanding of the consequences of doing so. I agree with James (2002) when he argues that, to some extent, Heidegger's views on human relationships with technology have become embedded in the modern moral of "sustainability," the driver of many policies aimed at leaving our descendants with the same wealth of nature and physical resources that we enjoy today. To my mind, Heidegger's argument can also be applied to the release of any artificially created entity that might affect the environment in the widest sense, implying that we should proceed only when we have sufficient knowledge to be able to predict the consequences, if any. This may be viewed as a risk-averse strategy, but it is very similar to the precautionary principle, a process of risk assessment used in the European Union (European Commission, 2000).

Civil society may also have moral concerns about the emerging fields of synthetic biology and self-replicating nanotechnology, in which researchers are trying to assemble new forms of life from chemical and biological components. Some will argue that creating new life is the province of evolution and the deity, not of humans. But we have been creating new life forms for thousands of years, starting with the extraordinary creation of what Amman (personal communication, 2000) refers to as "genetic monsters" such as maize and wheat. These were created by our ancestors from either deliberate or spontaneous hybrids and are in effect new life forms with novel genomes that have founded whole civilizations. Many of our fruits, vegetables, and ornamental plants have been created in similar ways. The moral argument against synthesizing life may be weak; however, we cannot escape from the issues of safety and regulation to ensure that the entities created by researchers do not threaten the integrity and functionality of Earth systems and are safe for the environment and for humans. As discussed earlier in this chapter, risk assessment will be needed, and ethical codes may develop to provide an ethical framework with defined boundaries for research in these areas.

Ethical Codes and Protocell Research

Ethical codes that capture both morals and best practice are now common in biological research, for example, governing the development and testing of drugs, laboratory animal welfare, and some types of microbiology (including transgenics).

These codes are usually developed by researchers themselves, working together with representatives from civil society such as NGOs, government, and industry. Not only do they ensure that researchers follow best practice, but they also give the public confidence that what is being done (often with public money) is both safe and morally sound. Topical examples are in stem cell and cloning research, for which an ethical code governs the use of embryonic tissue in Europe, and sets boundaries for cloning experiments. It is likely that research into synthetic biology and self-replicating nanotechnology such as that being developed by the PACE project (PACE, 2004) will need ethical codes at some point in the future, perhaps to minimize the risks of synthesizing novel pathogens and to guard against inadvertent release into the environment of self-replicating synthetic entities.

Regulation will almost certainly be needed, especially if self-replicating entities are intended to be released into either industrial or natural environments. Synthetic biological organisms would probably fall into the realms of existing regulations that govern the release of novel organisms such as transgenics, but entities such as protocells, based on chemistries that differ from synthetic biology, may need new approaches for risk assessment. To some extent, the issues are familiar; like transgenic and pathogenic microbes, protocells could be microscopic "invisible" entities that cannot be recalled once released. But we have no experience with how these "organisms" might behave in the wider environment, so risk assessment will be a challenge. There are also the familiar questions of who controls such potentially powerful technologies, who benefits, and who takes the risks? As the pace of research in these areas accelerates, regulatory authorities and civil society need to address these issues in the very near future, or they will be left floundering in the wake of invention. At some point there will almost certainly be a debate about this science; it would be wise for all those involved in its research and regulation to be prepared. Researchers and technologists might wish to promote the debate in order to resolve societal and safety issues as early as possible. However, choosing the most appropriate time to do so is notoriously difficult; if such a debate starts too early, researchers are unlikely to be able to provide cogent answers to legitimate questions of ethics and risk. I propose that a European task force of researchers, ethicists, sociologists, representatives of civil society, and regulators be assembled and charged with developing a strategy for engaging with the public, the media, and the political community on the subject of self-replicating cellular technologies. The aim would be to promote a mature and rational debate, at an appropriate pace and time, about the scientific, environmental, and societal implications of such technologies without necessarily including the stage of exaggerated claims and counter claims that has been the hallmark of such debates in the past. Involving representatives of civil society would be an attempt at *upstream engagement*, a principle supporting public engagement with science put forward by the UK think tank Demos (Demos, 2004). Public engagement in this potentially controversial area of science and technology would be more likely to be successful if civil society is allowed to frame the issues, perhaps using methods similar to those used by the UK GM Science Review (2003), a process aimed at engaging with public concerns about the science and technology of gene transfer. At the start of that process, the review panel commissioned a survey of public concerns about gene transfer as a framework for their review. The review was published on the Web, and received global attention and political approval for its rigor and transparency. If there is to be a deep debate about initiatives such as the PACE project, early public involvement of this kind would be better than the pattern we have witnessed during the GM debate, where the public were suddenly and unexpectedly presented with the products of biotechnology, with little if any prior knowledge of either the science involved in producing them or the health and environmental implications of using them.

References

Amman, K. (2000). Director of Botanical Garden, Bern, Switzerland. Personal communication.

Berg, P. (2004). Asilomar and recombinant DNA. Available online at: http://nobelprize.org/nobel_prizes/chemistry/articles/berg/index.html (accessed February 2007).

Carson, R. (1962). *Silent spring*. New York: Houghton Mifflin Company.

Chapman, G. (2005). A brief history of road safety. Available online at: http://www.chapmancentral.co.uk/web/public.nsf/documents/history?opendocument (accessed February 2007).

Coghlan, A. (2003). Europe backs stem cell research. *New Scientist.com* news service, available online at: http://www.newscientist.com/article.ns?id=dn4404 (accessed February 2007).

Crashtest.com (2006). Available online at: http://www.crashtest.com/explanations/stats/index.htm (accessed February 2007).

Demos. (2004). See-through science: Why public engagement needs to move upstream. Available online at: http://www.demos.co.uk/publications/paddlingupstream (accessed February 2007).

European Commission. (2000). *Communication on the precautionary principle*. February 2, Brussels. Available online at: http://ec.europa.eu/environment/docum/20001_en.htm (accessed February 2007).

European Union Joint Research Centre. Parma, Italy. Available online at: http://gmoinfo.jrc.it/gmc_browse.aspx?DossClass=0 (accessed February 2007).

Gray, M. W., Burger, G., & Lang, B. F. (2001). The origin and early evolution of mitochondria. *Genome Biology*, 2 (6), reviews 1018.1–1018.5.

Heidegger, M. (1977). *The question concerning technology*. New York: Harper & Row.

Hoban, T. J. (1998). Trends in consumer attitudes about agricultural biotechnology. *AgBioForum*, 1 (1), 3–7. Available online at: http://www.agbioforum.org (accessed February 2007).

House, L., Jaeger, S., Lusk, J., Moore, M., Morrow J. L., Traill, W. B., et al. (2004). Categories of GM risk-benefit perceptions and their antecedents. *AgBioForum*, 7 (4), 176–186. Available online at: http://www.agbioforum.org (accessed February 2007).

James, S. P. (2002). Heidegger and the role of the body in environmental virtue. *The Trumpeter*, 18 (1). Available online at: http://trumpeter.athabascau.ca/index.php/trumpet/issue/view/15 (accessed February 2007).

Krebs, J. R. (2000). GM foods in the UK between 1996 and 1999: Comments on "Genetically modified crops: Risks and promise" by Gordon Conway. *Conservation Ecology*, 4 (1), 11.

Mieth, D. (2000). Going to the roots of the stem cell debate: The ethical problems of using embryos for research. *EMBO Report*, 1 (1), 4–6.

Observer. (2003). Genetically modified crops? Not in my backfield. Available online at: http://observer.guardian.co.uk/foodmonthly/story/0,971026,00.html (accessed February 2007).

PACE (Programmable Artificial Cell Evolution). (2004). Homepage at: http://www.istpace.org/ (accessed February 2007).

Stewart, C. N. Jr., & Wheaton, S. K. (2001). GM crop data—agronomy and ecology in tandem. *Nature Biotechnology, 19,* 3.

Trewavas, A. J., & Leaver, C. J. (2001). Is opposition to GM crops science or politics? *EMBO Reports, 2* (6), 455–459. Available online at: http://www.nature.com/embor/journal/v2/n6/full/embor393.html (accessed February 2007).

UK GM Science Review. (2003). Reports available online at: http://www.gmsciencedebate.org.uk (accessed February 2007).

UK Statutory Instrument No. 1366. (1993). The crop residues (burning) regulations, 1993, HMSO, London.

Villarreal, L. P., & DeFilippis, V. R. (2000). A hypothesis for DNA viruses as the origin of eukaryotic replication proteins. *Journal of Virology, 74* (15), 7079–7084.

World Health Organization Report. (1999). *Injury: A leading cause of the global burden of disease.* Geneva, Switzerland: World Health Organization.

Wolverhampton Electricity Supply. (2005). The history of electricity supply in the area. Available online at: http://www.localhistory.scit.wlv.ac.uk/articles/electricity/history3.htm (accessed February 2007).

3

Social and Ethical Implications of Creating Artificial Cells

Mark A. Bedau and Mark Triant

A striking biotechnology research program has been quietly making incremental progress for the past generation, but it will soon become public knowledge. One small sign of this is a 2002 article in the widely distributed Sunday supplement *Parade Magazine*, in which one could read the following prediction: "Tiny robots may crawl through your arteries, cutting away atherosclerotic plaque; powerful drugs will be delivered to individual cancer cells, leaving other cells undamaged; teeth will be self-cleaning. Cosmetically, you will change your hair color with an injection of nanomachines that circulate through your body, moving melanocytes in hair follicles" (Crichton, 2002a, p. 6). This may sound incredible and it is certainly science fiction today, but scientists working in the field believe that within the next decade or so the basic technology underlying this prediction will exist. That technology could be called *artificial cells*.

Artificial cells are microscopic, self-organizing and self-replicating autonomous entities created artificially from simple organic and inorganic substances. The quoted *Parade* article was written by Michael Crichton to promote his best-seller, *Prey* (Crichton, 2002b). Crichton's book imagines the disastrous consequences of artificial cell commercialization gone awry (swarms of artificial cells preying on humans). Although one can question many scientific presuppositions behind Crichton's story (Dyson, 2003), research to create artificial cells is proceeding apace, and it is fair to say that the potential risks and benefits to society are enormous. When pictures of Dolly, the Scottish sheep cloned from an adult udder cell, splashed across the front pages of newspapers around the world, society was caught unprepared. President Clinton immediately halted all federally funded cloning research in the United States (Brannigan, 2001), and polls revealed that ninety percent of the public favored a ban on human cloning (Silver, 1998). To prevent the future announcement of the first artificial cells from provoking similar knee-jerk reactions, we should start thinking through the implications today.

This chapter reviews many of the main strategies for deciding whether to create artificial cells. One set of considerations focuses on intrinsic features of artificial cells. These include the suggestions that creating artificial cells is unnatural, that it treats life as a commodity, that it promotes a mistaken reductionism, and that it is playing God. We find all these considerations unconvincing, for reasons we explain here. The alternative strategies focus on weighing the consequences of creating artificial cells. Utilitarianism and decision theory promise scientifically objective and pragmatic methods for deciding what course to chart. Although we agree that consequences have central importance, we are skeptical whether utilitarianism and decision theory can provide much practical help because the consequences of creating artificial cells are so uncertain. The critical problem is to find some method for choosing the best course of action in the face of this uncertainty. In this setting, some people advocate the doomsday principle, but we explain why we find this principle incoherent. An increasing number of decision makers are turning for guidance in such situations to the precautionary principle, but we explain why we find this principle also unattractive. We conclude that making decisions about artificial cells requires being courageous about accepting uncertain risks when warranted by the potential gains.

The Intrinsic Value of Life

Arguments about whether it is right or wrong to develop a new technology can take either of two forms (Reiss & Straughan, 1996, ch. 3; Comstock, 2000, chs. 5–6). Extrinsic arguments are driven by the technology's consequences. A technology's consequences often depend on how it is implemented, so extrinsic arguments do not usually produce blanket evaluations of all possible implementations of a technology. Presumably, any decision about creating a new technology should weigh its consequences, perhaps along with other considerations. Evaluating extrinsic approaches to decisions about artificial cells is the subject of the two subsequent sections. In this section, we focus on intrinsic arguments for or against a new technology. Such arguments are driven by the nature of the technology itself, yielding conclusions pertinent to any implementation of it. The advances in biochemical pharmacology of the early twentieth century and more recent developments in genetic engineering and cloning have been criticized on intrinsic grounds, for example. Such criticisms include the injunctions against playing God, tampering with forces beyond our control, or violating nature's sanctity; the prospect of creating artificial cells raises many of the same kinds of intrinsic concerns. This section addresses arguments about whether creating artificial life forms is intrinsically objectionable.

Reactions to the prospect of synthesizing new forms of life range from fascination to skepticism and even horror. Everyone should agree that the first artificial cell will herald a scientific and cultural event of great significance, one that will force us to reconsider our place in the cosmos. But what some would hail as a technological milestone, others would decry as a new height of scientific hubris. The "Franken-cell" tag attached to Venter's minimal genome project reveals the uneasiness this prospect generates. So it is natural to ask whether this big step would cross some forbidden line. In this section, we examine four kinds of intrinsic objections to the creation of artificial cells, all of which frequently arise in debates over genetic engineering and cloning. These arguments all stem from the notion that life has a certain privileged status and should in some respect remain off limits from human intervention and manipulation.

One objection against creating artificial cells is simply that doing so would be unnatural and, hence, unethical. The force of such arguments depends on what is meant by "unnatural," and why unnatural is wrong. At one extreme, one could view all human activity and its products as natural since we are part of the natural world. But then creating artificial cells would be natural, and this objection would have no force. At the other extreme, one could consider all human activity and its products as unnatural, defining the natural as what is independent of human influence. But then the objection would deem all human activities to be unethical, which is absurd. So the objection has any force only if "natural" is interpreted in such a way that we can engage in both natural and unnatural acts and the unnatural acts are intuitively wrong. But what could that sense of "natural" be? One might consider it "unnatural" to intervene in the workings of other life forms. But then the unnatural is not in general wrong; far from it. For example, it is surely not immoral to hybridize vegetable species or to engage in animal husbandry. And the stricture against interfering in life forms does not arise particularly regarding humans, for vaccinating one's children is not generally thought to be wrong. So there is no evident sense of "unnatural" in which artificial cells are unnatural and the unnatural is intrinsically wrong.

Another objection is that to create artificial life forms would lead to the commodification of life, which is immoral.[1] Underlying this objection is the notion that living things have a certain sanctity or otherwise demand our respect, and that creating them undermines this respect. The commodification of life is seen as analogous to the commodification of persons, a practice most of us would find appalling. By producing living artifacts, one might argue, we would come to regard life forms as one among our many products, and thus valuable only insofar as they are useful to us. This argument is easy to sympathize with, but is implausible when followed to its conclusion. Life is after all one of our most abundant commodities. Produce, livestock, vaccines, and pets are all examples of life forms that are bought and sold every day. Anyone who objects to the commodification of an artificial single-celled organism should also object to the commodification of a tomato. Furthermore, creating, buying, and selling life forms does not prevent one from respecting those life forms. Family farmers, for example, are often among those with the greatest reverence for life.

The commodification argument reflects a commonly held sentiment that life is special somehow, that it is wrong to treat it with no more respect than we treat the rest of the material world. It can be argued that, though it is not inherently wrong to commodify living things, it is still wrong to create life from nonliving matter because doing so would foster a reductionistic attitude toward life, which undermines the sense of awe, reverence, and respect we owe it.[2] This objection does not exactly require that biological reductionism be false, but merely that it be bad for us to view life reductionistically. Of course, it seems somewhat absurd to admit the truth of some form of biological reductionism while advocating an antireductionist worldview on moral grounds. If living things are really irreducible to purely physical systems (at least in some minimal sense), then creating life from nonliving chemicals would presumably be impossible, so the argument is moot. By the same token, if living things are reducible to physical systems, it is hard to see why fostering reductionistic beliefs would be unethical. It is by no means obvious that life per se is the type of thing that demands the sense of awe and respect this objection is premised on, but even if we grant that life deserves our reverence, there is no reason to assume that this is incompatible with biological reductionism. Many who study the workings of life in a reductionistic framework come away from the experience with a sense of wonder and an enhanced appreciation and respect for their object of study. Life is no less amazing by virtue of being an elaborate chemical process. In fact, only after we began studying life in naturalistic terms have we come to appreciate how staggeringly complex it really is.

Inevitably, the proponents and eventual creators of artificial cells will have to face up to the accusation that what they are doing is *playing God*.[3] The playing-God argument can be fleshed out in two ways: It could be the observation that creating life from scratch brings new dangers that we simply are not prepared to handle, or it could be the claim that, for whatever reason, creating life from scratch crosses a line that humans simply should never cross. The former construal concerns the potential bad consequences of artificial cells, so it will be discussed in a subsequent section.

If creating artificial cells is crossing some line, we must ask exactly where the line is and why we should not cross it. What exactly would be so horrible about creating new forms of life from scratch? If we set aside the *consequences* of doing this, we are left with little to explain why crossing that line should be forbidden.

The term *playing God* was popularized in the early twentieth century by Christian Scientists in reaction to the variety of advances in medical science taking place at the time. With the help of new surgical techniques, vaccines, antibiotics, and other pharmaceuticals, the human lifespan began to extend, and many fatal or otherwise untreatable ailments could now be easily and reliably cured. Christian Scientists opted out of medical treatment on the grounds that it is wrong to "play God"— healing the ill was God's business, not ours. Yet if a person living today were to deny her ailing child medical attention on the grounds that medical science is playing God, we would be rightly appalled. So, if saving a life through modern medicine is playing God, then playing God is morally required.

Questions surrounding the playing-God argument are related to the more general question of how religious authority should influence scientific policy decisions. Though religious doctrine will surely be invoked in future discussions of artificial cell science, and though it might help guide policy makers, in modern nonsectarian democratic states, religious doctrine itself has a decreasing role in shaping public policy.

All of the intrinsic objections to the creation of artificial cells canvassed in this section turn out to be vague, simplistic, or ill-conceived. So, in the next section we examine extrinsic objections that turn on the *consequences* of artificial cells.

Evaluating the Consequences

Policy makers will soon have to face questions about whether and under what conditions to create artificial cells. Should artificial cells be developed for commercial, industrial, or military applications? How strictly should they be regulated? The reasons for pursuing artificial cell research and development are matched by concerns about their safety (see the introduction to this volume). Especially this early in the decision-making process, there is no obvious way of knowing whether the speculative benefits are worth pursuing, given the speculative risks. Ruling out any intrinsic ethical qualms against creating artificial cells, the choices we make will be for the most part a matter of how we weigh the risks and benefits of artificial cells, and what strategies and principles we employ when making decisions in the face of uncertain consequences.

The utilitarian calculus is one obvious tool we might employ in assessing a course of action in light of its consequences: Possible actions are measured according to the overall "utility" they would produce (e.g., the net aggregate happiness or well-being that would result), where the course of action we *should* pursue is the one that would produce the greatest utility. There are, of course, problems with any utilitarian calculus, one being the question of how to quantify a given outcome. Is it worse for one person to die or for a corporation to be forced to lay off ten thousand employees? At what point do benefits to public health outweigh ecological risks? The answers to these questions become even less clear when realistically complex situations are taken into account, but this is not a problem peculiar to utilitarianism. Normal acts of deliberation often require us to evaluate and compare the possible outcomes of our actions, no matter how different the objects of comparison. Though one may balk at the notion of assigning a monetary equivalent to the value of a human life, practical social institutions face this task every day. Whether or not the value of money is comparable to the value of a human life in any objective sense, certain acts of deliberation require that their values be compared insofar as they enter into deciding between mutually exclusive courses of action.

This problem of comparing apples to oranges is not the only obstacle the utilitarian approach faces. In its simplest forms, utilitarianism is insensitive to the distribution of risks and benefits resulting from a course of action. One outcome is considered better than another only if it possesses the greater *aggregate* utility. Imagine, for instance, that an experimental batch of artificial cells could be developed in the field in rural Asia that eventually saves tens of thousands of people throughout Asia, but also imagine that during their development the artificial cells would accidentally infect and kill ten people. If our choice were limited to whether or not to release the experimental artificial cells, a simple utilitarian calculus would require the new drug to be developed. Now imagine that a commercial artificial cell manufacturer makes a fortune off one of its products, but in doing so causes significant damage to the environment. Depending on the details of the situation and the relative value of things like corporate wealth and environmental health, the utilitarian calculus may require the commercialization of artificial cells. If this consequence seems unjust, the utilitarian rule could be adjusted to take proper account of the distribution of harms and benefits, along with their quantity and magnitude.

Beyond the question of distribution, we might want other factors to influence how harms and benefits are weighted. For instance, many consider the ongoing process of scientific discovery to be crucial for society's continued benefit and enrichment, so that when in doubt about the consequences of a given research program, we should err on the side of free inquiry. One might also bias one's assessment of risks and benefits by erring on the side of avoiding harm. We normally feel obligated not to inflict harm on others, even if we think human welfare overall would increase as a result. A surgeon should never deliberately kill one of her patients, even if doing so would mean supplying life-saving organ transplants to five other dying patients. In fact, the Hippocratic oath for doctors says that, first, one should do no harm. So if, all else being equal, we should avoid doing harm, it is sensible to bias one's assessment of harms and benefits in favor of playing it safe.

Though it may be true that we have a special obligation against doing harm, a harm-weighted principle of risk assessment is unhelpful in deciding how to proceed with a program as long-term and uncertain as artificial cell science. As Stephen Stich observes, "The distinction between doing good and doing harm presupposes a notion of the normal or expected course of events" (1978, p. 201). In the surgeon's case, killing a patient to harvest his organs is clearly an instance of a harm inflicted, whereas allowing the other five patients to die due to a scarcity of spare organs is not. But when deciding the fate of artificial cell science, neither pursuing nor banning this research program could be described as inflicting a harm, because there is no way of knowing (or even of making an educated guess about) the kind of scenario to compare the outcome against. In the end, artificial cells might prove to be too difficult for humanity to control, and their creation might result in disaster. But it could also be that by banning artificial cell technology we rob ourselves of the capability to withstand some other kind of catastrophe (such as crop failure due to global warming or an antibiotic-resistant plague). The ethical dilemma artificial cells pose is not one that involves deciding between alternative *outcomes*; it requires that we choose between alternative *standards of conduct* where the outcome of any particular course of action is at best a matter of conjecture. So in the present case, the imperative against doing harm is inapplicable.

The ethical problem posed by artificial cells is fundamentally speculative, and cannot be solved by simply weighing good against bad and picking the choice that comes out on top. We can, at best, weigh *hypothetical* risks against *hypothetical* benefits and decide on the most prudent means of navigating this uncertainty.

This still leaves room for something like a utilitarian calculus, however. Decision theory formulates principles for choosing among alternative courses of action with uncertain consequences, by appropriately weighing risks and benefits. Given a particular decision to make, one constructs a "decision tree" with a branch for each candidate decision, and then subbranches for the possible outcomes of each decision (the set of outcomes must be mutually exclusive and exhaustive). A utility value is then assigned to each possible outcome (subbranch). If the probabilities of each possible outcome are known or can be guessed, they are assigned to each subbranch and the situation is called a decision *under risk*. Decisions under risk are typically analyzed by calculating the expected value of each candidate decision (averaging the products of the probabilities of each possible outcome and their utilities), and then recommending the choice with the highest expected value.

If some or all of the probabilities are unknown, then the situation is called a decision *under uncertainty*. Various strategies for analyzing decisions under uncertainty have been proposed. For example, the risk-averse strategy called *minimax* recommends choosing whatever leads to the best of the alternative worst-case scenarios. The proper analysis of decisions under uncertainty is not without controversy, but plausible strategies can often be found for specific kinds of decision contexts (Resnick, 1987). In every case, decisions under both risk and uncertainty rely on comparing the utilities of possible outcomes of the candidate choices.

Decisions under risk or uncertainty should be contrasted with a third kind of decision—what we will term *decisions in the dark*—that are typically ignored by decision theory. Decisions in the dark arise when those facing a decision are substantially ignorant about the consequences of their candidate choices. This ignorance has two forms. One concerns the set of possible outcomes of the candidate choices; this prevents us from identifying the subbranches of the decision tree. The other is ignorance about the utility of the possible outcomes; this prevents us from comparing the utilities of different subbranches. In either case, decision theory gets no traction and has little if any advice to offer on how to make the decision.

New and revolutionary technologies like genetic engineering and nanotechnology typically present us with decisions in the dark. The unprecedented nature of these innovations makes their future implications extremely difficult to forecast. The social and economic promise is so huge that many public and private entities have bet vast stakes on the bio-nano future, but at the same time, the imagined risks are generating growing alarm (Joy, 2000). Even though we are substantially ignorant about their likely consequences, we face choices today about whether and how to support, develop, and regulate them. We have to make these decisions in the dark.

The same holds for decisions about artificial cells. We can and should speculate about the possible benefits and risks of artificial cell technology, but the fact remains that we now are substantially ignorant about their consequences. Statistical analyses of probabilities are consequently of little use. So, decisions about artificial cells are typically decisions in the dark. Thus, utilitarianism and decision theory and other algorithmic decision support methods have little if any practical value. Any decision-theoretic calculus we attempt will be limited by our current guesses about the shape of the space of consequences, and in all likelihood our picture of this shape will substantially change as we learn more.

This does not mean that we cannot make wise decisions; rather, it means that deciding will require the exercise of good judgment. Most of us have more or less well-developed abilities to identify relevant factors and take them into account, to discount factors likely to appeal to our own self-interest, and the like. These methods are fallible and inconclusive, but when deciding in the dark they generally are all we have available. It might be nice if we could foist the responsibility for making wise choices onto some decision algorithm such as utilitarianism or decision theory, but that is a vain hope today.

Deciding in the Dark

Even though the consequences of creating artificial cells will remain uncertain for some time, scientific leaders and policy makers still will have to face decisions about whether to allow them to be created, and under what circumstances. And as the science and technology behind artificial cells progresses, the range of these decisions will grow.[4] The decisions will include whether to permit various lines of research in the laboratory, where to allow various kinds of field trials, whether to permit development of various commercial applications, how to assign liability for harms of these commercial products, whether to restrict access to research results that could be used for nefarious purposes, and so on. The uncertainty about the possible outcomes of these decisions does not remove the responsibility for taking some course of action, and the stakes involved could be very high. So, how should one meet this responsibility to make decisions about artificial cells in the dark?

When contemplating a course of action that could lead to a catastrophe, many people conclude that it is not worth the risk and instinctively pull back. This form of reasoning illustrates what we call the *doomsday principle*, which says: *Do not pursue a course of action if it might lead to a catastrophe.*[5] Many people in the nanotechnology community employ something like this principle. For example, Merkle (1992) thinks that the potential risks posed by nanomachines that replicate themselves and evolve in a natural setting are so great that not only should they not be constructed; they should not even be designed. He concludes that achieving compliance with this goal will involve enculturating people to the idea that "[t]here are certain things that you just *do not do*" (Merkle, 1992, p. 292, emphasis in original). This illustrates doomsday reasoning that absolutely forbids crossing a certain line because it might lead to a disaster.

A little reflection shows that the doomsday principle is implausible as a general rule, because it would generate all sorts of implausible prohibitions. Almost any new technology could, under some circumstances, lead to a catastrophe, but we presumably do not want to ban development of technology in general. To dramatize the point, notice there is always at least some risk that getting out of bed on any given morning *could* lead to a catastrophe. Maybe you will be hit by a truck and thereby be prevented from discovering a critical breakthrough for a cure for cancer; maybe a switch triggering the world's nuclear arsenal has been surreptitiously left beside your bed. These consequences are completely fanciful, of course, but they are still possible. The same kind of consideration shows that virtually every action could lead to a catastrophe and so would be prohibited by the doomsday principle. *Not* creating artificial cells could lead to a disaster because there could be some catastrophic consequence that society could avert only by developing artificial cells, if for example, artificial cells could be used to cure heart disease. Since the doomsday principle prohibits your action no matter what you do, the principle is incoherent.

The likelihood of triggering a nuclear reaction by getting out of bed is negligible, of course, while the likelihood of self-replicating nanomachines wreaking havoc might be higher. With this in mind, one might try to resuscitate the doomsday principle by modifying it so that it is triggered only when the likelihood of catastrophe is non-negligible. But there are two problems with implementing such a principle. First, the threshold of negligible likelihood is vague and could be applied only after being converted into some precise threshold (for example, probability 0.001). But any such precise threshold would be arbitrary and hard to justify. Second, it will often be impossible to ascertain the probability of an action causing a catastrophe with anything like the requisite precision. For example, we have no way at present of even estimating if the likelihood of self-replicating nanomachines causing a catastrophe is above or below 0.001. Estimates of risks are typically based on three kinds of evidence: toxicological studies of harms to laboratory animals, epidemiological studies of correlations in existing populations and environments, and statistical analyses of morbidity and mortality data (Ropeik & Gray, 2002). We lack even a shred of any of these kinds of evidence concerning self-replicating nanomachines, because they do not yet exist.

When someone proposes to engage in a new kind of activity or to develop a new technology today, typically this is permitted unless and until it has been shown that some serious harm would result.[6] Think of the use of cell phones, the genetic modification of foods, or the feeding of offal to cattle. In other words, a new activity is innocent until proven guilty. Anyone engaged in the new activity need not first prove that it is safe; rather, whoever questions its safety typically bears the burden of proof for showing that it really is unsafe. Furthermore, this burden of proof can be met only with scientifically credible evidence that establishes a causal connection between the new activity and the supposed harm. It is insufficient if someone suspects or worries that there might be such a connection, or even if there is scientific evidence that there *could* be such a connection. The causal connection must be credibly established before the new activity can be curtailed.

This approach to societal decision making has, in the eyes of many, led to serious problems. New activities have sometimes caused great damage to human health or the environment before sufficient evidence of the cause of these damages had accumulated. One notorious case is thalidomide, which was introduced in the 1950s as a sleeping pill and to combat morning sickness, and was withdrawn from the market in the early 1960s when it was discovered to cause severe birth defects (Stephens & Brynner, 2001). Support has been growing for shifting the burden of proof and exercising more caution before new and untested activities are allowed, that is, treating them as guilty until proven innocent. This approach to decision making is now widely known as the *precautionary principle*: *Do not pursue a course of action that might cause significant harm even if it is uncertain whether the risk is genuine.* Different formulations of the precautionary principle can have significantly different pros and cons (Parke & Bedau, ch. 5, this volume). Here, we concentrate just on the following elaboration (Geiser, 1999, p. xxiii), which is representative of many of the best known statements of the principle:[7]

The Precautionary Principle asserts that parties should take measures to protect public health and the environment, even in the absence of clear, scientific evidence of harm. It provides for two conditions. First, in the face of scientific uncertainties, parties should refrain from actions that might harm the environment, and, second, that the burden or proof for assuring the safety of an action falls on those who propose it.

The precautionary principle is playing an increasing role in decision making around the world. For example, the contract creating the European Union appeals to the principle, and it governs international legal arrangements such as the United Nations Biosafety Protocol.[8] The precautionary principle is also causing a growing controversy.[9]

The authors of this chapter are skeptical of the precautionary principle. It is only common sense to exercise due caution when developing new technologies, but other considerations are also relevant. We find the precautionary principle to be too insensitive to the complexities presented by deciding in the dark. One can think of the precautionary principle as a principle of inaction, recommending that, when in doubt, leave well enough alone. It is sensible to leave well enough alone only if things are well at present and they will remain well by preserving the status quo. But these presumptions are often false, and this causes two problems for the precautionary principle.

Leaving well enough alone might make sense if the world were unchanging, but this is manifestly false. The world's population is continuing to grow, especially in poor and relatively underdeveloped countries, and this is creating problems that will not be solved simply by being ignored. In the developed world, average longevity has been steadily increasing; over the last hundred years in the United States, for example, life expectancy has increased more than 50% (Wilson & Crouch, 2001). Today heart disease and cancer are far and away the two leading causes of death in the United States (Ropeik & Gray, 2002). Pollution of drinking water is another growing problem. There are an estimated hundred thousand leaking underground fuel storage tanks in the United States, and a fifth of these are known to have contaminated the groundwater; a third of the wells in California's San Joaquin Valley have been shown to contain ten times the allowable level of the pesticide DBCP (Ropeik & Gray, 2002). These few examples illustrate that the key issues society must confront are continually evolving. The precautionary principle does not require us to stand immobile in the fact of such problems. But it prevents us from using a method that has not been shown to be safe. So the precautionary principle ties our hands when we face new challenges.

This leads to a second, deeper problem with the precautionary principle. New procedures and technologies often offer significant benefits to society, many of which are new and unique. Cell phones free long-distance communication from the tether of land lines, and genetic engineering opens the door to biological opportunities that would never occur without human intervention. Whether or not the benefits of these technologies outweigh the risks they pose, they do have benefits. But the precautionary principle ignores such benefits. To forgo these benefits causes harm— what one might call a "harm of inaction." These harms of inaction are opportunity costs, created by the lost opportunities to bring about certain new kinds of benefits. Whether or not these opportunity costs outweigh other considerations, the precautionary principle prevents them from being considered at all. That is a mistake.

These considerations surfaced at the birth of genetic engineering. The biologists who were developing recombinant DNA methods suspected that their new technology might pose various new kinds of risks to society, so the National Academy of Sciences, USA, empaneled some experts to examine the matter. They quickly published their findings in *Science* in what has come to be known as the *Moratorium letter*, and recommended suspending all recombinant DNA studies "until the potential hazards . . . have been better evaluated" (Berg et al., 1974, p. 303). This is an early example of precautionary reasoning. Recombinant DNA studies were suspended even though no specific risks had been scientifically documented. Rather, it was thought that there *might* be such risks, and that was enough to justify a moratorium even on recombinant DNA *research*.

The Moratorium letter provoked the National Academy of Sciences to organize a conference at Asilomar the following year, with the aim of determining under what conditions various kinds of recombinant DNA research could be safely conducted. James Watson, who signed the Moratorium letter and participated in the Asilomar conference, reports having had serious misgivings about the excessive precaution being advocated, and writes that he "now felt that it was more irresponsible to defer research on the basis of unknown and unquantifiable dangers. There were desperately sick people out there, people with cancer or cystic fibrosis—what gave us the right to deny them perhaps their only hope?" (Watson, 2003, p. 98). Watson is here criticizing excessive precautionary actions because they caused harms of inaction.

Some harms of inaction are real and broad in scope, as the threat of rampant antibiotic resistance can illustrate. The overuse and misapplication of antibiotics during the last century has undermined their effectiveness today. By the year 2000 as many as 70% of pneumonia samples were found to be resistant to at least one first-line antibiotic, and multidrug-resistant strains of *Salmonella typhi* (the bacterium that causes cholera and typhoid) have become endemic throughout South America and Africa, as well as many parts of South and East Asia (World Health Organization, 2000). Hospital-acquired *Staphylococcus aureus* infections have already become widely resistant to antibiotics, even in the wealthiest countries. Many strains remain susceptible only to the last-resort antibiotic vancomycin, and even this drug is now diminishing in effectiveness (Enright, Robinson, & Randle, 2002; World Health Organization, 2000). The longer we go on applying the same antibiotics, the less effective they become. It takes on average between 12 and 24 years for a new antibiotic to be developed and approved for human use. Pathogens begin developing resistances to these drugs in just a fraction of this time (World Health Organization, 2000). So, our current antibiotics will become ineffective in the long run.

Weaning ourselves from antibiotics requires effective preventive medicine. One such program that has begun to gain popularity among nutritionists, immunologists, and pathologists is probiotics, the practice of cultivating our own natural microbial flora. Microbes live on virtually every external surface of our bodies, including the surface of our gastrointestinal tract and on all of our mucous membranes, so they are natural first lines of defense against disease. The cultivation of health-promoting bacterial symbiotes has been shown to enhance the immune system, provide essential nutrition to the host, and decrease the likelihood of colonization by hostile microbes (Bocci, 1992; Erickson & Hubbard, 2000; Mai & Morris, 2004). People hosting healthy microbial populations have been shown to have a decreased risk of acquiring HIV (Miller, 2000).

Our natural microbial flora have no effect against many diseases, but wherever a colony of pathogenic organisms could thrive, so conceivably could the right innocuous species. So an innocuous microbe could successfully compete against the pathogen and prevent it from thriving in the human environment. Genetic engineering and artificial cell technologies offer two ways we could develop novel probiotics to compete against specific pathogens. These solutions, moreover, would be viable in the long term and more beneficial than present techniques, as a method of curing as well as preventing disease.

However, the precautionary principle would bar us from taking advantage of these new long-term weapons against disease. It is impossible to be certain that a new probiotic would cause no problems in the future, especially when human testing is out of the question (again, because of the precautionary principle). Though releasing new probiotics into human microbial ecosystems is undoubtedly risky and must be done cautiously, it may prove the only way to prevent deadly global epidemics in the future. So it is not at all implausible that the consequences of inaction far outweigh the potential risks of these technologies. Thus, excessively narrow and restrictive versions of the precautionary principle leave us trapped like a deer in the headlights.

Society's initial decisions concerning artificial cells will be made in the dark. The potential benefits of artificial cells seem enormous, but so do their potential harms. Without gathering a lot more basic knowledge, we will remain unable to say anything much more precise about those benefits and risks. We will be unable to determine with any confidence even the main alternative kinds of consequences that might ensue, much less the probabilities of their occurrence. Given this lack of concrete knowledge about risks, an optimist would counsel us to pursue the benefits and have confidence that science will handle any negative consequences that arise. The precautionary principle is a reaction against precisely this kind of blind optimism. Where the optimist sees ignorance about risks as opening the door to action, the precautionary thinker sees that same ignorance as closing the door.

As an alternative to the precautionary principle and to traditional risk assessment, we propose dropping the quest for universal ethical principles, and instead cultivating the virtues we will need for deciding in the dark. Wise decision making will no doubt require balancing a variety of virtues. One obvious virtue is caution. The positive lesson of the precautionary principle is to call attention to this virtue, and its main flaw is ignoring other virtues that could help lead to wise decisions. Another obvious virtue is wisdom; giving proper weight to different kinds of evidence is obviously important when deciding in the dark.

We want to call special attention to a third virtue that is relevant to making proper decisions: *courage.* That is, we advocate carefully but proactively pursuing new courses of action and new technologies when the potential benefits warrant it, even if the risks are uncertain. Deciding and acting virtuously requires more than courage. It also involves the exercise of proper caution; we should pursue new technologies only if we have vigilantly identified and understood the risks involved. But the world is complex and the nature and severity of those risks will remain somewhat uncertain. This is where courage becomes relevant. Uncertainty about outcomes and possible risks should not invariably block action. We should weigh the risks and benefits of various alternative courses of action (including the "action" of doing nothing), and have the courage to make a leap in the dark when on balance that seems most sensible. Not to do so would be cowardly.

We are not saying that courage is an overriding virtue and that other virtues such as caution are secondary. Rather, we are saying that courage is one important virtue for deciding in the dark, and it should be given due weight. Precautionary thinking tends to undervalue courage.

Our exhortation to courage is vague, of course; we provide no mechanical algorithm for generating courageous decisions. We do not view this as a criticism of our council for courage. For the reasons outlined in the previous section, we think no sensible mechanical algorithm for deciding in the dark exists. If we are right, then responsible and virtuous agents must exercise judgment, which involves adjudicating conflicting principles and weighing competing interests. Because such decisions are deeply context dependent, any sensible advice will be conditional.

New technologies give us new powers, and these powers make us confront new choices about exercising the powers. The new responsibility to make these choices wisely calls on a variety of virtues, including being courageous when deciding in the dark. We should be prepared to take some risks if the possible benefits are significant enough and the alternatives unattractive enough.

Conclusions

Artificial cells are in our future, and that future could arrive within a decade. By harnessing the automatic regeneration and spontaneous adaptation of life, artificial cells promise society a wide variety of social and economic benefits. But their ability to self-replicate and unpredictably evolve creates unforeseeable risks to human health and the environment. So it behooves us to start thinking through the implications of our impending future with artificial cells now.

From the public discussion on genetic engineering and nanotechnology one can predict the outline of much of the debate that will ensue. One can expect the objections that creating artificial cells is unnatural, that it commodifies life, that it fosters a reductionistic perspective, and that it is playing God; we have explained why these kinds of considerations are all unpersuasive.

Utilitarianism and decision theory offer scientifically objective and pragmatic methods for deciding what course to chart, but they are inapplicable when deciding in the dark. No algorithm will guarantee sound policies as long as society is largely ignorant of the potential consequences of its actions. The precautionary principle is increasingly being applied to important decisions in the dark, but the principle fails to give due weight to potential benefits lost through inaction (what we called "harms of inaction"). We suggest that appropriately balancing the virtues of courage and caution would preserve the attractions of the precautionary principle while avoiding its weaknesses.

Acknowledgments

This research was supported by a Ruby Grant from Reed College. For helpful discussion, thanks to Steve Arkonovich, Hugo Bedau, Todd Grantham, Scott Jenkins, Leighton Reed, and to audiences at the LANL-SFI Workshop on Bridging Non-Living and Living Matter (September 2003), at the ECAL'03 Workshop on Artificial Cells (September 2003), at a Philosophy Colloquium at Reed College (October 2003), and at a philosophy seminar at the College of Charleston (November 2003).

Notes

1. For discussions of this argument as applied to other forms of biotechnology, see Kass (2002, chapter 6), and Comstock (2000, pp. 196–198).

2. See Dobson (1995), Kass (2002), and chapter 10 of this volume for discussions of this objection in other contexts.

3. For discussions of the playing-God objection as it has entered into the genetic engineering controversy, see Comstock (2000), pp. 184–185 and Reiss and Straughan (1996), pp. 79–80, 121.

4. For one attempt to identify the key triggers for stages of ethical action regarding artificial cells, see Bedau, Parke, Tangen & Hantsche-Tangen (2008).

5. This principle to our knowledge was first discussed by Stich (1978).

6. One notable exception to this pattern is the development of new drugs, which must be proven to be safe and effective before being allowed into the public market.

7. The formulation of the precautionary principle adopted at the Wingspread Conference (Raffensperger & Tickner, 1999, pp. 353f) and the ETC Group's formulation (2003, p. 72) are very similar to the formulation quoted in the text.

8. Appendix B in Raffensperger and Tickner (1999) is a useful compilation of how the precautionary principle is used in international and U.S. legislation.

9. Attempts to defend the precautionary principle and make it applicable in practice are collected in Raffensperger and Tickner (1999), while Morris (2000) gathers skeptical voices.

References

Bedau, M. A., Parke, E. C., Tangen, U., & Hantsche-Tangen, B. (2008). Ethical guidelines concerning artificial cells. Available online through the PACE site, at http://www.istpace.org (accessed August 2008).

Berg, P., Baltimore, D., Boyer, H. W., Cohen, S. N., Davis, R. W., Hogness, D. S., et al. (1974). Potential biohazards of recombinant DNA molecules. *Science, 185,* 303.

Bocci, V. (1992). The neglected organ: Bacterial flora has a crucial immunostimulatory role. *Perspectives in Biology and Medicine, 35* (2), 251–60.

Brannigan, M. C. (2001). Introduction. In M. C. Brannigan (Ed.), *Ethical issues in human cloning: Cross-disciplinary perspectives* (pp. 1–4). New York: Seven Bridges Press.

Comstock, G. L. (2000). *Vexing nature? On the ethical case against agricultural biotechnology.* Boston: Kluwer Academic Publishers.

Crichton, M. (2002a). How nanotechnology is changing our world. *Parade,* November 24, 6–8.

Crichton, M. (2002b). *Prey.* New York: HarperCollins.

Dobson, A. (1995). Biocentrism and genetic engineering. *Environmental Values, 4,* 227.

Dyson, F. J. (2003). The future needs us! *New York Review of Books, 50* (13 February), 11–13.

Enright, M. C., Robinson, D. A., & Randle, G. (2002). The evolutionary history of methicillin-resistant *Staphylococcus aureus* (MSRA). *The Proceedings of the National Academy of Sciences of the United States of America, 99* (11), 7687–7692.

Erickson, K. L., & Hubbard, N. E. (2000). Probiotic immunomodulation in health and disease. *The Journal of Nutrition, 130* (2), 403S–409S.

ETC Group (2003). *The big down: From genomes to atoms.* Available online at: http://www .etcgroup.org (accessed June 2003).

Geiser, K. (1999). Establishing a general duty of precaution in environmental protection policies in the United States: A proposal. In Raffensberger & Tickner (Eds.), *Protecting public health and the environment: Implementing the precautionary principle* (pp. xxi–xxvi). Washington, DC: Island Press.

Joy, B. (2000). Why the future doesn't need us. *Wired 8* (April). Available online at: http:// www.wired.com/wired/archive/8.04/joy.html (accessed September 2007).

Kass, L. R. (2002). *Life, liberty, and the defense of dignity: The challenge for bioethics.* San Francisco: Encounter Books.

Mai, V., & Morris, J. G. Jr. (2004). Colonic bacterial flora: Changing understandings in a molecular age. *The Journal of Nutrition, 134* (2), 459–64.

Merkle, R. (1992). The risks of nanotechnology. In B. Crandall & J. Lewis (Eds.), *Nanotechnology research and perspectives* (pp. 287–294). Cambridge, MA: MIT Press.

Miller, K. E. (2000). Can vaginal lactobacilli reduce the risk of STDs? *American Family Physician, 61* (10), 3139–3140.

Morris, J. (Ed.) (2000). *Rethinking risk and the precautionary principle.* Oxford: Butterworth-Heinemann.

Raffensperger, C., & Tickner, J. (Eds.) (1999). *Protecting public health and the environment: Implementing the precautionary principle.* Washington, DC: Island Press.

Reiss, M. J., & Straughan, R. (1996). *Improving nature? The science and ethics of genetic engineering.* Cambridge: Cambridge University Press.

Resnick, M. (1987) *Choices: An introduction to decision theory.* Minneapolis: University of Minnesota Press.

Ropeik, D., & Gray, G. (2002). *Risk: A practical guide to deciding what's really safe and what's really dangerous in the world around you.* Boston: Houghton Mifflin.

Silver, L. M. (1998). Cloning, ethics, and religion. *Cambridge Quarterly of Healthcare Ethics,* 7, 168–172. Reprinted in M. C. Brannigan (Ed.) (2001), *Ethical issues in human cloning: Cross-disciplinary perspectives* (pp. 100–105). New York: Seven Bridges Press.

Stephens, T. D., & Brynner, R. (2001). *Dark remedy: The impact of thalidomide and its revival as a vital medicine.* New York: Perseus.

Stich, S. P. (1978). The recombinant DNA debate. *Philosophy and Public Affairs,* 7, 187–205.

Watson, J. D. (2003). *DNA: The secret of life.* New York: Alfred A. Knopf.

Wilson, R., & Crouch, E. A. C. (2001). *Risk-benefit analysis.* Cambridge, MA: Harvard University Press.

World Health Organization (2000). World Health Organization report on infectious diseases 2000: Overcoming antimicrobial resistance. Available online at: http://www.who.int/ infectious-disease-report/2000/ (accessed May 2004).

4

The Acceptability of the Risks of Protocells

Carl Cranor

Protocells

With the advent of a variety of DNA techniques, better understanding of cells, their genetics, and how they function in organisms, scientists are beginning to go in the microscopic world where humans have not gone before. Modified living cells hold the promise of medical advances, of the creation of animals for experimental purposes (such as cancer-prone rodents), and even of environmental cleanup. Nanotechnology advocates often suggest similar beneficial uses for the products of the very tiniest particles, constructed molecule by molecule. Much of this research is basic—scientists simply trying to better understand the world or some part of it. However, the possibility of creating products is frequently not far from researchers' minds, since their successful creation can bring considerable payoffs, amply rewarding a scientific career when funding has become increasingly difficult to obtain. Moreover, when the promise is high enough (for example, stem cell research to repair life-rending injuries or to restore those with physically debilitating diseases to normal functioning), the public can yearn for the benefits that drive research.

Indeed, we are living in an exciting time for technological advances, but also possibly in a brave new scientific world in which a wide range of consequences of interesting and potentially beneficial scientific creations is not well understood. Whatever risks attend the potential benefits of biotechnology and nanotechnology, protocells appear to raise the stakes on risks from technological innovations. What are "protocells?" Some that are under consideration would be "theoretical minimal cell[s] that [incorporate] all properties of the living state, including growth and evolution" (Pohorille & Deamer, 2002, p. 123). Constructed from the "bottom up," from component parts, they would resemble bacteria with a boundary membrane, a replicating molecular system, and a system for translating genetic information "to direct the synthesis of specific proteins" (Pohorille & Deamer, 2002, p. 123); they could also be somewhat simpler (Rasmussen et al., 2004). Thus, much as with genetically modified organisms, the creators of protocells aim for bacterialike substances that can replicate, migrate, and mutate to respond to a changing microecosystem, and perform useful tasks. They might perform "pharmacological and medical diagnostic functions," act as biosensors, metabolize environmental toxicants, and even cure heart disease. Unlike existing bacteria or genetically modified organisms, however, protocells appear to have little relationship to existing entities in the natural world; science is entering "uncharted waters" (see introduction to this volume). In this sense, protocells are "alien life forms."

How should we think about the possibility of creating such microorganisms? In this chapter, I take some steps toward addressing this question. First, I provide some psychometric evidence that the public takes into account when assessing the acceptability of risks. This result is strengthened by philosophical reasons for their judgments. Second, I argue that any risks posed by protocells are likely to be among the most unacceptable risks to the public. Finally, I discuss some trustee institutions and their properties that should be considered for regulating any risks. If protocell projects go forward, these institutions *might* provide trustees that could control the risks despite great concerns about protocells. That is, even if protocell risks generically are among the most unacceptable of risks, all things considered, the risks might be less unacceptable, and perhaps even cross the threshold of acceptability, if well-designed institutions functioned sufficiently to provide an ample margin of safety to protect the public. I close with a cautionary note. Even well-designed institutions might not always function well, in which case advocates of protocells have a personal responsibility to take the lead in ensuring the safety and containment of their products, because they pose such radical potential risks.

Risks and Their Acceptability

Over a 25-year period researchers in the social sciences, well exemplified by Paul Slovic, have shown that there are a variety of dimensions to the public's perceptions of risks (Slovic, 1987). This research suggests that the public perceives the risks with properties in the second column of table 4.1 as worse than those with properties that fall in the first column. What "worse than" means is open to several interpretations. (1) The public might judge the risks with properties in the second column as more probable and more serious in magnitude than risks in the first class. (This interpretation seems unlikely, and would be a seriously mistaken attribution.) (2) The public might find features of risks in the second group to be *less acceptable* to members of the public from *their personal points of view* than risks in the first group (at least somewhat independent of the probabilities involved). In this view,

Table 4.1
Public perception of risk

More Acceptable Features	Less Acceptable Features
Dimensions of Risk Perception	
Controllable	Not controllable
Merely injurious	Fatal
Equitable	Not equitable
Low risk to future generations	High risk to future generations
Easily reduced	Not easily reduced
Voluntary	Involuntary
Does not affect me	Affects me
Not dreaded	Dreaded
Epistemic Considerations	
Observable	Not observable
Known	Unknown to those exposed
Effect immediate	Effect delayed
Old risk	New risk
Risk known	Risk not known to science

Source: Slovic, 1987 (organization author's).

one would find risks in the second group less acceptable to oneself than risks with properties in the first group. Or, (3), "worse than" might mean that features of risks in the second group tend to be *less* acceptable to the public from an impartial or moral point of view. Interpretations 2 and 3 seem more plausible, but it is difficult to tell from the studies because the researchers tend not to distinguish these possible interpretations. For purposes of this chapter, I believe the second interpretation is probably the more accurate characterization of the public's risk judgments. However, as I argue in the following, for the most part individuals have good reasons (beyond mere perceptions) to judge risks with properties in the second group to be less acceptable than risks with properties in the first group. In addition, as I have argued elsewhere, these *personal judgments* of acceptability provide important background for constructing moral arguments for their acceptability (Cranor, 2007). Thus, two reason-based assessments of risks are not *mere perceptions*: reasoned personal assessments (not utilizing moral considerations) and more obvious reasoned assessments explicitly utilizing moral considerations. I have argued elsewhere that reason-based assessments from the personal point of view can be important considerations for moral assessments from a quasi-Kantian theory or a contractualist point of view like Scanlon's (Cranor, 2007).

The psychometric research has been read by some as revealing that the public merely has certain "perceptions" of risks. Although some risk theorists have seemed to argue precisely that, and to dismiss the public's judgments about risks as "mere mistaken perceptions," such judgments should not be treated as mistakes. As I have argued elsewhere, these judgments about the acceptability of risks are not just perceptions that can be dismissed as mirages or some other kind of perceptual or cognitive mistakes on the part of the observer. The issue is the *acceptability* of the risks, not whether the perception is a correct view of an independent reality. There are good reasons, from a personal point of view, for agents to regard most of these properties of risks as well grounded (Cranor, 1995, 2007). In short, the differences between the first, more acceptable group, and the second, less acceptable group of properties need not be mere perceptions.

One can begin to understand the plausibility of the acceptability of risks from a personal point of view by considering what generic reasons risk-bearers might have for finding risks acceptable or unacceptable (Scanlon, 1998, pp. 195, 204, 230). The kinds of generic reasons I have in mind include various ways *individual* persons are generically burdened or benefited by exposure to risks (this is not an impartial moral assessment), and what kinds of things persons generically would have reason to avoid or embrace from a personal, rather than a moral point of view. For example, they would have reasons to avoid threats to their life or health, but might embrace risks that are central to personal projects they might have (Cranor, 2007; Scanlon, 1998). They would, I argue, also have good personal reasons to regard risks that they could (easily) detect by their natural senses as more personally acceptable, *ceteris paribus*, than those they could not detect by their natural senses. When risks are comparatively easy for a person to detect, he or she is likely to be able to respond to the risk and, unless it is not too overwhelming, take some protective action if he or she is personally threatened.

Generic personal reasons point to some good normative rationales for members of the general public to have more complex assessments of risks, compared with technical experts, many of whom base their views merely on the probability and magnitudes of the risks. I believe their reasons can be fairly well grounded in philosophical considerations. That is, risks have *normative features* that are more or less independent of their probability of occurring and the magnitude of damage they might do, and these normative features affect our judgments of the risks' acceptability.

In considering risks and their acceptability, for clarity I use a standard conception of risk: A risk is the chance, or the probability, of some loss or harm, that is, the chance of a mishap. More precisely one might say that a risk is represented by the probability of a loss or harm times its severity. Moreover, a risk in this sense should be distinguished from the personal (and moral) acceptability of the risk in question. Thus, all might agree that there is a risk of being harmed by running a chain saw, or of contracting cancer from exposure to benzene, but disagree both personally and from a moral point of view about the degree of acceptability of the two risks.

Moreover, we should distinguish between "taking a risk," "risk-taking," or being a "risk taker" and "being exposed to a risk," "risk-exposure," or being a "risk bearer." Not all risks are "taken;" often people are exposed to risks to which they might not want to be or to which they have not explicitly chosen to be exposed. Any clarity on this topic requires making such basic distinctions. Moreover, for some activities, one person or group might take or create the risks, but others might find those risks or related ones *imposed* on them. A company might take a financial risk in creating a product, and take workplace risks in manufacturing it, but *impose* health risks on the workforce or those in the larger environment.

The source of risks is a further general point in understanding their acceptability. To the extent that risks are naturally caused, several questions about their acceptability could be posed: How serious are the risks, how dangerous are they if they materialize, and what should be done to minimize them and their consequences, or reduce their chances of materializing and causing harm? By contrast, risks caused by human activities pose additional questions about acceptability: To what extent should they be permitted in the community? Could they be reduced or eliminated by conducting the activities differently? Collectively, since in principle we have control over humanly created risks (and typically greater control than we have over at least the occurrence of more visible naturally created risks from hurricanes, earthquakes, and the like), we should ask whether the activities are really needed or whether they should be conducted in the manner that poses the risks in question, and so on.[1] Are there good reasons for some persons in the community to bear such risks? Humanly created risks seem much more plastic, malleable, and controllable (at least by the community as a collective whole) than many of the more obvious naturally caused risks. Given these differences, there are some generic reasons for supposing that persons would judge as less acceptable humanly created risks of the same magnitude and probability as naturally caused risks. Humanly created risks could easily be seen as much more imposed on or invasive to persons than naturally caused risks, simply because the activities need not be done (or need not be done in the way that creates the risks in question). Moreover, beyond personal judgments of acceptability, humanly created risks pose issues of distribution. Since the risks are not internalized within the activity and are imposed on fellow citizens, are they rightly or justly imposed? Should they be distributed as they are? Such questions are more properly questions of morality and justice, but they are largely inappropriate for naturally caused risks. One can plausibly ask moral and justice questions about naturally caused risks, but for the most part they will be different.

The magnitudes of potential benefits and potential harms and the probabilities of each materializing are pertinent (but not decisive) considerations in individual judgments of the acceptability of risks. If the benefits are minor or quite trivial while the potential harms are substantial or life-threatening, this asymmetry is likely to lead a person to judge that from a personal point of view the risks are less rather than more acceptable.[2]

There is another cautionary note as well. An asymmetry often exists between what is known about a chemical substance's benefits (i.e., why it is created and developed commercially) and what is known about the risks of any harms it may cause. Acquiring the relevant scientific information about harmful effects of chemical substances can be difficult, unless these effects are quite dramatic. For example, diseases that have high relative risks or leave signature effects are relatively easy to detect, but these tend to be the exception, not the rule. By contrast, long latency periods for disease, obscure causal mechanisms, causal overdetermination of adverse effects, lack of "signature" effects, and subtlety of effects compared with background conditions, *inter alia,* all increase the difficulties of acquiring the adverse effect data (Cranor, 2006). In addition, advocates for new technologies appear to be motivated by the potential benefits of their creations rather than by their potential downsides.

More important for the acceptability of risks is their relationship to a person's life. Any risks associated with benefits central to one's life plan are likely to be more acceptable than if similar benefits were less central. By contrast, if the benefits from an activity are peripheral to one's life plan, but its associated risks are serious or life-threatening, the activity is likely to be judged much less acceptable. For some activities, risks may be not only a constituent part of the activity, but part of what makes it attractive, such as mountaineering. Consequently, the risks of being a rock climber, scuba diver, lifeguard, or stuntman that are central to one's life will be embraced, but equivalent risks of suffering disease or death from exposure to substances, such as carcinogens in the water supply or altered genes in plants, are likely to be judged much less acceptable, or even quite unacceptable.

In addition, less obvious benefits from risky activities tend to make risks more acceptable. For example, if an activity is one's personal project (e.g., mountaineering, or home furniture construction), any risks connected to it are likely to be much more acceptable, simply because the personal project is part of what makes life worth living. Other activities might require substantial expertise, in which one takes pride, such as skill-based activities like rock-climbing or scuba diving. Still others can be morally rewarding, such as being a firefighter or emergency medical technician.

By contrast, risks to which a person is subjected that are peripheral to one's life and not part of personal projects are legitimately seen, *ceteris paribus*, as less acceptable. This remains true, I believe, even if the numerical chances of imposed risks are identical in probability and magnitude to other risks that are central to one's life plan and personal projects. Thus, for example, if there are low-level risks from genetically modified foods, but any benefits from them are peripheral to one's life, they are likely judged from a personal point of view to be much less acceptable than identically low-level risks of harm to a home furniture builder (such as being injured by saws or other machinery).

Who creates a risk bears on its personal and moral acceptability. For example, if a person creates a risk to which she herself is exposed, she is likely to know of the risk, understand it better than a risk others have created and imposed on her, and ultimately find it more acceptable. Going beyond personal reasons, from the moral point of view, one party creating a risk and imposing it on others raises distributive issues about the activity's benefits and risks. These would need to be addressed for the justice of the resulting distribution.

Slovic's research reveals that people regard more controllable risks as more acceptable than less controllable ones. This is also endorsed by philosophic considerations. Risks can be subject to a variety of kinds and degrees of control. For example, individuals operating chain saws or other equipment know they can obviously cause harm but, in most cases, have considerable control over these risks and whether or not they materialize. That is, one can choose to engage in the activity or not; one can choose to stop the activity at any time; one can stop the activity if it becomes particularly dangerous, and so on. Other risks are largely outside one's personal control, such as exposure to harmful air pollutants. This suggests that the more dimensions of risks over which one has control (for example, how one becomes exposed to them, how the risks materialize, how much damage they do) the more acceptable such risks will be. By contrast, the less control one has over a range of dimensions of risks, the less acceptable the risks are likely to be. When one is subject to risks largely or wholly out of one's control, one has good reasons to feel more vulnerable, and will likely be more vulnerable, because by definition one cannot control many aspects of the risk. Various degrees of control provide some measure of self-protection, and from a moral point of view, for at least some major risks over which individuals have considerable control, individuals can reasonably be left to protect themselves, with a lesser need for social or legal protections (many details would need to be discussed to work out a more fine-grained analysis).

What I have labeled *epistemic features* here are important additional features of risks that bear on their acceptability. For example, transparent risks from large, dangerous objects, such as motor vehicles or hazardous equipment, will tend to be judged more acceptable by those exposed to them simply because they manifest dangers that normal adults can readily detect. Consequently, it is not surprising that Slovic's research shows that risks that are observable or known tend to be judged more acceptable by the person at risk simply because they can detect the risks, perhaps estimate the scope, extent, and degree of danger they present, and have a sense of how to conduct themselves to avoid these risks. Risks whose effects are immediate (as opposed to delayed) again provide a kind of feedback to those affected so that typically they can judge, assess, and estimate the nature and extent of the risks. Old, familiar risks may be more acceptable than new, unfamiliar ones simply because those exposed have had some experience with the risks, and perhaps have had opportunities to assess and measure them, as it were. Trial and error with risks provides opportunities to assess their scope, as well as clues to the chances they will materialize and produce adverse effects. As one acquires such information, depending on what is revealed, the risks' acceptability may increase, but it could also decrease (if they are too unpredictable). Finally, risks known to science may tend to be judged more acceptable because members of the public are prepared to place some degree of trust in scientists' knowledge and understanding of the risks, compared with risks that are not known to science.

By contrast, exposures to risks that are difficult or virtually impossible to detect by means of one's senses, such as chemical carcinogens or reproductive toxicants, leave one without any sensory input to inform one of the risks, their scope and consequences.[3] If the genetic transformation of plants poses risks because of allergenic features or viruses used to carry the genetic material, typically these would be undetectable until adverse reactions occurred (if one could recognize and trace the causal path). Consequently, there seem to be good epistemic reasons for ordinary persons without special sensory devices that extend the natural senses to judge that risks that are more epistemically accessible are more acceptable than those that are less epistemically accessible. Of course, epistemically accessible risks might still be unacceptable. But in general, as far as the knowledge dimension is concerned, if a risk, its properties, and its consequences are more easily known through the senses, normal adult agents can detect them and perhaps do more to protect themselves. Moreover, persons can exercise caution, care, and perhaps some degree of control with risks that are more epistemically accessible than with those that are not. This is not necessarily so, however. Awareness of a risk (e.g., an incoming hurricane), does not necessarily mean that people can do much to reduce its effect on them; *knowledge about* and *control of* a risk are clearly different. Nonetheless, even for risks that are epistemically quite knowable and over which persons have some degree of control, we may still want to be sure that children and adults fully understand the nature of the risk, so they can guide their activities accordingly to the extent they can control exposure.

Voluntarily incurred risks tend to make the exposure or activity more acceptable than if the risks are not voluntarily embraced. However, there are some conditions on fully voluntary risks that may be difficult to satisfy for a number of risk exposures: One must be aware of the risks one is incurring and understand and appreciate a good deal about them, be competent to make decisions about the risks, and in some important sense have agreed or consented to them (explicitly or implicitly). These are all important features of medical informed consent, which we should bear in mind as a kind of model for voluntarily accepted risks. However, the appreciation condition is not easily satisfied (Slovic, 2006). For risks that are epistemically undetectable, it would be difficult to argue that risks are voluntarily accepted, because the knowledge condition on voluntarily incurred risks is not satisfied. In some circumstances persons may be informed of the presence of invisible risks (for example, by use of Geiger counters or warning devices), which might mitigate some of these concerns. The circumstances under which one might judge that such exposures are voluntary need to be considered in more detail. By contrast, some risks are clearly voluntarily chosen, such as those from mountaineering or scuba diving, and a person can easily avoid them simply by refraining from the activity in question.

Chemicals, Genetically Modified Plants, and Protocells

Protocells are quintessentially humanly created products, thus any associated risks are similarly humanly created (not natural). In many respects, any risks they pose will tend to resemble those from chemicals and some risks from genetically modified food products. At present it is likely that a large percentage of the public will not find any welfare, personal project, experiential, expertise, or morally rewarding benefits from the products. A few will, of course. For the creators of protocells, their creation is a personal project through which they can exemplify their considerable scientific expertise. They might even have modest community pride in a substantial scientific achievement, such as landing astronauts on the moon. If some of the promised benefits to the general public (for example, relieving suffering from certain diseases) materialize, they may also feel morally rewarded. Any obvious welfare benefit to reasonable numbers of the larger public (e.g., cures for cancer) would substantially influence the acceptability judgment. At present, however, it appears that for the most part any risks from protocell products will have substantial externalities for the public and the environment and raise distributive issues that must be addressed.

If protocells escape and expose the broader public to risks, this exposure is unlikely to be voluntary, and thus on this dimension would be judged quite unacceptable. Of course, they would typically be epistemically undetectable, so individuals could likely do little or nothing to avoid exposure to the risks such tiny products carried. If the public is unaware of their presence, avoiding them will be difficult. Moreover, their undetectability will give the public little or no personal control over the materialization of and exposure to the risks, to say nothing of the kind of continuing control one might have over, for example, risks from machinery. How acceptable or unacceptable any such risks might be will depend on what and how serious they are, as well as their other properties. If protocells are robust, that is, capable of living and reproducing readily outside a laboratory, they will be more worrisome than if they are weak, difficult to keep alive, and much less robust than existing bacteria or other forms of microscopic life that have survived for long periods of time.

Public Trustees

According to the risk perception literature, any risks from protocell products appear to have a number of properties that presumably would lead the public to judge them as quite unacceptable compared with other risks in their lives. This risk perception is reinforced by philosophic considerations. Nonetheless, even if the public would tend to regard the risks from a technology as unacceptable and have good reasons for doing so, it might judge the same risks as more acceptable if a governmental agency were to act as a trustee for citizens to ensure an appropriate level of acceptability. What are some of the conditions institutional trustees should satisfy in order to achieve this?

We can begin to address this issue by reviewing some of the shortcomings of existing or previous institutions in regulating similar risks. The design and management of future institutions for guiding the development of and regulating risks of new products should be changed to prevent similar problems from arising. Because protocell products resemble both chemicals and genetically modified plants (or animals for that matter), other regulatory experience in related areas provides some ideas about how institutional trustees might approach protocells to better assess their safety and prevent any risks they pose.

In many respects institutions regulating chemicals have failed to ensure the production of data about their potential adverse effects on public health or the environment. The regulation of chemicals (with the potential for carcinogenic or adverse reproductive effects), other than drugs and to some degree pesticides, did not begin in earnest until the early 1970s. Despite, or perhaps because of, the very recent attempt to regulate exposures to chemicals, scientists and regulatory agencies tend to know little about them. Approximately 100,000 chemical substances or their derivatives are registered for commerce, but most have not been well assessed for health effects (Huff & Hoel, 1992; U.S. Congress OTA, 1995). Moreover, about 1,500 to 2,000 substances are added yearly (IOM/NRC, 2005). In 1984, seventy-eight percent of the 3,000 top-volume chemicals in commerce lacked the most basic toxicity data, and this was little improved by 1998 when the U.S. Environmental Protection Agency (EPA), industry, and environmental groups entered into a voluntary agreement to conduct a full battery of tests on the approximately 3,000 chemicals produced in the highest volumes (NRC, 1984; EDF, 1998).

More specifically, the U.S. National Research Council (NRC, 1984) found the data shown in table 4.2 for different categories of chemical substances in 1984. The

Table 4.2
Commercial chemical substances

Category of Substance	Number of Substances	Percent for Which No Toxicity Data Are Available	Percent for Which Minimal Toxicity Data Are Available	Percent with Enough Data for Partial Health Hazard Assessment	Percent with Enough Data for Complete Health Hazard Assessment
Chemicals in commerce: ≥1 million lbs produced/year	12,860	78	11	11	0
Chemicals in commerce: <1 million lbs produced/year	13,911	76	12	12	0
Food additives	8,627	46	1	14	5
Drugs*	1,815	25	3	18	18
Cosmetics	3,410	56	10	14	2
Pesticides*	3,350	36	2	24	10
Chemicals in commerce: Production volume unknown	21,752	82	8	10	0

Source: Data from NRC (1984, p. 12, figure 2).
*Substances subject to premarket screening laws.

greatest amount of safety data are available for products regulated under premarket screening statutes (i.e., pharmaceuticals and pesticides), and even these data are not especially impressive (NRC, 1984). (Premarket screening laws are those that require the manufacturer of a product to conduct certain tests on products and submit the results to a regulatory agency. The agency reviews the studies and ensures the safety of the product as required by the applicable law, and then permits the product into commerce only if it is safe, or poses no unacceptable risks in accordance with the law; U.S. Congress, 1987.) Much less is known about substances subject to post-market laws. In table 4.2, the vast majority of commercial substances were subject to postmarket regulation; an overwhelming number of these had no toxicity data and only about 11 to 12 percent had enough data for even a partial hazard assessment. (Postmarket laws permit substances into commerce and subject them to regulation only if they pose harms or risks of harm in violation of a statute.) Of course, much more is likely to be known about products whose manufacturers have been required to submit test results and data to a governmental agency than products whose manufacturers are not required to conduct any tests before the substances enter commerce. However, existing data suggest that we should not be overly optimistic even for premarket screening laws. That is, even for pesticides and drugs, two categories of substances for which premarket screening is required, 25 percent of drugs and 36 percent of pesticides have no toxicity data in the public record, and complete risk assessments could be conducted for only 18 percent of drugs and 10 percent of pesticides. Thus, premarket laws have not provided very thorough data for assessing the risks from such substances. (The record may have improved for new products, since data requirements are better and more consistently used now than they used to be.)

To date it appears that the creation of products is outstripping the legal and scientific ability to provide basic health and safety data about them, and to assess any risks they might pose. This is not surprising, since typically there are asymmetries between the knowledge of products' benefits and of their risks. That is, firms have substantial incentive to test and refine the benefits of their products before they enter commerce, so that they can extract the maximum commercial potential from them. While research on risks is needed to protect firms' self-interest and avoid problems with regulation or personal injury law, however, incentive appears to be insufficient to eliminate or greatly minimize risks. In fact, firms have substantial incentive to take chances with some of the risks from their products. For instance, the probability that injured parties can trace their injuries to a particular exposure is quite remote; the chances are also quite low that someone who has been wrongfully injured from exposure will actually pursue a legal remedy (Gillette & Krier, 1990; Saks, 1992).

In addition, there appears to be little or no analysis of life-cycle effects of products and little or no information about long-term environmental or health effects (Cranor, 2003). An NRC committee has pointed out that understanding of the ecological sciences is in its infancy (NRC, 2002). Once products are in commerce, with a few exceptions, such as drugs and pesticides, there is little sensitivity and credible response to warnings of adverse effects, that is, systematic monitoring of possible adverse health effects from products in the market. And, even for drugs, it is not clear to what extent manufacturers are legally obligated to monitor or report adverse environmental and human health effects from their products (U.S. Department of Health and Human Services, 1994).

For products that can have adverse environmental effects, the risk issues are heightened. For one thing, we no longer appear to live in a "frontier environment" (with limitless possibilities, resources and absorption capacity). A more appropriate description of our environment is a crowded, significantly polluted, and fragile fish bowl. If this is the case, any affected ecosystems likely have less capacity to absorb, dilute, neutralize, and detoxify products without suffering adverse environmental impacts than they once did. In addition, the NRC has raised "special concerns about ecological effects because so little is understood about ecological systems" (NRC, 2002, pp. 184, 188). When a new technology is introduced, such as transgenic modification of plants or other species that have the potential to adversely affect the environment, these issues are greatly heightened (NRC, 2002). As the NRC committee argued, introducing a poorly understood technology into poorly understood ecosystems with which it will interact should be done with the utmost caution.

Institutional Guides to Preventing Unacceptable Risks

Given the record of existing institutions in generating knowledge about chemical risks, what are some structural improvements that could lead to better performance in the future? One plausible model for better protecting public health and the environment would have several features, many of which would increase the information generated by those who produce or distribute products (Cranor, 2003). First, it would require greater affirmative knowledge generation by manufacturers of potentially harmful substances than postmarket laws typically do. That is, it would require appropriate premarket product testing to ensure that any risks to health or the environment would be identified as early as possible. In this way, it would resemble features of U.S. premarket drug screening laws. Second, it would also require appropriate premarket review and assessment of the safety of substances or products before they were permitted into commerce. A manufacturer would have a burden of proof to persuade a governmental agency that the product was appropriately "safe" or posed "no unacceptable risk."[4] Third, a health-protective regime would require ongoing monitoring of a product's effects on public and workforce health with legally required reporting to an agency of adverse effects of products. Much too frequently this does not occur. Consequently, the public and even regulatory agencies are often surprised by a product's adverse effects. In addition, post-commercial monitoring of products provides feedback for the accuracy of premarket approval procedures. Fourth, it should provide for more expeditious and less legally burdensome responses to early warnings of adverse health effects in order to reduce or eliminate exposure to toxic products faster rather than slower. At present, this can take considerable time. For example, even under a premarket screening law it recently took the Food and Drug Administration (FDA) about five years to require a pharmaceutical company to put warning labels on a breast milk lactation suppression drug that appeared to cause seizures, heart attacks, and possibly strokes. Once the FDA became even more convinced of the serious adverse effects from this product, it took an additional five years to force the product from the market (FDA, 1994).

Where the full model could not be implemented or full-fledged premarket laws were impractical or unworkable, if postmarket laws were used, they would aim to replicate as many of these features as possible within a postmarket framework (e.g., better monitoring of public health and products that could cause adverse effects, as well as more rapid responses to reduce exposures or to remove products when early warnings were received).

This model premarket law (or postmarket laws with analogous features) has several attractive elements. It is knowledge generating. Firms are required to generate greater health and safety knowledge about their products before they enter commerce than under postmarket statutes. It also requires the generation of health and safety knowledge about the product once it is in commerce. Premarket screening provisions provide for neutral reviews by an impartial body to ensure that the product meets legal safety standards and restricts exposure until the legal degree of safety is ensured. By imposing on the manufacturer an affirmative duty to monitor and report adverse effect data and by requiring comparatively quick withdrawal, it places a considerable premium on protecting public health and the environment.

Beyond Laws and Structures

The preceding discussion concerns only legal requirements and suggests some institutional structures that would improve the regulation of risks from new technologies. However, institutions are more than laws and procedures. Some additional human elements and attitudes would improve the function of institutions. The NRC has critiqued agencies responsible for assessing the risks of transgenic plants, from which we can also learn.

Within a regulatory institution, what other features might facilitate its functioning?

1. In some circumstances tiered premarket screening of products can be done before they enter commerce. Products might merit more or less scrutiny, depending on what is known about their adverse effects. In the chemical area, very large chemicals that do not have certain obviously toxic chemical subgroups attached to them may not need as much scrutiny as smaller chemicals with known toxic subgroups attached. However, making these discriminations requires considerable knowledge and experience with different kinds of chemical products, and may not be readily apparent in the case of a new technology. If there were any meaningful distinctions to be made on these grounds for protocell products, they would likely occur only after considerable experience with them.

2. People who work within an agency must develop a culture that considers seriously the possibility of adverse public health and environmental effects that can arise from the regulated products (NRC, 2002).

3. Moreover, the agency must seek considerable independent scientific input about potential environmental and human health problems from deliberate or accidental releases of the product. The NRC found that there is currently insufficient independent scientific input to the agencies regulating the environmental effects of transgenic plants (NRC, 2002). Scientific isolation, or, worse, agency coziness with the regulated companies, leads to risk of scientific mistakes during the review process.

4. Agency personnel must be successful in identifying the adverse effects from products. That is, an agency must be "scientifically reliable" so that its decisions are accurate over time; this is a possible shortcoming of existing institutions that regulate genetically modified plants. This reliability must be sufficiently high (and reassuring) that the public can properly place confidence in the institution.

5. Where agencies must assess environmental risks from test plots, they need to conduct adequate sampling to ensure that the results are appropriately representative of the distribution of modified organisms (if that is the issue, as it is for genetically modified organisms). The generalization of this point is that they must conduct, or insist that manufacturers use, scientifically sound and appropriate tests for identifying any adverse effects.

6. In addition to reliability, a trustee agency must have sufficient credibility with the general public, perhaps even through substantial public input to their procedures, so that their decisions have social acceptability. This is likely to be especially important with respect to radical new technologies like protocell products. The NRC notes that it is important for the internal culture of an agency to foster an

attitude of welcoming greater public discussions of the issues under consideration and public contributions to regulatory decision making (NRC, 2002).

7. For environmental effects from genetically modified plants, the NRC has noted that inventory and monitoring of the farming and natural environments is insufficient at present for agencies to rest comfortably with current procedures for reviewing genetically modified plants. Too little is known about the current farming or natural environment to judge well adverse departures from the norm, when they are invaded by outside substances that might upset balances. Clearly, a better inventory would be appropriate for new technologies like protocells, if there is a chance they will enter such environments.

The material in the last two sections suggests some of the institutional features that would assist in identifying any risks from protocell products, and ideally prevent risks from arising that could cause harm to public health and the environment. However, institutional arrangements are not foolproof. They must be applied by personnel with a strong commitment to protecting public health and the environment. Laws must be implemented well, and institutional structures must function well even after they have been in place for a long time. However, that still might not be enough.

Much of my research has been on science and the law: what scientific evidence has been used in regulatory and personal injury law, and how it has been and should be used by private parties and adjudicated by courts and regulatory agencies (Cranor, 1993, 2006). That research has clarified an important social issue: Too often an adversarial relationship develops in which regulated businesses try to minimize, downplay, or in the most egregious cases, hide unfavorable evidence that jeopardizes their products, or fabricate studies trying to show greater safety for their products (Michaels & Monforton, 2005). The law-regulated industry relationship has become an elaborate ritual played out in agency hearings or before personal injury courts, because of its adversarial nature. As a consequence, the public and the environment lose because risky, sometimes quite dangerous products (such as asbestos, PCBs, DES, or Vioxx) continue to be present in commerce and cause considerable harm to public health, the environment, or both. This may continue until the harms are so manifest that agencies and the public will not tolerate them any longer.

If the proponents of protocell products approach the environmental and public health safety reviews of their products in the same way that some other regulated industries have done, no institutional safeguards will provide sufficient protections from any risks and harms the products pose. Protocells could resemble bacteria that sense, respond to, and adapt to their environments; they could evolve in response to changing ecological niches, reproduce themselves, and presumably even migrate (to achieve beneficial effects). If substantial risks are associated with protocells, but they otherwise resemble bacteria in their invisibility, mobility, and reproducibility, they would be worrisome indeed. They will be a new life form, heretofore unseen on Earth, likely without natural predators or other neutralizers, with the ability to procreate, migrate, and evolve, which might well cause considerable damage much as rabbits have in Australia only with more far-reaching consequences.[5] If these concerns (or some reasonably close to them) are correct, the creators of protocells would be well-advised to adopt humble, cooperative attitudes toward society's likely concerns about their products and toward their products' potential downsides. In the first instance, they should not go forward with their products until they have a high degree of understanding of the products' properties and potential risks and the environments in which they will be embedded. Even then, as they approach commercialization advocates should work with regulatory agencies, and not frustrate their efforts, hide data, minimize risks, or "spin" the issues in their favor, but rather give a full, honest assessment of their products. Too much appears to be at stake to do otherwise.

Notes

1. Of course, we should also recall that human decisions with respect to naturally caused risks can make a substantial difference in any harm they cause. Human decisions with respect to hurricane Katrina (both before and during its occurrence) made the harms caused much worse than they otherwise would have been with different decisions.

2. In discussing risks of a product or activity, one should distinguish between (1) risks, where possible outcomes are known and probabilities can be assigned, (2) uncertainties, where there is no adequate basis for assigning probabilities to outcomes, and (3) ignorance, where there is lack of certainty about possible outcomes and no basis for assigning probabilities. Often these are not separated, which leads persons to treat them all as risks, which are much more easily known, than as circumstances of considerable uncertainty or great ignorance. Moreover, one should probably have considerable humility concerning a *new* technology's benefits, harms, and associated probabilities because its features are often poorly understood, especially in early stages. For new and poorly understood technologies, an assessment of their downsides is more likely to be an ignorance or uncertainty assessment (in the preceding terms) than a risk assessment (narrowly construed).

3. Of course, various technological devices can extend one's senses, such as microscopes or Geiger counters, but I am largely interested in persons' natural sensory abilities to detect risks, since that is the circumstance in which most of us find ourselves. We do not for the most part have sensory-extending devices available to us. Moreover, if one has some general idea that invisible risks exist in the area, one can simply avoid the area if one wishes to avoid the risks. (I owe these points to Mark Bedau.)

4. Some substances might have features that would require greater testing and scrutiny while others might require less. Thus, there could be a "tiered" review system, with greater or lesser degrees of scrutiny.

5. As indicated earlier, they might lack such robust properties. If they do and if they are comparatively weak and unlikely to survive outside a laboratory, this would lessen some concerns about them. How potentially harmful they might be within a lab would also need to be assessed.

References

Cranor, C. F. (1993). *Regulating toxic substances: A philosophy of science and the law.* New York: Oxford University Press.

Cranor, C. F. (1995). The use of comparative risk judgments in risk management. In A. Fan & L. Chang (Eds.), *Toxicology and risk assessment: Principles, methods and applications* (pp. 817–833). New York: Marcel Dekker, Inc.

Cranor, C. F. (2003). Some legal implications of the precautionary principle: Improving information generation and legal protections. *The European Journal of Oncology, Library* 2, 31–51.

Cranor, C. F. (2006). *Toxic torts: Science, law and the possibility of justice.* Cambridge, UK, New York: Cambridge University Press.

Cranor, C. F. (2007). Toward a non-consequentialist approach to acceptable risks. In T. Lewens (Ed.), *Risk: Philosophical perspectives* (pp. 36–53). London: Routledge.

Environmental Defense Fund (EDF) (1998). *Toxic ignorance.* Oakland: Environmental Defense Fund.

Gillette, C. P., & Krier, J. E. (1990). Risk, courts and agencies. *University of Pennsylvania Law Review, 38,* 1077–1109.

Huff, J., & Hoel, D. (1992). Perspective and overview of the concepts and value of hazard identification as the initial phase of risk assessment for cancer and human health. *Scandinavian Journal of Work Environment and Health, 18,* 83–89.

Institute of Medicine and National Research Council (IOM/NRC), Committee on the Framework for Evaluating the Safety of Dietary Supplements (2005). *Dietary supplements: A framework for evaluating safety.* Washington, DC: National Academy Press.

Michaels, D., & Monforton, C. (2005). Manufacturing uncertainty: Contested science and the protection of the public's health and environment. *American Journal of Public Health, 95* (Suppl. 1), S39–S48.

National Research Council (NRC) (1984). *Toxicity testing: Strategies to determine needs and priorities.* Washington DC: US Government Printing Office.

National Research Council (NRC) (2002). *Environmental effects of transgenic plants.* Washington, DC: National Academy Press.

Pohorille, A., & Deamer, D. (2002). Artificial cells: Prospects for biotechnology. *Trends in Biotechnology, 20* (3), 123–128.

Rasmussen, S., Chen, L., Deamer, D., Krakauer, D., Packard, N. H., Stadler, P. F., & Bedau, M. A. (2004). Transitions from nonliving to living matter. *Science, 303,* 963–965.

Saks, M. J. (1992). Do we really know anything about the behavior of the tort litigation system—And why not? *Pennsylvania Law Review, 140,* 1147–1289.

Scanlon, T. M. (1998). *What we owe to each other*. Cambridge, MA: Harvard University Press.

Slovic, P. (1987). Perception of risks. *Science, 236,* 280–285.

Slovic, P. (2006). Written direct examination of Paul Slovic, Ph.D. Submitted by the United States Pursuant to Order 471, United States of America v. Phillip Morris, Civil Action No. 99-CV-02496 (GK).

U. S. Congress, Office of Technology Assessment (OTA) (1987). *Identifying and regulating carcinogens*. Washington, DC: U.S. Government Printing Office.

U.S. Department of Health and Human Services, Food and Drug Administration (FDA) (1994). Notice on opportunity for a hearing on a proposal to withdraw approval of the indication of bromocriptine mesylate (Parlodel) for the prevention of physiological lactation. *59 Federal Register.* 43347 (Aug. 23).

5

The Precautionary Principle and Its Critics

Emily C. Parke and Mark A. Bedau

The precautionary principle is an important guideline for making decisions about new technologies with weighty and uncertain consequences. It was established to handle situations in which a new technology might pose risks to human health or the environment, but significant scientific uncertainty about those risks exists. The principle arises in contemporary discussions of the social, ethical, and regulatory implications of things like genetically modified (GM) crops and organisms (GMOs), stem cell technology, and the products of synthetic biology, including protocells. The principle has been adopted in many important national and international treaties and policy statements.[1] However, in spite of its widespread and increasing presence in legislation and policy, there is no consensus as to its exact definition, status, or implications. This chapter aims to clarify the nature and proper use of the precautionary principle.

Evaluating the social and ethical implications of protocells is a clear case in which the precautionary principle becomes relevant. In addition to having highly uncertain and potentially irreversible consequences, both beneficial and harmful, the creation from scratch of new life forms raises challenging philosophical questions such as what it means to be alive, and whether life forms have qualitatively unique ethical features. So, when making difficult decisions about our future with protocells, one needs good guidance, and some form of the precautionary principle might help.

The variability of the precautionary principle is notable, and has been a subject of debate in ethics and policy-related literature for nearly a decade. The principle is understood, formulated, and practically interpreted in many ways. It is neither stable nor well-defined, and it has been very controversial. This chapter will explain its variability, evaluate its most serious threats, and identify what is required for its future use to be constructive and compelling.

Variability of the Precautionary Principle

While some sources claim to describe *the* precautionary principle with one formulation (Barrett & Flora, 2000; Saunders, 2000; Lieberman & Kwon, 2004), it is really a vague principle with many realizations. However, a common structure unifies this variability.[2] This structure has three elements: a triggering factor, an uncertainty condition, and a directive for action.[3] These respectively express what potential threats trigger precautionary action, what degree of scientific uncertainty clouds understanding of the threats, and what precautionary action is called for. These three elements exist in every formulation of the principle, in the following form:

If [the triggering factor holds] and [the uncertainty condition holds], then [the directive for action holds],

or, if we let T, U, and A stand for the triggering factor, uncertainty condition, and directive for action:

If T and U, then A.

The principle is often expressed as something like this: In situations where potential threats are perceived and there is insufficient scientific evidence regarding the potential harm involved, measures should be taken to prevent harm. Here, the T is "potential threats are perceived," U is "insufficient scientific evidence regarding the potential harm," and A is "measures should be taken to prevent harm."

Each of the three elements is subject to a particular kind of variability. Triggering factors vary in *scope* from narrow to broad. For example, "scientific evidence of damage" applies to a relatively narrow range of situations compared to "lightly founded rumor"; broader factors trigger precautionary action in a broader range of situations. Uncertainty conditions vary in *strength*. "Great uncertainty about potential harms" is much stronger than "minimal uncertainty." Directives for action vary in the *severity* of the constraints they place on our behavior. Allowing a new technology to be used conditionally while assessing potential alternatives is not very severe, whereas banning the new technology is (Parke, 2004).

Once T, U, and A in a particular context are determined, their content can be more or less specific. In the preceding example, none of the parameters is very specific. It is not specified what kind of perception of potential threat triggers precaution, nor how much scientific evidence is sufficient, nor what specific precautions should be taken. In general, the less specific the principle, the less clear it is how to interpret its elements in application.

Thus, we can think of each element of the principle as having a two-dimensional variability: scope-specificity for T, strength-specificity for U, and severity-specificity for A. This way of conceptualizing the variability of the principle serves as a framework for evaluating it as a decision-making tool, in the face of various criticisms that have been voiced in the literature.

Arguments Against the Precautionary Principle

Many criticisms have been raised against the precautionary principle, some directed at practical considerations and others at the principle's fundamental premises or motives. Some arguments are more damaging than others, and many pertain only to certain versions of the principle. Here we evaluate eleven of the principle's serious criticisms and indicate how the principle could be defended, with the hopes of setting the scene for further clarification of the principle as a practical decision-making tool.

Superfluousness
Critics have charged that the precautionary principle is unnecessary because current regulatory establishments (such as the Food and Drug Administration (FDA) and the system of tort law) are already filling the role it aims to fill. This line of argument makes two assumptions about current establishments. The first is that they are adequately preventing harm to health and the environment. Durodié (2000), for example, argues that in the face of significant public hype and concern over an extremely rare hazard—choking deaths—preventive regulatory measures already more than adequately addressed the hazard, given its statistical unlikelihood. The second assumption is that current regulatory establishments are already equipped to handle exactly the types of situations to which the precautionary principle purports to apply, that is, those involving perceived risk combined with scientific uncertainty. Morris (2000) illustrates this argument when he dismisses a "widespread misconception" on behalf of the principle's proponents regarding the insufficiency of established tort law and regulatory systems as mechanisms for preventing environmental damage, concluding that the precautionary principle is unnecessary.

To respond to this criticism, a proponent of the principle should point out that the precautionary principle and current regulatory establishments are trying to accomplish different goals. In general, established regulatory institutions tend to address the issue of *acceptable levels* of hazards; that is, they assume that safe levels of hazards (for example, chemicals in a workplace) can be identified and used to determine appropriate regulatory standards. The precautionary principle becomes relevant when uncertainty and complexity make it impossible to determine safe levels, and so one can only try to minimize potential harms. By emphasizing the limitations of scientific knowledge, the precautionary principle applies when we are too ignorant about risks to quantify them and determine appropriate responses. So, the principle cannot be written off as superfluous given existing regulatory institutions.

Disguised Protectionism

Critics have targeted the motives behind the precautionary principle by accusing it of being a disguised form of trade protectionism. Many parties and nations that have dedicated themselves to the precautionary principle (under national and international agreements such as the Rio Declaration, Cartagena Protocol on Biosafety, and World Charter for Nature) are also bound to agreements of the World Trade Organization (WTO), which checks for disguised obstacles to trade. The WTO supports the idea that science, rather than rumor or unfounded beliefs, should be behind trade regulatory decisions. Critics of the precautionary principle can claim it threatens free trade, arguing that it establishes an easy gateway for politically biased, unscientific justification of regulatory action—and thus conceals ulterior motives like trade protectionism (see, e.g., Jones, 2000; Morris, 2000; Yandle, 2000).

There are concrete cases in which the principle has been accused of applying a form of disguised protectionism (e.g., the European Commission's ban on beef using growth hormones; see Kastner, 2002, for further discussion). These apprehensions are valid and merit attention, but they are grounds for arguing against the principle's *misapplication*, not against its existence. Proponents should find practical ways to implement the precautionary principle while still complying with obligations arising from WTO agreements. Important work has already been done toward this end. The European Commission, for example, aims to "avoid unwarranted recourse to the precautionary principle, which in certain cases could serve as a justification for disguised protectionism . . . far from being a way of evading obligations arising from the WTO agreements, the envisaged use of the precautionary principle complies with these obligations" (2000, p. 9). The point that the principle can be (and arguably has been) misused does not damage it *per se*. Rather, it is an incentive for closer attention to the complex interactions between precautionary thinking and the goals of trade globalization. An analogous point holds for *any* decision-making principle; any principle can justify bad decisions if it is misinterpreted and misapplied. The existence of such potential is not a special objection against the precautionary principle.

Excessive Regulation

Critics have claimed that the precautionary principle fails to acknowledge the fact that "the dose makes the poison." The principle is accused of disregarding basic principles of toxicology by implying that a substance dangerous at some level is dangerous at any level, no matter how small. Several critics cite cases in which unreasonable inferences have been made. Durodié (2000), for example, writes of the "plastic panic," in which infant teethers containing polyvinyl chloride (PVC) were removed from the market after public hype over potential toxicity, even though the only definitive evidence against PVC was that consuming 50 grams of the plastic each day would have potentially carcinogenic effects. Lieberman and Kwon (2004) describe another case in which trace benzene levels found in Perrier mineral water resulted in halted production worldwide, despite the FDA's conclusion that, given the level of benzene (which is 2,000 times less than that found in a pack of cigarettes), lifetime consumption of 16 ounces of Perrier per day *might* increase lifetime cancer risk by one in one million. These and other examples of "unfounded health scares" are used to show that the precautionary principle leads to regulatory constraints that are out of proportion with genuine threats (see also Morris, 2000).

This criticism maintains the unreasonableness of inferences from the fact that X is found to be harmful at an extreme level to the conclusion that X is harmful in general, at any level. Wildavsky dismisses this so-called "no safe dose" theory for carcinogens as a completely irrational premise for policy decisions, pointing out that "virtually everything human beings breathe or eat or drink brings them into contact with carcinogens" at extremely minute levels (2000, p. 30).

But this criticism is easily deflected. The precautionary principle need not assume that any level of a harmful substance is harmful. Some *adherents* of the principle might choose to make that assumption, but other adherents might choose otherwise. In particular, when the dose really does make the poison, applications of the precautionary principle should not assume the reverse. So the excessive regulation criticism applies only to overzealous applications of the principle. Any decision-making principle has the potential to be applied excessively, even principles that are sound. This possibility is not an inherent flaw in the precautionary principle.

Worst-Case Bias

Another argument against the precautionary principle is that it always assumes worst-case scenarios (Rubin, 2000; Miller & Conko, 2000). It accuses the principle of having a kind of tunnel vision, focusing only on the worst possible thing that could happen and, in doing so, ignoring the entire range of other possible outcomes.

Advocates of the precautionary principle could respond to this criticism by claiming that the weight they give to worst-case scenarios is merited. Risk aversion is one of the main concepts underlying precautionary thinking; the precautionary principle involves considering that the worst might happen, and taking steps to avoid this possibility. But giving due weight to worst-case scenarios should be distinguished from assuming, when confronted with uncertainty and potential harm, that the worst possible scenario *will* happen. To whatever extent this criticism takes the precautionary principle to assume worst-case scenarios are most likely to happen, it is based on a misunderstanding.

This criticism comes down to the question of how risk averse we should be. By arguing that the principle gives undue weight to worst-case scenarios, critics think that we should not be so risk averse. Proponents of the principle, on the other hand, think that we *should* be so risk averse. The risk averseness of a given formulation of the precautionary principle depends on the scope of its triggering factor, or the severity of its directive for action. A formulation with a broad triggering factor, such as calling for precaution in response to *mere rumor*, is very risk averse. Similarly, a formulation with a severe directive for action that calls for banning everything that poses any threat is more risk averse than one that calls for just proceeding with caution. Thus, the more risk averse the formulation of the principle, the stronger this criticism. The degree to which someone sympathizes with the precautionary principle's focus on worst-case situations depends on that person's aversion to risk.

Unsound Science

Many critics have argued that the precautionary principle is inconsistent with the norms of sound science, because it calls for precautionary action prior to definitive scientific proof of harm. This line of argument generally holds that the principle threatens the established integrity of science-based risk assessment by promoting value-laden decision making that fails to put reason and practicality at the forefront. These critics often prefer quantitative risk assessment to the precautionary principle. The crux of this criticism is that application of the precautionary principle is based on science that is dubious or nonexistent (Gray, 1990; Bewers, 1995; Whelen, 1996).[4]

A good deal of literature has defended the principle from this criticism. Most defenses maintain that the precautionary principle and sound science, when properly understood, are not just consistent but complementary. Critics charging the principle with unsound science equate sound science with conventional risk assessment, which assumes that unambiguous data and scientific inferences about possible outcomes will enable us to make wise decisions. From this perspective, sound science allows us to calculate the probabilities and expected consequences of possible identified risks. It deals with complexity and uncertainty by narrowing the focus of research until empirical data implies clear quantitative risk assessment, isolated from value-laden and qualitative social and political considerations, and basing decisions on statistical considerations (Gray, 1990).

Defenders of the principle emphasize the need for a *different kind* of science, but one that they claim is still perfectly sound.[5] This perspective emphasizes the inability of contemporary risk assessment to assess complex decision contexts properly. A risk is not equal to the sum of its parts; we cannot understand the risk of a new industrial chemical, for example, by separately analyzing the effects of its components in isolation. The principle's advocates believe that sound decisions must be based on a sound scientific assessment of the larger ramifications of the new technology.[6]

The fundamental difference between critics and defenders of the precautionary principle in this case turns on decision-making values. "Unsound science" critics value decisions based narrowly on scientific risk assessment, whereas the defenders value decisions that rest on a broader range of considerations. Emphasizing fragmentation, quantitative analysis, and analysis of highly specific, narrowly defined components of risks as "sound science" signals adherence to one set of values. Likewise, emphasizing scientific inclusion, complexity, nonlinearity, emergent properties, and acknowledgment of uncertainty implies adherence to a different set of values. Defenders of the precautionary principle argue that the principle does not appeal to science that is less sound than that of scientific risk assessment. The scientific foundation of the principle includes contributions from such disciplines as population biology, marine ecology, chemistry, microbiology, public health, and economics, and it uses this information to build a broad scientific basis for making complex decisions. In the eyes of the principle's defenders, anything less would be scientifically unsound.

Thus, the unsound science criticism at bottom is about the values that guide our norms for making decisions about new technologies with uncertain risks. Those values determine which scientific questions one seeks to answer. Once the questions have been identified, though, presumably everyone can agree that one should seek their answers using the soundest possible science.

Appealing to Emotions

The precautionary principle has been accused of valuing "the politics of emotionalism over reasoned debate" (Durodié, 2000, p. 140). The worry is that the emotional appeal of the precautionary principle produces irrational or unreasonable decisions that lead to unwise regulation.

Defenders of the principle should acknowledge that our emotions are inevitably raised by certain kinds of decisions, and point out that this is not necessarily a bad thing. Tickner and associates point out, for example, that "[d]ecision-making about health is not value-neutral" (1998, p. 16). It is natural to react emotionally when our health or environment is threatened. We take these things personally.

If the precautionary principle's application is rooted in blind emotion and irrationality, this could lead to objectionable results. It arguably has been so applied in Durodié's "plastic panic" example (Durodié, 2000). Nonetheless, one can still deflect the charge of irrational emotionalism. What proponents of the precautionary principle should do is work toward clearing up misconceptions. We saw that to avoid the charge of unsound science, one must be clear about which scientific questions gate sound decisions. Similarly, in this case, the principle should be construed in such a way as to clarify that, though it focuses on emotionally charged issues, it will avoid inappropriate emotional appeals. Instead, it will acknowledge the emotional and value-laden issues but promote decisions based on the highest standards of rational analysis.

Self-Contradiction

There is a conceptual argument that the precautionary principle is inconsistent. This argument points out that the principle halts potentially harmful actions with uncertain consequences. But for many such actions, it is the case that *not* doing those actions could also involve potential harms and uncertain consequences. Comstock's (2000) discussion of GM crops illustrates this point. The precautionary principle was used to ban GM crops in some parts of Europe; however, because of the uncertainty of their environmental effects, there are many reasons to believe GM crops could *prevent* more environmental degradation than they cause, for example, by making food production and distribution more stable. In this case, the precautionary principle would simultaneously ban both developing and not developing GM crops. So, when confronted with uncertainty and potential harm, the precautionary principle becomes a self-contradictory principle of inaction (see, e.g., Sandin, ch. 6, this volume; Bedau & Triant, ch. 3, this volume).

At this point, we should recall that actions that might be harmful are permitted by many formulations of the precautionary principle. In particular, mild directives for action can allow some new technology to be used with restrictions. Thus, only very severe directives for action, such as phaseouts, bans, and moratoriums, are subject to self-contradiction.

Certain severe formulations avoid self-contradiction, because they are triggered by a narrow range of conditions (recall the spectrum from narrow to broad triggers discussed earlier). Narrower triggers require precaution in response to fewer potential threats, and broader triggers are tripped by more potential threats. Very narrow triggers require precaution *only* given weighty evidence of potential threats; they are narrower because they apply to fewer situations. Narrow triggers have a higher epistemic hurdle and require stronger evidence; contradictory ones are less likely to lead to the same precautionary action. Thus, self-contradiction threatens only for-mulations with broad or vague triggering factors combined with severe directives for action.

Consider a formulation of the precautionary principle that bans everything with uncertain consequences that could threaten human health or the environment, but which also allows things that could dramatically improve human health or the environment. These formulations leave the door open for situations (perhaps like GM crops) that require two contradictory actions. This problem affects all formula-tions that specify severe directives for action and require (or allow, through vague-ness) broad triggering factors. Proponents of the principle should avoid those self-contradictory formulations.

Demanding the Unknowable

Critics have argued that the precautionary principle is unreasonable because it establishes an epistemologically impossible burden of proof, that is, expecting zero uncertainty about risk before going forward (Morris, 2000). Implicit in this sort of claim is the worry that the principle will be far too exclusive in application and will prevent the introduction of any new technologies (because they will not be able to meet this unrealistic standard of proof).

Like the contradiction argument, this criticism is valid against only those versions of the principle with severe directives for action and broad (or vague) triggering factors. Formulations calling for absolute proof of safety prior to allowing some-thing to proceed can fairly be accused of establishing an infinitely high standard of proof.

Of this set of formulations with strong directives for action, a particular subset can deflect the charge that generally applying the precautionary principle would necessitate the regulation of every single new technology. That subset includes those formulations with specific, narrow triggers. If *only* the cases in which concrete sci-entific evidence of harm is presented lead to demanding proof of safety and precau-tionary action, it is clearly not the case that every new technology would be regulated, and their proponents faced with unreasonable expectations. If formula-tions with strong directives for action and strong or vague triggering factors were adopted as a screening process for all new technologies, this would indeed be estab-lishing unreasonable expectations. However, many formulations of the precaution-ary principle avoid this.

So, the same formulations that did not stand up to the contradiction argument do not stand up to this one. In order to respond to the criticism, the principle should be refined to clarify that it is not mandating proof beyond a shadow of doubt.

Rather, it should mandate that promoters of an activity or technology that is perceived as possibly dangerous look more closely and inclusively at its potential effects than regulatory institutions currently require.

But clearly not all formulations require that anything suspected to be harmful be proved entirely safe (European Commission, 2000). That is, the precautionary principle does not assume that we can demand zero risk. It does, however, assume that we can demand safer alternatives to processes or products that might threaten health or the environment.

Variability and Vagueness

Another criticism of the precautionary principle holds that it can never be taken seriously or successfully implemented because of its vagueness and variability. This involves two related but separable claims. The first is that because of variability *among* formulations (i.e., the variety of ways the principle has been and can be formulated), a single universal definition of the principle could never be reached. The second claim is that because of variability *within* formulations (i.e., the many different ways one can practically interpret the variables within a given formulation), the principle is too vague to ever offer workable guidelines for implementation. The variability and vagueness worry leads people to question how any version of the principle could ever be embraced as an international standard (Foster et al., 2000; Jones, 2000). Variability among formulations would also be a practical problem if two (or more) different formulations gave contradictory advice about the same situation.

The criticism about variability within formulations argues that the precautionary principle expresses an intuitive sentiment that offers little guidance for practical application to policy decisions. Critics point out the lack of concrete guidance for determining the weight of evidence needed to trigger the principle, and for deciding which of many possible precautionary measures should be applied under given circumstances. This makes the principle seem too vague to serve as an operable regulatory standard (Bodansky, 1991; Morris, 2000).

Variability *within* formulations seems even more worrisome than variability *among* formulations. It is true that the vast majority of formulations are extremely vague. Advocates of the principle offer little operational guidance, in official documents or in the secondary literature. It seems to be assumed that appealing to precaution will lead to open, informed, well-thought-out decisions. This is not necessarily true; more clarity about how to interpret the principle is needed if it is to be a coherent *operational* guideline rather than just "a token theoretical idea that may be acknowledged and subsequently ignored" (Santillo, Johnston & Stringer, 1999, p. 41).

One instance of the precautionary principle offers especially detailed operational guidelines for its application—the UN's Straddling Fish Stocks and Highly Migratory Fish Stocks agreement (United Nations, 1995). Annex II of the agreement outlines detailed procedures for determining "precautionary reference points," boundaries intended to constrain harvesting of fish stocks within safe biological limits. This agreement takes an unprecedented step among international policy documents that vaguely call on the principle, by specifying the *particular kinds of values* that should be plugged into the variables (within the very specific realm of decision making it governs). It calls on decision makers to take precautionary "measures," but then specifies that this should involve allowing the potentially harmful activity in question (i.e., harvesting of fish stocks) to continue under the provision that restrictions (in the form of "precautionary reference points") are determined, implemented, and upheld. This level of specificity and concrete guidance concerning the directive for action sets an important precedent for context-sensitive application of the precautionary principle, offering concrete practical direction to supplement invocation of a vague guideline. This kind of guidance is just what is needed to dispel the worries posed by the argument about variability within formulations.

Alternatively, the precautionary principle can also be viewed not as one single entity but as an overarching principle manifested in a variety of ways and consistent with a variety of specific implementations. Rather than as an operational algorithm issuing concrete directions, it could be viewed more as a guideline that requires decision makers to exercise judgment depending on the details of the situation. From this perspective, variability and vagueness are welcome and expected; in fact, some would argue that vagueness is *good* because it provides flexibility of application. The corresponding danger to avoid, in this case, is so much flexibility that scant guidance is provided and abuse is permitted.

Forgone Benefits and Stifled Innovations

The precautionary principle is sometimes charged with being harmful because it leads to forgone benefits or negative consequences (Goklany, 2000; Hammitt, 2000; Miller & Conko, 2000).[7] Any compelling version of the precautionary principle must confront this issue. Before a new technology's potentially harmful effects trigger precautions, one should make sure that those precautions will not cause more harm than good. One harmful consequence often attributed to precautionary action is stifled innovation. The worry that precaution stifles innovation arises especially when directives for action are excessive, such as banning any new technologies with any possibility of harmful consequences (Wildavsky, 2000; Bedau & Triant, ch. 3, this volume).[8] Requiring absolute proof of safety before implementing any new technology would certainly stifle innovations, but less severe directives (such as allowing activities to proceed with some restrictions) would fix this.

The problem of forgone benefits can be lessened by attending to a certain distinction. Of those formulations with severe directives, those with narrow triggers will stifle innovation less. Any severe precautionary principle that bans every technological innovation pending proof of safety clearly can be charged with stifling innovation. Formulations with narrow triggers require more evidence of harm, however, and thus can be expected to lead to fewer cases of precautionary action. Therefore, the stifling innovation argument has varying levels of force against different formulations of the precautionary principle.

It was mentioned earlier that the general argument about forgone benefits (in contrast with the specific point about stifling innovation) applies to the entire landscape of formulations of the principle. This charge—that the principle suffers from tunnel vision, focusing on potential harm and ignoring harm that could come from imposing regulations—is by no means trivial. To address the concerns behind this criticism, proponents must appropriately reframe the principle. As it presently stands, the principle can be generally interpreted as being biased toward single-mindedly preventing harm, in advance of scientific certainty. It makes sense to worry that in focusing on addressing that harm, decision makers might fail to address any harm or forgone benefits that could follow from the regulatory action adopted.

The stifling innovation argument shows that the precautionary principle should avoid extremely severe directives such as categorical bans or denials. Even further, it should not demand that *any one* particular course of action be taken in every situation. Appropriate responses to potential risks are case sensitive. Advocates of the principle should clarify that the principle does not automatically ban everything that is potentially harmful.

Mistaken Priorities

Another criticism charges that the precautionary principle leads to mistaken priorities because it distracts attention from known, proven threats to health and the environment. This criticism stresses that we all face countless risks, from minor to major, from hypothetical to actual, and we have limited resources to handle them. Focusing time, energy, and resources on *hypothetical* risks of any magnitude causes harm by diverting resources from significant *actual* risks.[9] This charge cannot be dismissed as a misunderstanding; it requires a different response.

The charge of mistaken priorities rests on the claim that uncertain hypothetical threats are less important to address than definite known threats, such as cancer, AIDS, and ozone depletion. However, different principles can address different problems, and the precautionary principle is not responsible for addressing and weighing all priorities. Social institutions generally are established to address *particular* issues. The FDA addresses harms (both hypothetical and definite) from foods and drugs. The American Cancer Society addresses the threat of cancer. The Clean Air Act addresses harms from pollution. And so on. It would be silly to condemn the American Cancer Society for focusing only on cancer and thus detracting resources that could be applied to other serious problems, such as pollution. So why should one condemn the precautionary principle for choosing to allocate resources toward some serious problems rather than others?

Nevertheless, we still must decide where to direct resources, so setting proper priorities is vital. Users of the precautionary principle, especially on national and international levels, should check whether it would divert resources that would be better applied elsewhere. The European Commission (2000) has stressed that applications of the principle should be *proportional* to the potential threats (p. 18), and determining proportionality involves weighing the costs and benefits of action and inaction (p. 19).

It is an open question whether cost-benefit analysis is antithetical to the precautionary principle. Precaution is especially attractive to those who find the potential harms of new technologies more threatening than many actual familiar harms, even if this is not reflected in typical contemporary cost-benefit calculations. Proponents of using the precautionary principle should at least clarify what role they give to cost-benefit analysis, and how they would determine priorities and allocate resources.

Conclusion

One measure of the importance of the precautionary principle is the amount of critical attention it has received. This chapter has reviewed eleven criticisms that are commonly raised against it. Some of the criticisms can be dismissed because they are based on misunderstanding or otherwise misdirected; the superfluity and variability and vagueness criticisms fall into this category. Some other worries can be classified as legitimate but directed at *misuses* of the principle rather than the principle itself. This applies to the arguments of disguised protectionism, excessive regulation, and appealing to emotions.

The worst-case bias and unsound science arguments have another kind of force. They fundamentally hinge on honest differences of opinion and fundamental values. Whether one sympathizes with the precautionary principle's focus on worst-case scenarios is largely determined by one's level of risk aversion. Those who are more risk averse will be more sympathetic with precaution. Similarly, the plausibility of the charge of unsound science depends on one's opinions about what kinds of considerations should be given weight when making decisions.

Most of the remaining criticisms of the precautionary principle are serious problems only for certain forms of the principle. In particular, the criticisms based on self-contradiction, demanding the unknowable, forgone benefits, or stifled innovation are serious worries only for those versions of the principle with severe directives for action or broad or vague triggering conditions. Anyone who wants to defend such a formulation of the precautionary principle must be prepared to face and overcome these criticisms. Milder, narrower, or more specific principles can more easily deflect these worries. Thus, these criticisms block excessive precaution, but allow precaution that is moderate.

This leaves one more criticism: the charge that the precautionary principle leads to mistaken priorities. The natural response is to subject precautionary directives to a cost-benefit analysis, to ensure that they pass a test of proportionality, as the European Commission has suggested. Traditional scientific risk assessment might not offer much help for this kind of cost-benefit analysis, for risk assessment requires one to map out the possible consequences of different courses of action, but the precautionary principle arises precisely when those consequences are especially uncertain (Bedau & Triant, ch. 3, this volume). There is no accepted recipe for appropriately weighing costs and benefits in contexts of great uncertainty. Compelling appeals to the precautionary principle will require wisdom and good judgment.

New technologies like protocells will have highly uncertain consequences and the potential to create both significant harms and significant benefits. The prospect of protocells provides good reason to seek practical versions of this intuitively appealing but vague and controversial decision-making principle. Applications of the precautionary principle will be persuasive only if combined with enough specificity about the triggering conditions, acceptable uncertainty, and directives for action. Plausible principles must not be too vague, be triggered by conditions that are too broad, or require actions that are too severe. Appeals to precaution can be compelling, provided appropriate care and good judgment are devoted to weighing the costs and benefits of proposed directives, and the fundamental values driving precaution are clarified and justified.

Notes

1. See, e.g., Conference of the Parties to the Convention on Biological Diversity (2000), United Nations (1982, 1992a, 1992b, 1992c), Ministry of the Environment (1984, 1987), and OSPAR Commission (1992), to name just a few. See the Municipal Code Corporation (2006) for a recent example from municipal policy in California.

2. It is possible to make further distinctions and analyses of the general formulation of the precautionary principle (Sandin, 1999; Applegate, 2000; European Commission, 2000; Jones, 2000; Foster et al., 2000; Barrett, 2001; Raffensperger & Barrett, 2001; Parke, 2004).

3. Our analysis of the precautionary principle follows Sandin's analysis (1999), except that we find it more convenient to combine two of Sandin's four "dimensions" of the precautionary principle in the directive for action.

4. See, in particular, Gilland (2000) for a discussion of citing the precautionary principle as rationale for a moratorium on genetically modified crops in the UK, with next to no scientific evidence to justify concerns about harm they might cause.

5. For example, Quijano (2003) argues that there are irreconcilable differences between the precautionary principle and conventional notions of science that normally guide policy, and that this should be solved by rejecting the latter in favor of the former. Others have argued that precautionary policy, though departing from the conventional models of scientific regulation, by no means abandons sound science. The principle may conflict with what some critics consider "sound" science, but it aspires to be guided by a broader, more inclusive view of which scientific information is relevant to a decision. Precautionary policy creates opportunities for scientists to rethink how they conduct their studies and communicate their results (Tickner, 2003).

6. Barrett and Raffensperger (1999) outline one useful framework for thinking about these two kinds of science, and argue that the science advocated by the precautionary principle is an improvement on, rather than an antagonist of, the conventional risk-based approach critics advocate. They distinguish between "mechanistic science," which is characterized by specialization, reductionism, separation from social issues, quantitative experimental data, and "precautionary science," which is characterized by a multidisciplinary and inclusive approach, looking to qualitative and experiential as well as quantitative experimental data. The latter sort of science, they argue, incorporates a broader perspective and confronts uncertainty and complexity head-on.

7. Cross (1996) has extensively developed this argument about harm from forgone benefits and stifled innovation. He maintains that the precautionary principle's "truly fatal flaw" is the unsupported presumption that actions aimed at public health and environmental protection will not have negative effects on either (pp. 859–860). He identifies three categories of risk associated with application of the principle: (1) risks from alternatives wherein, for example, the decision to ban a product based on suspicion results in the introduction of an even more harmful alternative; (2) forgone benefits; and (3) risks of remediation, that is, unforeseen negative effects of precautionary action. See also Goklany (2000), who argues that applying the principle to ban or stall production of GM crops based on concerns about the uncertainty involved and potential risk to health or the environment would in fact increase overall risk to both.

8. Critics who have argued that the principle stifles innovation typically take the precautionary principle to be extremely severe (see, e.g., Holm & Harris, 1999, p. 398; Miller & Conko, 2000, p. 84; Wildavsky, 2000, p. 22f).

9. Cross (1996) distinguishes two types of indirect risk. The first is resource misallocation, for example, taking resources away from things we know are integral to well-being, like vaccinations and day care, in favor of precautionary action against risks that are only

potential. The second type is economic risk, for example, the costs of regulating a material ultimately borne not by the corporations producing it, but by consumers (in the form of increasing prices) and workers (in the form of reduced wages). See also Goklany (2000) for discussion of negative economic effects of precautionary actions to reduce greenhouse gas emissions.

References

Applegate, J. S. (2000). The precautionary preference: An American perspective on the precautionary principle. *Human and Ecological Risk Assessment, 6* (3), 413–443.

Barrett, K. (2001). Applying the precautionary approach to living modified organisms. Presented at the Intergovernmental Committee for the Cartagena Protocol on Biosafety, Montpellier, France (December 2000). Available online at: http://www.sehn.org/biotech.html (accessed September 2007).

Barrett, K., & Flora, G. (2000). Genetic engineering and the precautionary principle: Information for extension. Science and Environmental Health Network publication. Available online at: http://www.sehn.org/biotech.html (accessed September 2007).

Barrett, K., & Raffensperger, C. (1999). Precautionary science. In C. Raffensperger and J. Tickner (eds.), *Protecting public health and the environment: Implementing the precautionary principle* (pp. 106–22). Washington, DC: Island Press.

Bewers, J. M. (1995). The declining influence of science on marine environmental policy. *Chemistry and Ecology, 10,* 9–23.

Bodansky, D. (1991). Scientific uncertainty and the precautionary principle. *Environment, 33* (7), 4–5, 4344.

Comstock, G. (2000). Are the policy implications of the precautionary principle coherent? Available online at: http://www.lifesciencesnetwork.com/repository/defining_pp.pdf (accessed August 2008).

Conference of the Parties to the Convention on Biological Diversity (2000). *Cartagena protocol on biosafety.* Supplementary agreement to the Convention, adopted January 29. Available online at: http://www.biodiv.org/biosafety/protocol.asp (accessed September 2007).

Cross, F. B. (1996). Paradoxical perils of the precautionary principle. *Washington and Lee Law Review, 53,* 851–925.

Durodié, B. (2000). Plastic panics: European risk regulation in the aftermath of BSE. In J. Morris (Ed.), *Rethinking risk and the precautionary principle* (pp. 140–166). Oxford: Butterworth-Heinemann.

European Commission (2000). *Communication from the commission on the precautionary principle.* Available online at: http://portal.unesco.org/shs/en/ev.php-URL_ID=6615&URL_DO=DO_TOPIC&URL_SECTION=201.html (accessed September 2007).

Foster, K., Vecchia, P., & Repacholi, M. H. (2000). Science and the precautionary principle. *Science, 12,* 979–981.

Gilland, T. (2000). Precaution, GM crops and farmland birds. In J. Morris (Ed.), *Rethinking risk and the precautionary principle* (pp. 60–83). Oxford: Butterworth-Heinemann.

Goklany, I. M. (2000). Applying the precautionary principle in a broader context. In J. Morris (Ed.), *Rethinking risk and the precautionary principle* (pp. 189–228). Oxford: Butterworth-Heinemann

Gray, J. S. (1990). Statistics and the precautionary principle. *Marine Pollution Bulletin, 21,* 174–176.

Hammit, J. K. (2000). Global climate change: Benefit-cost analysis vs. the precautionary principle. *Human and Ecological Risk Assessment, 6,* 387–398.

Holm, S., & Harris, J. (1999). Precautionary principle stifles discovery. *Nature, 400* (July), 398.

Jones, P. B. C. (2000). Implementing the precautionary principle. Science, Technology and Innovation Program, Harvard Center for International Development. Available online at: http://www.biotech-info.net/BPCJ_viewpoint.html (accessed September 2007).

Kastner, J. (2002). General possibilities: The precautionary principle again. Commentary from the Food Safety Network. Available online at: http://www.foodsafetynetwork.ca/en/article-details.php?a=3&c=17&sc=133&id=493 (accessed September 2007).

Lieberman, A., & Kwon, S. (2004). *Facts versus fears: A review of the greatest unfounded health scares of recent times.* Available online at: http://www.acsh.org/publications/pubID.154/pub_detail.asp (accessed September 2007).

Miller, H., & Conko, G. (2000). Genetically modified fear and the international regulation of biotechnology. In J. Morris (Ed.), *Rethinking risk and the precautionary principle* (pp. 84–104). Oxford: Butterworth-Heinemann.

Ministry of the Environment (1984). *First International Conference on Protection of the North Sea: Bremen declaration.* Bremen, October 31–November 1. Available online at: http://www.seas-at-risk.org/n2_archive.php?page=9&PHPSESSID=9a9e58a56400b09659375ab8395d618a (accessed September 2007).

Ministry of the Environment (1987). *Second International Conference on the Protection of the North Sea: London declaration.* London, November 24–25. Available online at: http://www.seas-at-risk.org/n2_archive.php?page=9&PHPSESSID=9a9e58a56400b09659375ab8395d618a (accessed September 2007).

Morris, J. (Ed.) (2000). *Rethinking risk and the precautionary principle.* Oxford: Butterworth-Heinemann.

Municipal Code Corporation (2006). *City and county of San Francisco municipal code.* Available online at: http://www.municode.com/Resources/gateway.asp?pid=14134&sid=5 (accessed September 2007).

OSPAR Commission for the Protection of the Marine Environment of the North-East Atlantic (1992). *1992 OSPAR convention.* Ministerial meeting of the Oslo and Paris Commissions, Paris, September 21–22. Available online at: http://www.ospar.org/eng/html/convention/welcome.html (accessed September 2007).

Parke, E. (2004). *An analysis and evaluation of the precautionary principle.* Portland, OR: Reed College, BA Thesis.

Quijano, R. F. (2003). Elements of the precautionary principle. In J. Tickner (Ed.), *Precaution, environmental science, and preventive public policy (*pp. 21–27). Washington, DC: Island Press.

Raffensperger, C., & Barrett, K. (2001). In defense of the precautionary principle. *Nature Biotechnology, 19* (9), 811–812.

Rubin, C. T. (2000). Asteroid collisions and precautionary thinking. In J. Morris (Ed.), *Rethinking risk and the precautionary principle* (pp. 105–126). Oxford: Butterworth-Heinemann.

Sandin, P. (1999). Dimensions of the precautionary principle. *Human and Ecological Risk Assessment, 5,* 889–907.

Santillo, D., Johnston, P., & Stringer, R. (1999). The precautionary principle in practice: A mandate for anticipatory preventative action. In C. Raffensperger and J. Tickner (Eds.), *Protecting public health and the environment: Implementing the precautionary principle* (pp. 36–50). Washington, DC: Island Press.

Saunders, P. (2000). Use and abuse of the precautionary principle. Available online at: http://www.biotech-info.net/precautionary_use-and-abuse.html (accessed September 2007).

Tickner, J. (Ed.) (2003). *Precaution, environmental science, and preventive public policy.* Washington, DC: Island Press.

Tickner, J., Raffensperger, C., & Myers, N. (1998). *The precautionary principle in action: A handbook.* Science and Environmental Health Network. Available online at: http://www.biotech-info.net/handbook.pdf (accessed August 2008).

United Nations (1982). *World Charter for Nature.* October 28. Available online at: http://www.un.org/documents/ga/res/37/a37r007.htm (accessed September 2007).

United Nations (1992a). *Convention on Biological Diversity.* Product of the United Nations Conference on Environment and Development, Rio de Janeiro, June 3–14. Available online at: http://www.biodiv.org/convention/articles.asp?lg=0&a=cbd-00 (accessed September 2007).

United Nations (1992b). *Convention on the Protection and Use of Transboundary Watercourses and International Lakes.* Product of the Helsinki Conference, March 17. Available online at: http://www.unece.org/env/water/text/text.htm (accessed September 2007).

United Nations (1992c). *Rio Declaration on Environment and Development.* Product of the United Nations Conference on Environment and Development, Rio de Janeiro, June 3–14. Available online at: http://www.unep.org/Documents/Default.asp?DocumentID=78& ArticleID=1163 (accessed September 2007).

United Nations (1995). *Agreement for the implementation of the provisions of the United Nations Convention on the Law of the Sea of 10 December 1982 relating to the conservation and management of straddling fish stocks and highly migratory fish stocks.* Product of the United Nations Conference on Straddling Fish Stocks and Highly Migratory Fish Stocks, sixth session, July 24–August 4. Available online at: http://www.oceanlaw.net/texts/unfsa.htm (accessed September 2007).

Whelen, E. M. (1996). Our "stolen future" and the precautionary principle. *Priorities for Health, 8,* 3.

Wildavsky, A. (2000). Trial and error versus trial without error. In J. Morris (Ed.), *Rethinking risk and the precautionary principle* (pp. 22–45). Oxford: Butterworth-Heinemann.

World Trade Organization (2000). *The WTO Agreement on the application of sanitary and phytosanitary measures.* Available online at: http://www.wto.org/English/tratop_e/sps_e/spsagr_e.htm (accessed September 2007).

Yandle, B. (2000). The precautionary principle as a force for global political centralization: A case-study of the Kyoto protocol. In J. Morris (Ed.), *Rethinking risk and the precautionary principle* (pp. 167–188). Oxford: Butterworth-Heinemann.

6

A New Virtue-Based Understanding of the Precautionary Principle

Per Sandin

In recent years, the idea of artificial life has become less fiction and more science (Rasmussen et al., 2004). By *artificial life*, I mean artificial entities that continually regenerate and replicate themselves, and are capable of undergoing evolution. The phenomenon of life is more complicated than this (Deamer, 2005), but I will not discuss that issue here. Simple cell-like entities produced in the laboratory are called protocells. Protocell research, that is, research working toward understanding and producing protocells, is currently at the forefront of artificial life research. However, protocells have not yet been created. The discussion in this chapter applies to all forms of artificial life research; thus, in this context, *protocell research* might be understood as shorthand for "research into artificial life in general."

Like gene technology, protocell research raises a host of ethical questions. What benefits and risks are involved, and how should they be distributed? Are scientists "playing God," and is that a reason for banning the research? Is there a possibility that protocells could, in some way, cause a major health or environmental disaster? Should there be a moratorium on protocell research?

In discussions on the ethics of such new technologies with uncertain but potentially frightening consequences, the precautionary principle is often invoked. This is true of gene technology (Myhr & Traavik, 2003) and a number of other technologies as well. Protocell research is a case in point.

However, the precautionary principle is notoriously controversial. Some of this controversy is due to the fact that it has been interpreted in a number of ways. Is the precautionary principle an intuitively simple, commonsense idea of "when in doubt, don't" or "better safe than sorry" transformed into regulation or policy? Or is it rather "a neologism coined by opponents of technology who wish to rationalize banning or over-regulating things they don't like" (Miller & Conko, 2000, p. 95)? This confusion has led some to conclude that "there is no such thing as 'the' precautionary principle" (Graham, 2000, p. 383).

Does the precautionary principle tell us whether we should proceed with protocell research or not? Or is it merely a useful tool to be employed by people who dislike protocell technology and wish to rationalize banning it? In this chapter, I discuss three types of precautionary principles presented in the literature, and note that there are problems with each version. Thereafter I present a different, and to my knowledge new, way of understanding the precautionary principle that avoids many of the problems of the standard versions. This new way of understanding the precautionary principle is based on precaution conceived of as a virtue, and the corresponding principle is thus *a virtue principle of precaution*.

Types of Precautionary Principles

Consider the following two versions of the precautionary principle:

In order to protect the environment, the precautionary approach shall be widely applied by States according to their capabilities. Where there are threats of serious or irreversible damage, lack of full scientific certainty shall not be used as a reason for postponing cost-effective measures to prevent environmental degradation. (UNCED, 1993)

[T]he precautionary principle, that is to take action to avoid potentially damaging impacts of substances that are persistent, toxic and liable to bioaccumulate even when there is no scientific evidence to prove a causal link between emissions and effects. (cited in Haigh, 1994, p. 244)

The first example is from Principle 15 of the 1992 Rio Declaration, and the second is from the Third Conference on the North Sea, The Hague, in 1990. The Rio Declaration applies to the environment in general; the North Sea Conference version addresses toxic substances in particular. Versions of the precautionary principle that are similar in wording to the ones quoted have been included in a number of legal and policy documents over the years. The Rio version occurs almost verbatim in several other documents, sometimes called *precautionary principle*, sometimes *precautionary approach*. One example is the Cartagena Protocol on Biosafety (Secretariat of the Convention on Biological Diversity, 2003).

Let us return to the two versions of the precautionary principle. They are notably different. The Rio version states what types of reasons for postponing action should be considered legitimate. The North Sea Conference version, on the other hand, talks of actions to be carried out rather than reasons. There are numerous other versions as well, and despite extensive discussions over the last fifteen years or so, disagreement still exists about what the precautionary principle means, should mean, or might reasonably mean. Critics of the precautionary principle have indeed noted this, and a common complaint has been that it is ill-defined or vague, perhaps so vague that it is useless (Sandin et al., 2002; chapter 5, this volume).

Thus, unsurprisingly, philosophers and other philosophically minded commentators have attempted to clarify the matter. One approach has been lexicographical. Such approaches are partly empirical, and aim at identifying a "core" in different formulations of the precautionary principle. Examples are Sandin (1999), Arcuri (2004), and Sunstein (2005, pp. 119f). There is much to be said of such approaches. One author ridicules what she terms "precaution spotting" and "mining" of examples of precaution (Fisher, 2002).

Nevertheless, the attempts at clarifying the precautionary principle have yielded suggestions that can be categorized in at least three groups. According to the first, the precautionary principle is a *rule of choice* (Sunstein, 2005), that is, an algorithm for choosing a particular action or set of actions in a given situation. The Wingspread version of the precautionary principle cited in the next section is a good example of this. According to the second group, it is an *epistemic rule or principle,* that is, a principle about knowledge, concerning what it is reasonable to believe, or what is to qualify as knowledge (Peterson, 2007). According to the third group, the precautionary principle is a *procedural requirement* (Fisher, 2002; Arcuri, 2004). Such a precautionary principle would not be an algorithm for choosing particular courses of action, but rather a requirement for how such decisions are to be made. One example of such a procedural version of the precautionary principle is the Rio Declaration version cited earlier.

These three groups are not entirely distinct, and it may be possible to combine them. Nevertheless, the distinctions matter. This is because all three of them are problematic, and it has been argued that we should replace one version of the precautionary principle with another in order to avoid problems with the former version. For instance, Peterson (2007) suggests that we should abandon the precautionary principle as a rule of choice in favor of an epistemic version.

I will treat these versions of the precautionary principle in turn.

Rules of Choice

Interpreted as a rule of choice, the precautionary principle states that in particular types of situations, certain courses of action should be chosen. A rule of choice, here, is an algorithm for choosing a particular action or set of actions in a given situation—for instance, a rule that prescribes or prohibits certain actions.

The Wingspread Statement is an example of a version of the precautionary principle that is reasonably interpreted as a rule of choice:

When an activity raises threats of harm to human health or the environment, precautionary measures should be taken even if some cause-and-effect relationships are not fully established scientifically. (Raffensperger & Tickner, 1999, pp. 354–355)

Many authors have taken such versions of the precautionary principle to be saying something like the following:

All actions that might have harmful consequences are forbidden.

This interpretation makes the precautionary principle a consequentialist principle demanding action on the mere possibility of a consequence occurring. Such rules of choice are problematic. There are always some cause-and-effect relationships that are not fully established scientifically. Every action *might* raise threats to human health and the environment, if scenarios that are far-fetched enough are taken into consideration. Such a principle will prohibit every action, including the action of taking precautionary measures. Perhaps, through an unlikely but possible causal chain, I will cause a global disaster by picking my nose. Decision makers will thus be paralyzed. According to this argument, the precautionary principle is self-refuting or incoherent: "The real problem is that the principle offers no guidance—not that it is wrong, but that it forbids all courses of action, including regulation" (Sunstein, 2005, p. 26).

Suppose now that someone argues, with reference to the precautionary principle, that protocell research should be stopped because such research might, through some possible but extremely unlikely causal chain, result in the end of all natural life on Earth. Call this the *Mutant Protocell Scenario*. It might then be countered that there is an alternative scenario, also possible, in which *not* proceeding with protocell research leads to the extinction of all natural life on Earth. Artificial life forms may, possibly, provide the only defense against the *Alien Space Invaders*. Thus, given that both the Alien Space Invader Scenario and the Mutant Protocell Scenario are possible, we should, according to the precautionary principle, both proceed and not proceed with protocell research.

It should be noted that a similar problem remains even if the principle is given a less extreme formulation than in this example. Consider the following:

It is recommended that we abstain from all actions that might have harmful consequences.

Here, all actions are not prohibited, we are merely recommended to abstain from them. But the problem here is that we are offered no action guidance at all.

This problem has been noted by a number of commentators over the years. In fact, the argument is older than the phrase "precautionary principle." In one example, from the early days of DNA technology, Stich (1978) criticizes what he calls the "doomsday scenario argument" (p. 189). Stich himself, like other authors, mentions Pascal's Wager and the Many Gods Objection (Manson, 2002). Pascal's Wager is the argument for belief in God given by the French seventeenth-century philosopher and mathematician Blaise Pascal. According to Pascal's argument, God either exists or does not exist. Furthermore, we can choose either to believe in God or not to believe in God. If we believe in God and He exists, we will go to heaven. This is, supposedly, infinitely valuable. On the other hand, if we believe in God, and he does not exist, we will have lost very little, if anything. Given that going to heaven is so valuable, the mere *possibility* that He exists justifies belief in Him.

However, there is a problem with the argument. The situation in the Wager is insufficiently described. Consider not only the Christian God, but also some vengeful and jealous god (say, Baal). Suppose that God does not exist, but Baal does, and is set on throwing all who believe in the *Christian* God into Hell, which supposedly is infinitely bad, just as going to heaven is infinitely good. What do we do? Given that there might be different gods with different preferences regarding our belief in the Christian God, Pascal's argument would urge us both to believe and not to believe in God. This is the so-called Many Gods Objection to Pascal.

The precautionary principle poses a parallel problem to that of Pascal's Wager. If we are to act on mere possibilities, albeit of very great harm, we will have to face the Many Gods Objection, or rather the Many Threats Objection, as in the case of the Mutant Protocell Scenario versus the Alien Space Invaders.

Thus, the problem is a very real one for proponents of the precautionary principle as a rule of choice. There may be ways of circumventing the problem, in order to save the principle as a rule of choice, for instance, through restricting its sphere of application. I have argued so elsewhere in response to this criticism of the precautionary principle (Sandin, 2006; cf. Harris & Holm, 2002; Holm, 2006). However, I will not pursue the issue further here. Instead I will mention the two other ways of making sense of the precautionary principle. First, I consider it as an *epistemic* principle.

Epistemic Rules or Principles

Epistemic principles are principles about knowledge, with *epistemology* being the theory of knowledge. Epistemic principles thus concern what it is reasonable to believe, or what is to qualify as knowledge.

It has been suggested that the precautionary principle can be interpreted as an epistemic principle. There are different ways of doing this, one of which Harris and Holm (2002) discuss and dismiss. According to Harris and Holm, an epistemic precautionary principle requires that evidence suggesting a causal link between an activity and possible harm should be given greater weight than it would otherwise have been given. However, as they write, "systematic discounting of evidence would systematically distort our beliefs about the world, and would necessarily, over time, lead us to include a large number of false beliefs in our belief system" (p. 362). I think they are mainly correct in this. Thus, I do not think that the precautionary principle is reasonably interpreted as an epistemic principle.

An additional reason for rejecting epistemic versions of the precautionary principle is that it would be strange to prescribe what to believe on the basis of nonepistemic consequences. It would be more reasonable to prescribe that we should act *as if* certain things were true, pending further information. This is especially the case in policy. Policy, fortunately, rarely tells us what to believe.

One such "as if" version of the precautionary principle is the precautionary default approach discussed in Sandin et al. (2004). A precautionary default is a cautious or pessimistic assumption that is used in the absence of adequate information, and is to be replaced when such information is obtained. But that is not an epistemic principle. It is, in fact, more like the third group, namely, the precautionary principle interpreted as a procedural requirement.

Procedural Versions

Procedural versions of the precautionary principle are not algorithms for choosing particular courses of action. Rather, they are requirements for what procedures to use when making decisions: What arguments are to be considered legitimate? Who is to be heard in the debate? On whom does the burden of proof fall? And so on.

This might be termed procedural precaution, reflected in procedural versions of the precautionary principle (Arcuri, 2004). For example, Fisher (2002) points out that the precautionary principle is a legal principle, in Dworkin's sense of a principle that "states a reason, that argues in one direction, but does not necessitate a particular decision" (Dworkin, 1978, p. 26). It is not a "rule that dictates a particular outcome in a certain set of circumstances" (Fisher, 2002, p. 15). One example of a procedural version of the precautionary principle is the version from the Rio Declaration cited previously.

Procedural versions of the precautionary principle constitute a motley group. There seems to be precious little that unites different versions of precaution, and any version is difficult to isolate. Procedural versions of the precautionary principle are intermingled with other legal principles, rules, questions of legitimate authority, and other procedural requirements. As Fisher (2002, p. 15) points out, "the discussion so far has tended to characterize the precautionary principle in a way that is at odds with the actual legal nature of the precautionary principle and how it operates in particular contexts."

The problem with procedural versions of the precautionary principle is that there are so many possible varieties. Sunstein suggests that "[f]or every regulatory tool, there is a corresponding precautionary principle" (2005, p. 120). Examples mentioned are a "Funding More Research Precautionary Principle," an "Economic Incentives Precautionary Principle," and an "Information Disclosure Precautionary Principle." From this plethora of options, it is hard to see that anything like *the* precautionary principle can be isolated. There are, at best, family resemblances among all different procedural requirements that have been thought to be instances of *the* precautionary principle. This does not mean, of course, that such principles cannot be rationally discussed and evaluated. But this has to be done on a case-by-case basis.

The Virtue Approach

As we have seen, all three types of precautionary principle have problems. I do not think that the criticisms are necessarily decisive. Perhaps the various precautionary principles can be amended to circumvent the problem. For instance, some of the objections might be met by restricting the sphere of application of the precautionary principle.

However, such discussions have been underway for a number of years now, and controversy remains. I will therefore consider a different possible approach to precaution and the precautionary principle—the virtue approach. The hope is that by conceiving of precaution as a virtue, we can construct a virtue principle of precaution. This is, compared to previous discussions on precaution, working backwards. Let me explain.

At one time I argued that it is reasonable to analyze precaution in terms of what it means for an *action* to be precautionary (Sandin, 2004, 2007a). I argued, first, that precautionary actions are precautionary only with respect to something, and, second, that three criteria have to be fulfilled for an action to be classified as precautionary (as opposed to, e.g., preventive): intentionality, uncertainty, and epistemic reasonableness. Taken together, these criteria state that

an action A is precautionary with respect to something undesirable U, if and only if

(1) A is performed with the intention of preventing U,

(2) the agent does not believe it to be very likely that U will occur if A is not performed, and

(3) the agent has externally good epistemic reasons (a) for believing that U might occur, (b) for believing that A will in fact at least contribute to the prevention of U, and (c) for not believing it to be certain or very likely that U will occur if A is not performed.

But why analyze precautionary actions? Is it not at least as interesting to know what makes a decision precautionary or a person cautious?

My primary reason for choosing precautionary actions as the subject of analysis was what might be called *analytical neatness*. A person's being cautious and a decision's being precautionary can be analyzed in terms of precautionary actions: A decision is precautionary if, and only if, it consists in choosing a precautionary action, and a person is cautious if, and only if, he or she typically performs precautionary actions. Thus, an analysis of precautionary actions, arguably, gives us an analysis of precautionary decisions and cautious persons as well. But the analysis could be reversed. A precautionary action could be defined as the action that a cautious person would have carried out in the given circumstances. This is, in effect, the virtue approach. In the remainder of this chapter, I will sketch such an approach to the precautionary principle.

Cautiousness as a Virtue

How does one recognize a virtue? There are numerous conceptions of what a virtue is. A starting point, however, which may at least provide minimal criteria for identifying a virtue is Philippa Foot's seminal essay *Virtues and Vices* (Foot, 1997). She characterizes virtues roughly in the following way:

(i) Virtues are *beneficial characteristics* that a human being needs to have for his own sake and that of his fellows.

(ii) They have to actually engage *the will*, and are thus to be distinguished from *skills*.

(iii) They are *corrective*, in the sense that they are about what is difficult for humans in general (not necessarily for a particular individual).

These three characteristics can be operationalized into a "test" for whether a trait is a virtue. Let us call this the *Footian test*, and let us see whether cautiousness passes the Footian test.

In the Footian test, virtues are assumed to be characteristics of humans. In the following, I will be less specific, and leave open the possibility that virtues may be ascribed to other entities as well, such as corporations. This is a somewhat controversial assumption (cf. Schudt, 2000). I believe, however, that there are good reasons for adopting it, but this is not the place to argue this point (I do so elsewhere; see Sandin, 2007b; cf. also French, 1995, pp. 79ff.). We might also note that the term *cautious* is frequently used as a virtue term in everyday language, in the sense that it is predicated of persons (or agents).

To return to the Footian test, regarding (i), whether the character trait of cautiousness is beneficial, the answer is probably yes. Using most understandings of cautiousness, it would seem to be beneficial to the cautious agent and to his or her fellows. One observation that supports this is that critics of the precautionary principle sometimes seem to agree that precaution is a good thing. What they question is whether measures proposed by precautionary principle supporters really are precautions rather than, say, undercover trade protectionism. (I elaborate on this in Sandin, 2007a.)

Regarding (ii), we may apply Foot's test for distinguishing virtues from skills (Foot, 1997, p. 169). Her point is that someone can choose not to exercise a skill, without leading us to conclude that the skill is lacking. Consider the skill of performing an eskimo roll while kayaking. I possess that skill, but I may choose not to exercise it. I might, for instance, perform a failed roll to show onlookers a common error. This does not count against my rolling skill. If someone accuses me of being a poor eskimo roller, the statement that "I did it deliberately" rebuts the accusation. (Whether people believe me is another matter.) It is different with virtues. An accusation of lack of virtue cannot be rebutted in this way. Suppose someone accuses me of lacking the virtue of courage. In this case, claiming that I deliberately failed to be courageous does not neutralize the accusation.

And what about cautiousness? Suppose Henriette deliberately performs an action that is contrary to cautiousness, say, steps out into a busy street pushing a baby carriage without looking. (Here I am making the rather weak assumption that the action just described is not compatible with cautiousness.) I accuse Henriette of lacking cautiousness. She replies: "No, I did it deliberately." Would that rebut my accusation? I am strongly inclined to say no. Thus, it seems that cautiousness does in fact display the characteristic (ii) of engaging the will.

Let us turn to (iii), whether cautiousness is *corrective*. Is it about what is difficult for humans in general? There is no straightforward answer to this question. There is no obvious corresponding temptation. In what way could being cautious be difficult? There are at least two ways. First, being cautious does probably require some more thinking and planning than being incautious does. Arguably, a cautious person probably considers the available options and possible consequences more than someone not so disposed. Being cautious might be cumbersome. Second, cautiousness might mean that some possible rewards are forgone in the interest of safety. The answer to the question whether cautiousness is corrective is thus probably yes.

Thus, cautiousness passes the Footian test, though perhaps not by a large margin.

The Thickness of Precaution

There is another reason why it might be fruitful to think of precaution as a virtue. This reason is the *thickness* of the concept of precaution.

Traditionally, a distinction is made between a concept's descriptive (factual) content and its evaluative content. Some concepts are purely descriptive, that is, they have no evaluative content at all. An example is the concept "red." Redness, in itself, is not a value-laden concept, but only describes a fact. Some evaluative concepts contain both factual and evaluative elements. For instance, if someone applies the concept *greediness* to me, she says something factual (about my behavior and preferences toward money), but she also says something evaluative of me, namely, that there is something wrong with the way I behave and the attitudes I have.

Roughly speaking, an evaluative concept is thin if it has little descriptive content. The obvious example is the concept *good*, which seems purely evaluative. An evaluative concept is thick if it has a large descriptive content. Examples of such concepts are the ones referred to by virtue terms, such as *courageous, benevolent, honest,* and the like. The list of thick evaluative concepts could be extended almost indefinitely. And, I would argue, *cautious* should be added to it, at least tentatively. If an ethical concept is thick, then, as a rule of thumb, one could try to conceive of it as a property of an agent, that is, as a virtue or a vice, in order to see whether more sense can be made of it that way. This is not to say that all thick evaluative concepts are virtues or vices, of course. But conceiving of them as virtues or vices may be a good starting point for further inquiries, until we have reasons to believe they are not, in fact, virtues or vices.

The descriptive content of thick evaluative concepts is notoriously controversial. And if there is controversy regarding how an evaluative concept should be applied, this is a reason for suspecting it is a thick concept. This is the case for precaution. One author has even argued that "[p]recaution might join the class of essentially contested concepts" (Breyman, 1999). I believe this is an exaggeration, but the statement nevertheless highlights the ubiquitous controversy regarding the content of precaution, as reflected in the debates around the precautionary principle and the problem of its definition.

A Virtue Principle of Precaution

As we have seen, precaution is a thick concept and it passes the Footian test. Let us now ask whether the virtue conception of precaution could be used for formulating a precautionary principle. Tentatively, I think it is possible. Consider the following *virtue principle of precaution*:

Perform those, and only those, actions that a cautious agent would perform in the circumstances.

This is the precautionary principle as a rule of choice, but reinterpreted using cautiousness as a virtue. Quite possibly, epistemic and procedural versions of the precautionary principle could also be constructed in a similar fashion. For instance, a virtue version of an epistemic precautionary principle might be, "believe what a cautious agent would believe in the circumstances." A procedural version would be, perhaps, "admit only those arguments that a cautious agent would admit." I think epistemic versions should be rejected for a reason mentioned earlier: There is something strange about a principle prescribing what to believe on the basis of nonepistemic considerations. Procedural virtue-based precautionary principles might fare better. However, here I limit my discussion to the rule-of-choice version of the precautionary principle, not only for reasons of space, but also because rule-of-choice precautionary principles are the ones that have received the most attention in the recent debate.

Let us return to the virtue principle of precaution. Is it in any way better than the usual rule-of-choice versions of the precautionary principle? And can such a principle be tenable at all? Three immediate criticisms should be commented on.

First, the virtue principle of precaution can be criticized for not being action guiding. How are we to know what to do? This is a standard critique of virtue ethics (Hursthouse, 1999, ch. 1). However, in some cases, it will be quite obvious what a cautious agent would (and would not) do in the circumstances. In other cases it will be less obvious, but still possible to reach some sort of agreement. Thus, the virtue principle of precaution is not completely action guiding, but at least *partly* action guiding. And this, we might remember, is a lot better than what is the case for some versions of the precautionary principle as a rule of choice: "The real problem is that the principle offers no guidance—not that it is wrong, but that it forbids all courses of action, including regulation" (Sunstein, 2005, p. 26; cf. Peterson, 2006). It could be objected that a similar problem arises with the virtue principle of precaution: Could not a cautious agent reject action *A* as too risky, and at the same time reject action not-*A*, on the same grounds? So if we are to do as the cautious agent would, we are left paralyzed. This objection is mistaken, for two reasons. First, even if such situations could arise, there is nothing that says they will arise. And even if such dilemmas did arise, the virtue principle of precaution would still be partially action guiding, namely, in those situations in which such dilemmas do not arise. Second, and more important, the objection amounts to rejecting cautiousness as a virtue. Remember that one condition for a characteristic's being a virtue is that it is beneficial to the agent and to his or her fellows. If such a characteristic frequently puts the agent in situations where all options are deemed impermissible, the criteria of being beneficial would not be fulfilled. Incidentally, this is compatible with the idea of virtues as "golden means." Virtues are often conceived of as character traits occupying a middle ground between excess and deficit. Displaying the virtue of courage, for instance, means, among other things, proper balancing between deficit (cowardice) and excess (overboldness).

The second criticism of the virtue principle of precaution is that virtues and virtuous agents are unsuitable for policy making and regulation. What we need, it might be argued, are clear, unambiguous rules and principles, telling us what to do and not do, and allowing us to sanction those agents who fail to abide by the rules. Put differently, a virtue principle of precaution, being only partly action guiding, lacks transparency in implementation.

This is true. However, virtuous agents are already doing work in regulatory and policy contexts. One example is the *bonus pater familias* figure. This figure—the good family father—occurs, or at least occurred, in civil and family law. The general idea is to judge real people's actions against the standards of the *bonus pater familias*.

A perhaps better example is the concept of good seamanship. Every vessel, including a small pleasure boat, should be handled in accordance with good seamanship. But what is good seamanship? Certainly not something that can be completely enumerated in a set of clear and unambiguous principles, though such principles might play a *part* in good seamanship. Instead, good seamanship must be understood in terms of what a good seaman—or perhaps a good crew—would do, and would not do, in the circumstances. Alasdair Macintyre makes precisely this point: "The types of action required by a particular virtue can never be specified exhaustively by any list of rules. But failure to observe certain rules may be sufficient to show that one is defective in some important virtues" (1999, p. 109).

It may be argued that the good seaman and the good family father are highly dissimilar to the cautious agent introduced in my virtue version of the precautionary principle. First, family fathers and seamen or crews do have comparatively well-defined roles, against the background of which their virtues can be understood. Second, seamen and family fathers are entities with which we have experience. Most people have seen instances of better and worse family fathers, and also been exposed to numerous examples of evaluations of fathers' behavior. Regarding seamen, this is true of fewer people. Nevertheless, for those familiar with the seaman's tasks, they are in most cases able to judge whether a certain sailor's behavior was in accordance with good seamanship, that is, if he or she acted as a good seaman would have acted, in the given circumstances.

Against this I will argue that we *do* have ideas of what cautious people would and would not do in certain circumstances. We recognize a cautious person when we meet one. We may not always be sure, but that holds for good seamen and family fathers as well.

The third, related criticism of the virtue principle of precaution is that the precautionary principle is to apply in situations of which we have very little, if any, experience, and consequently, we are not able to say what a cautious agent would do in those particular circumstances, even if we know how cautious agents typically would act. For example, the virtue principle of precaution should be applicable to issues relating to new technologies. Take the example of protocell technology. We have no experience with introducing protocells into the world. We are not able to say what a cautious agent would (and would not) do in that particular case.

There are two replies to this. First, part of the problem regards the description of the situation. It might not be fair to describe it as one with which we have no experience. In fact, it may belong to a particular class of situations, and we may in fact have vast experience with that class. This, I believe, is the underlying idea behind a work such as Harremoës et al. (2002). They list a number of mostly historical examples where, as they claim, "early warnings" were present, and where the precautionary principle would have been applicable. Second, cautious agents would simply avoid, as far as possible, situations with which they have little experience. The very recognition that the situation is one with which there is little or no experience would prompt the cautious agent to stop or take a step back.

Conclusion

As we have seen, existing versions of the precautionary principle can be placed in at least three categories: rules of choice, epistemic rules or principles, and procedural requirements. The principles have problems in each category. The problems are perhaps not decisive, but nevertheless justify a search for alternative ways of interpreting the precautionary principle.

One such alternative way of understanding the precautionary principle, which may avoid some of the problems, is conceiving of precaution as a virtue. There are several reasons for adopting such an approach. Precaution is a thick evaluative concept, it is predicated of persons and agents in everyday language, and it passes the Footian virtue test.

With this as a basis, a virtue version of the precautionary principle as a rule of choice may be formulated, tentatively in the form of "perform those, and only those, actions that a cautious agent would perform in the given circumstances." Such a principle can be partly, if not completely, action guiding and thus fares better than many of the versions that have been discussed in the literature on the precautionary principle. And as I have suggested, we need not be as afraid of referring to virtuous agents in policy as we might think.

Where does this leave us in the protocell case? What would a cautious agent do in the circumstances? I cannot hope to answer this question completely, but I will provide at least the beginnings of an answer.

First, a cautious agent would certainly take great pains to collect as much high-quality information as possible when facing a decision about new technologies, such as protocell technology. In the case of protocells, there is time to do so. Unlike some other technologies, protocell technology is not being developed in response to a "ticking bomb" scenario. Consider a new technology for remedying some pressing environmental problem. In that case, "wait and see" is not an option. Something has to be done immediately. In the case of protocells, however, postponing decisions while searching for further options will not necessarily be problematic in this sense (cf. Hansson, 1996).

Second, a cautious agent would typically opt for small, reversible steps in introducing new technologies. Laboratory protocell research would probably be acceptable, at least in laboratories that are not located in rogue states and that have some degree of transparency and are subject to some sort of democratic supervision and accountability. On the other hand, a cautious agent would be reluctant to allow spreading of protocells in consumer products and other more widespread uses. A somewhat parallel case, nanotechnology, might help illustrate this. A cautious agent would not object to nanotechnology research, given reasonable constraints. However, he or she would certainly be more hesitant to allow inclusion of nanoparticles in consumer products. Today, some brands of car care products are advertised as containing nanoparticles (e.g., Eurochem, 2006). A cautious agent would be reluctant to allow this, particularly in a case where the benefits are so marginal. Car wash shampoo and "self-cleaning" kitchen sinks, other examples where nanoparticles are claimed to be included, are after all not particularly necessary items.

These examples show, I think, that "what a cautious agent would do" should be understood as a relational notion, for it may be objected that the preceding characterization of a cautious agent is trivial: *Every* reasonable agent, even one who is not particularly cautious, would try to collect all relevant information when facing a decision about a new technology. And every reasonable agent would, if possible, opt for reversible steps when introducing new technologies. However, experience shows that several new technologies have been introduced in a rather different fashion, without much consideration given to reversibility. Cell phones might be a case in point. Therefore, "what a cautious agent would do" should be understood as taking *more* pains to collect information and being *more* prone to opt for small, reversible steps than has been the case for many technology decisions in the past.

Where all this leaves us with regard to protocell technology policy is still an open question. But the virtue principle of precaution at least puts us in a situation to ponder the question from a somewhat different angle. Hopefully, it may help establish a middle ground where "precaution bashers" and the "opponents of technology who wish to rationalize banning or over-regulating things they don't like," referred to at the outset of this chapter, can meet.

Acknowledgments

A previous version of this chapter was presented at the Ethical Aspects of Risk conference in Delft, The Netherlands, June 14–16, 2006, where the participants provided helpful comments. I am particularly grateful to Mark Bedau and Emily

Parke, whose written comments on the penultimate draft have been especially helpful. The research leading up to this paper was partly funded by the Swedish Emergency Management Agency. All views expressed are those of the author.

References

Arcuri, A. (2004). *The case for a procedural version of the precautionary principle erring on the side of environmental preservation* (Global Law working paper 09/04). New York: Hauser Global Law School Program.

Breyman, S. (1999). The political economy of the precautionary principle. Paper prepared for presentation at the 1999 Annual meeting of the AAAS, Anaheim, California, January 21–25, 1999.

Deamer, D. (2005). A giant step towards artificial life? *TRENDS in Biotechnology, 28* (7), 336–338.

Dworkin, R. (1978). *Taking rights seriously.* London: Duckworth.

Eurochem. (2006). Available online at: http://www.eurochem.co.uk/products.php?categories_id=13 (accessed November 2006).

Fisher, E. (2002). Precaution, precaution everywhere: Developing a "common understanding" of the precautionary principle in the European Union. *Maastricht Journal of European and Comparative Law, 9* (1), 7–28.

Foot, P. (1997). Virtues and vices. In R. Crisp & M. Slote (Eds.), *Virtue ethics* (pp. 163–177). Oxford: Oxford University Press.

French, P. A. (1995). *Corporate ethics.* Fort Worth: Harcourt Brace College Publishers.

Graham, J. D. (2000). Perspectives on the precautionary principle. *Human and Ecological Risk Assessment, 6* (3), 383–385.

Haigh, N. (1994). The introduction of the precautionary principle into the UK. In T. O'Riordan & J. Cameron (Eds.), *Interpreting the precautionary principle* (pp. 229–251). London: Cameron May.

Hansson, S. O. (1996). Decision making under great uncertainty. *Philosophy of the Social Sciences, 26* (3), 369–386.

Harremoës, P., Gee, D., MacGarvin, M., Stirling, A., Keys, J., Wynne, B., & Guedes Vas, S. (Eds.) (2002). *The precautionary principle in the 20th century: Late lessons from early warnings.* London: Earthscan.

Harris, J., & Holm, S. (2002). Extending human lifespan and the precautionary paradox. *Journal of Medicine and Philosophy, 27* (3), 355–368.

Holm, S. (2006). Reply to Sandin: The paradox of precaution is not dispelled by attention to context. *Cambridge Quarterly of Healthcare Ethics, 15* (2), 184–187.

Hursthouse, R. (1999). *On virtue ethics.* Oxford: Oxford University Press.

Macintyre, A. (1999). *Dependent rational animals: Why human beings need the virtues.* Chicago and La Salle, IL: Open Court.

Manson, N. A. (2002). Formulating the precautionary principle. *Environmental Ethics, 24,* 263–274.

Miller, H. I., & Conko, G. (2000). Genetically modified fear and the international regulation of biotechnology. In J. Morris (Ed.), *Rethinking risk and the precautionary principle* (pp. 84–104). Oxford: Butterworth-Heinemann.

Myhr, A. I., & Traavik, T. (2003). Genetically modified (GM) crops: Precautionary science and conflicts of interest. *Journal of Agricultural and Environmental Ethics, 16* (3), 227–247.

Peterson, M. (2006). The precautionary principle is incoherent. *Risk Analysis, 28* (3), 595–601.

Peterson, M. (2007). Should the precautionary principle guide our actions or our beliefs? *Journal of Medical Ethics, 33,* 5–10.

Raffensperger, C., & Tickner, J. (Eds.) (1999). *Protecting public health and the environment: Implementing the precautionary principle.* Washington, DC: Island Press.

Rasmussen, S., Chen, L. H., Deamer, D., Krakauer., D. C., Packard, N. H., Stadler, P. F., & Bedau, M. (2004). Transitions from nonliving to living matter. *Science, 303* (5660), 963–965.

Sandin, P. (1999). Dimensions of the precautionary principle. *Human and Ecological Risk Assessment, 5* (5), 889–907.

Sandin, P. (2004). The precautionary principle and the concept of precaution. *Environmental Values, 13* (4), 461–475.

Sandin, P. (2006). A paradox out of context: Harris and Holm on the precautionary principle. *Cambridge Quarterly of Healthcare Ethics, 15* (2), 175–183.

Sandin, P. (2007a). Common-sense precaution and varieties of the precautionary principle. In T. Lewens (Ed.), *Risk: Philosophical perspectives* (pp. 99–112). New York: Routledge.

Sandin, P. (2007b). Collective military virtues. *Journal of Military Ethics, 6* (4), 303–314.

Sandin, P, Bengtsson, B. E., Bergman, A., Brandt, I., Dencker, L., Eriksson, P., et al. (2004). Precautionary defaults—a new strategy for chemical risk management. *Human and Ecological Risk Assessment, 10* (1), 1–18.

Sandin, P., Peterson, M., Hansson, S. O., Rudén, C., & Juthe, A. (2002). Five charges against the precautionary principle. *Journal of Risk Research, 5* (4), 287–299.

Schudt, K. (2000). Taming the corporate monster: An Aristotelian approach to corporate virtue. *Business Ethics Quarterly, 10* (3), 711–723.

Secretariat of the Convention on Biological Diversity (2003). The Cartagena protocol on biosafety: A record of the negotiations. Available online at: http://www.biodiv.org/doc/publications/bs-brochure-03-en.pdf (accessed June 2006).

Stich, S. P. (1978). The recombinant DNA debate. *Philosophy and Public Affairs, 7* (3), 187–205.

Sunstein, C. R. (2005). Laws of fear: Beyond the precautionary principle. Cambridge: Cambridge University Press.

UNCED (1993). *The earth summit: The United Nations conference on environment and development (1992: Rio De Janeiro).* Introduction and commentary by S. P. Johnson. London: Graham & Trotman.

7

Ethical Dialogue about Science in the Context of a Culture of Precaution

Bill Durodié

[W]henever society is in trouble it begins to moralize.
—Furedi (1997, p. 161)

Scientists hoping to develop the first protocells do so in a broader social and political climate that may not be wholly supportive of scientific innovation and technological deployment. The specific context varies a great deal across countries, and within countries across different themes. But the general contours are similar inasmuch as there are growing concerns in many quarters across the globe as to the ethics of deploying particular technologies, especially those emanating from the life sciences.

This chapter draws its examples from the United Kingdom, where a number of significant debates have occurred in recent years over related scientific matters from the safety of food, such as beef, to the risks posed by deploying genetically modified (GM) crops in the environment, to the administration of new vaccines and other technologies, such as cellular phones.

Increasingly, scientists and the scientific establishment are asked to adopt a more moral and dialogic stance to their work. It has been suggested that the old paternalistic formula labeled DAD (decide-announce-defend) be gradually replaced by a more inclusive approach that seeks to engage with the public on science and scientific decision-making issues at all levels.

Calls for greater inclusion of public views or "values" within the scientific process have come from many quarters, including, in the UK, the Royal Commission on Environmental Pollution (1998), the House of Lords (2000), the Parliamentary Office for Science and Technology (2001), and the authors of an influential Economic and Social Research Council publication (Hargreaves & Ferguson, 2001). Such inclusivity, it is held, will make for ideas and institutions that are more people centered and ethical in their outlook.

More recently members of Parliament on the Commons Science and Technology Committee announced that they were to examine the Royal Society over allegations that Britain's top scientific body is too "elitist" and "out of touch" (Meek, 2002). A plethora of new ethics committees, commissions, and codes of conduct has also been established to assess both the content of the science that is carried out and the purposes of those who undertake it.

Science, it appears, is breaking out of a reductionist paradigm to examine more global, holistic processes pertaining to the interface of science with society. Parenting, pollution, and public health now form as much a part of the content of scientific investigation as do genetic engineering, inorganic chemistry, or particle physics. These more social and ethical orientations are held by some to be good for both society and science.

But the consequences of this sea change in outlooks and attitudes have yet to be assessed. Some have questioned the purported effectiveness of negotiated dialogic processes (Coglianese, 1997). Others have argued that these changes have been driven in large part by fear and confusion rather than confidence and direction. If so, they may end up contributing to a more widespread disorientation and demoralization in science and society, rather than generating a new sense of purpose and trust. That can only be bad news for scientists—including those hoping to develop the first protocells.

Science and Society

Technological change, enhanced longevity, and social development are all testaments to the tremendous impact science has had on society. In addition, modern societies are necessarily dynamic and science is often at the forefront of upsetting the status quo. But even when its benefits are questioned, the emphasis usually placed on the importance of science in social change is one-sided (Gillott & Kumar, 1995).

Science, in addition to transforming society, is itself a product of society. Newton understood this when he wrote in his famous letter to Hooke in 1676: "If I have seen further it is by standing on the shoulders of giants" (Turnbull, 1959, p. 416). Science comes with a history. Its advances are limited by material reality, and circumscribed by the state of the society within which it develops—including its ambition and imagination (or lack of these).

The world of antiquity yielded many intellectual insights, but constrained by its social structures these proved to be of limited practical consequence (Kline, 1987). Then, from 400 to 1000 AD, Europe was, in scientific terms, a backwater. Some of the high points of Greek science were kept alive and developed in the Arab world, but the feudal order was largely static, positing a relationship between humanity and nature that was conceived as fixed for all eternity (Manchester, 1992).

It was the Italian Renaissance that first began to change and then challenge the old order. Built largely on the development of trade, it raised new demands on individuals and society, encouraging invention through the merger of intellectual activity with practical needs. With the discovery of America in 1492 trade routes began to shift to the Atlantic seaboard. England, Holland, and France now began to accelerate in development as important centers of innovation driven by their own commercial interests.

Within a few centuries, in addition to the development of perspective in art and the construction of Brunelleschi's Dome in Florence, the world had been circumnavigated, its largest continents discovered, the compass, telescope, and printing press invented. The world would never be the same again (Boas, 1970).

By 1660, when what was to become known as the Royal Society was founded in London, the ecclesiastical domination of the Holy See in Rome had been broken, while the trial and execution in 1649 of the monarch Charles I were fresh in people's minds. Accordingly, its founders adopted the Latin phrase *nullius in verba* (on the word of no one), from the Roman poet Horace—the son of a freed slave—as their motto.

This was a bold statement of intent, and reflected the political mood of the time. The champions of the new philosophy wished to emphasize the "experimental learning" central to their outlook, but also their reluctance to take any pronouncement on trust. Once the dogma of Pope and king had been dispensed with, acquired insight could henceforth truly aspire to replace received authority (Hampson, 1968).

Science now formally established itself as a new source of authority. As well as delivering remarkable achievements, it was to be a practical battering ram with which to challenge perception, prejudice, and power. But this was a reflection and pronouncement of faith in humanity itself, rather than merely in science. Social development had raised human expectations as to what was possible. It had given humanity confidence in the power of its own reason, a factor that then proved significant to the development of science.

The scientific revolution of the seventeenth century represented the triumph of rationality and experimentation over the superstition, speculation, dogma, and domination that had gone before. It was more than simply an advance in scientific knowledge; it was part of a wider shift in attitudes and beliefs. The scientific revolution was the product of dynamic social progress, as well as becoming an essential contributor to that progress. But just as the initial dynamic behind science was social change, so social change, or more particularly the lack of it, could circumscribe science.

The developing vision of nature and humanity was driven by aspirations for freedom and equality. These concepts represented the needs of a new elite—

the commercial, and later industrial, capitalist class. But as such, society would now encounter new constraints, both from the ongoing and vociferous rejection of the old religious and monarchical orders it had supplanted, and from the inherent limitations of this new social system and the particular world view of its proponents.

From 1789, at the time of the revolution in France, and later due to a growing threat from the dispossessed, promises of freedom, equality, and progress came to be seen as highly problematic, highlighting the failure of society to live up to them. The new establishment, in addition to social and political reformation, now needed to circumscribe the claims and effects on society of scientific enquiry, reason, and progress.

The nineteenth century saw the development of a model of science known as positivism, which consciously sought to facilitate the restoration of order (Pick, 1989). Reflecting the simple mechanical processes emerging in industry, it posited that science operates on objective, absolute, and ascertainable facts connected by rigid links of cause and effect (Hobsbawm, 1988). But this view of a clockwork universe with its uniform rules and truths being revealed by pristine individuals disinterestedly recording the underlying workings of invariable natural laws does not stand up to simple scrutiny.

It was a model of science still worthy of esteem, but robbed of any association with historical change and development. The link between the advances of science and society was lost. Many of today's confusions about science stem from the misapprehension that this approach, rather than being a limiting constraint, somehow continued the Enlightenment tradition.

Through the Victorian age, a compromise was effectively reached whereby science could still develop—quite rapidly at times—but it no longer systematically challenged the old authorities. Darwin's secular universe cohabited that of the bishops but did not seek to tread on their patch. Scientists were held in high regard, but science was now decoupled from the political aspiration to transform society, although its consequences continued to do so.

Over the course of the twentieth century, philosophers of science gradually placed greater emphasis on the uniqueness of individual experience. This corresponded intellectually to the tremendous changes, impasses, and uncertainties they found themselves caught up in. Two world wars, a depression, and continuing poverty and conflict in the developing world generated doubts as to the possibility of universal human progress and a "fear of the future" (Carr, 1990).

Accordingly, those seeking to defend science, including many in what we might now consider to be the scientific establishment, sought to separate it further from social and political transformation by increasingly placing it into a narrowly technological or reductionist straitjacket. Harnessed to the pursuit of American security through the Manhattan project and the Apollo missions, science also created opponents for itself among its old allies. The political left, which had traditionally supported the liberating potential of scientific advance, now came to view it with increased suspicion.

The left argued that aspiration itself, rather than its failure—as evidenced in the collapse of confidence in social progress—had turned nature into "mere objectivity" for humanity (Adorno & Horkheimer, 1989). This attitude could then be found reflected in the subordination of people and countries, and was increasingly facilitated through the use of instrumentalist technologies. Science was seen as the amoral steamroller of a dispassionate new modernity, crushing communities and tradition.

What is so poignant about the modern disenchantment with science is that it has emerged at a time when scientific achievements are without precedent. Mapping the human genome and nanotechnology, for example, offer the tantalizing possibilities of developments in science and society that could hardly have been dreamed of even a generation ago. But without social progress, the direction and purpose of science and society have become uncertain. Behind the current crisis of science lies a collapse of confidence in humanity and the possibility of social progress.

Risk and Morality

Clearly, science is far from being value free. It invariably reflects the dominant values of the historical period in which it finds itself. But if, as Marx would have it, "[t]he ideas of the ruling class are in every epoch the ruling ideas" (Marx, 1845, p. 64), then it is worth reflecting on what might happen to a society in which the establishment no longer holds its traditionally distinct ideas and values.

The unprecedented convergence, discussed in the previous section, of the political left's loss of faith in science and social transformation with the political right's traditional misgivings, has effectively encouraged a more pessimistic outlook about science to develop across the political spectrum, leading in part to what some have identified as the rise of a heightened risk consciousness (Beck, 1992).

Despite being two sides of the same coin, risk is now regularly emphasized over opportunity, and as a consequence safety and precaution have, in certain quarters, almost become new organizing principles. Scientists in the United Kingdom are now encouraged to highlight the uncertainties in their work, as if these were somehow new phenomena. But the uncertainties and unknowns of not deploying new technologies are rarely examined, let alone the missed opportunities and benefits of not experimenting.

The convergence of left and right and the ensuing depoliticization and demise of political debate have also coincided with and facilitated the breakdown of many forms of social organization. With the decline of families, neighborhoods, communities, religious congregations, informal associations, trade unions, political parties, or other institutions to be part of, it has become far easier for people's subjective impressions of the world to hold sway (Putnam, 2001).

Some have argued that old-style moral panics driven from the top down with a view to cohering society appear to have been replaced by more nebulous social anxieties involving a wider range of public interests and constituted by a vast number of free-floating threats, with new threats always lurking in the background (Ungar, 2001). Unsurprisingly, therefore, the authorities increasingly seek to provide assurances against those they believe to be self-serving or incautious, from profit-seeking multinationals down to feckless individuals.

It is commonly assumed that the media have a significant role to play in such matters by making us more aware than previous generations of the various hazards we face. Certainly, in the absence of political debate, the media do have a more prominent role. But what is often overlooked is the extent to which politicians, regulators, and even scientists themselves now have more ambiguous attitudes toward science and seek to encourage dissenting voices within the public (Royal Society, 2002a).

Our heightened awareness of risk not only latches on to new products and processes, but also reinterprets age-old activities that were once unquestioned. From eating beef to bullying in schools and from sex to sunscreen lotions, the sheer range and number of issues now perceived as risky suggest an underlying process beyond the intrinsic properties of each issue that we should seek to understand. It would appear that such problems are in abundant supply, limited only by our imagination.

The attitudinal and institutional changes prompted by these shifts have been celebrated by some (Hume, 2005). The challenges presented to traditional forms and sources of authority can be presented in an uncritically positive manner as a celebration of identity, choice, and personal preference. Clearly, it is good that patronage and conformity be consigned to the past. But we should be wary of replacing a culture of unthinking deference with an equally incapacitating culture of unnecessary fear.

Without the discipline of, and an active engagement in, broader concerns, individuals can be left feeling incredibly isolated. This social and political disengagement may then become reflected in, and further feed, public disenchantment with science. Such popular concerns, verging on cynicism, can in turn drive official policy in an unwarranted manner (Better Regulation Commission, 2006).

Rather than recognizing that a healthy skepticism about science is born of an active body politic, some are now consciously attempting to artificially restore trust in science and scientists through enhanced participation with a view to relegitimizing democratic processes across society (Durodié, 2004). Foremost among the new mechanisms proposed to regulate society and attenuate our fears has been the precautionary principle. The latter suggests that, in the absence of definitive scientific evidence to the contrary, measures to protect the environment or human health should be taken whenever any threat of serious or irreversible damage to either may be present (O'Riordan & Cameron, 1995).

Critics have countered that, because scientific certainty is never possible and irreversibility is inevitable, the application of the principle is a recipe for paralysis. Further, defining the extent of evidence necessary to justify concern, as well as what measures should be invoked and by whom, are considerations lending themselves to significant political, commercial, and nongovernmental manipulation (Durodié, 2000b). Nevertheless, because of the inflated perceptions of risk, the principle is set to play an ever-increasing role in scientific decision making.

Unsurprisingly perhaps, under permanent attack and held open to constant questioning, many institutions and experts now seem to lack self-belief, or even a clear vision or purpose. This has led many into overzealous reactions to events or perceived fears. Policy reversals and reactive regulation now appear increasingly commonplace, thereby sending confusing signals to an already sensitized public.

The Slovenian philosopher and psychoanalyst, Slavoj Zizek, has characterized "endless precautions" and "incessant procrastination" as "the subjective position of the obsessional neurotic." Far from indicating a respectable "fear of error," he suggests, this approach "conceals its opposite, the fear of truth" (Zizek, 1989, p. 191). But a pursuit of truth, however temporary, lies at the very heart of scientific inquiry. Scientists do not just record and measure; they assess, infer, and prioritize, as well as experiment and transform. It is these active and judgmental modes that are most at risk of being dissolved and lost today.

Ironically, to the extent that social life has increasingly become reorganized around risk, it has recreated a limited sense of moral purpose (Porritt, 2001). By using the technical language of risk assessment, this new morality does not announce itself as such. Though not preaching in an old-fashioned way, the new prescriptions for personal and professional conduct administered by unaccountable agencies and regulatory bodies are no less intrusive than the moral codes of previous generations.

Unlike scientists, however, these new bodies, including countless regulatory agencies, nongovernmental organizations, and appointed expert or lay panels, have a more direct relationship to the state, and by encouraging caution and self-limitation they set themselves against the very motive force of science—a desire to explore and experiment.

Equivocation and Inclusion

Nowadays, even when the scientific evidence is fairly categorical, scientists have learned to be much more equivocal about the outcomes of their research. Emphasis is increasingly placed on the uncertainties rather than the potential benefits of products and procedures. This has occurred because of the onslaught of calls emanating from the bodies identified in the opening section of this chapter, as well as many others, for scientists to show "more humility" than in the past (European Environment Agency, 2001). In the United Kingdom at least, there is now a perceived need to incorporate such so-called "lay and local knowledge" as well as "wider social interests and values" in scientific work (Wynne, 1997).

These developments had been evolving steadily over the previous decades, but were catalyzed to a new level by the BSE (bovine spongiform encephalopathy) debacle of the mid-1990s. They were then consolidated through the process of preparation and prompt endorsement of the government inquiry into the affair, known as the Phillips report (BSE Inquiry, 2000).

In the interim, a number of other major risk episodes in the United Kingdom achieved public prominence and notoriety, including the Stewart inquiry into the safety of mobile phones (IEGMP, 2000), the proposed release of genetically modified organisms into the environment (Royal Society, 2002b), and a furor over the supposed consequences for children of a new triple vaccine against measles, mumps, and rubella (Fitzpatrick, 2004).

The Phillips report marked the acceptance of the precautionary principle as a central tenet of future scientific policy making within the United Kingdom. Irrespective of which of its many formulations is used, the precautionary principle has the consequence of emphasizing worst-case scenarios, thereby encouraging a tendency to overreact to events and, more insidiously, elevating public opinion over professional expertise and subordinating science to prejudice. Accordingly, debates over "strong" or "weak" versions of precaution, or over whether it is a "principle" or merely an "approach," fall wide of the mark (Morris, 2000).

BSE is remarkable for acting as the basis and justification of much that has happened since, in many other, often unrelated areas. Yet, both the evidence of a significant problem, and the purported impact of the policies implemented for dealing with it, remain essentially inconclusive. In the history of the relationship between humanity and nature, this episode is unlikely to merit more than a footnote. Domesticated animals have been a potent source of infectious disease before, with measles, mumps, whooping cough, smallpox, and tuberculosis all crossing the species barrier at some stage with intermittently catastrophic consequences and mortality rates of around 90 percent (McNeill, 1976).

The link between BSE and variant Creutzfeldt-Jakob disease (vCJD), a degenerative brain disorder in humans, has yet to be conclusively proven or understood, and what little evidence there is suggests no clear connection. It is almost as if, desperate in their attempts to show the public their willingness to act, both the government and many leading scientists sought to pander to the popular mood in the belief that this would restore some kind of trust.

Thus, after neuropathologist Sir Bernard Tomlinson announced in December 1995 that he had stopped eating hamburgers and health secretary Stephen Dorrell announced a possible link between BSE and vCJD to the UK House of Commons in March 1996, concern about contaminated beef became rife.

Significantly, public concerns about BSE and its transmissibility to humans bore little relationship to its actual incidence. The Phillips report itself recognized that actions taken by ministers as early as 1988 had stemmed the epidemic, though they had not necessarily been comprehensive or completely enforceable. The ban on ruminant protein in cattle feed led to the number of BSE cases by year of birth falling from a peak of 36,861 in 1987 to 1 in 1996, the year of the panic (MAFF, 2000). Despite early predictions of as many as 500,000 cases of vCJD per year, to date there have been approximately 150 cases with evidence of a tailing-off. It is not evident that all of these can be directly attributed to eating beef, as cases of vegetarians developing vCJD were also reported.

At least one senior public health specialist has queried much of the prevailing orthodoxy (Venters, 2001). Using the standard epidemiological criteria of plausibility, strength of association, consistency, quality, and reversibility—analytical tools established by Austin Bradford Hill and Richard Doll's famous observations on the link between smoking and lung cancer in the 1960s—George Venters has questioned much of the evidence for a link between BSE and vCJD, noting that the incidence of vCJD would have been expected to rise anyway since systematic monitoring for it first started in 1990.

Venters has suggested that there was "a process of hypothesis confirmation rather than hypothesis testing" and further that "evidence that has been awkward or contrary, has either been played down or just outright ignored," accusing scientists and health experts of falling for "the belief that multiple pieces of suspect or weak evidence provide strong evidence when bundled together." "It is," he continues, "almost like they made up their minds about a link between BSE and nvCJD and so they set about confirming it" (O'Neill, 2001).

Irrespective of the evidence—or the lack of it—both government and scientists reorganized their operations according to the worst predictions. The Report of the BSE Inquiry is quite explicit about this, arguing that despite the lack of evidence for a link between BSE and vCJD, "[t]he importance of precautionary measures should not be played down on the grounds that the risk is unproved" (BSE Inquiry, 2000, p. 266). Certainly, BSE acted as the catalyst to a major restructuring and policy reorientation, both at the heart of the European Commission and within the UK, and the new approaches developed therefrom then began to encroach into other areas (Durodié, 2000b).

One of the other distinctive features of the BSE Inquiry was the prominent significance it gave to the relatives of the victims of vCJD (BSE Inquiry, 2000). Though this innovation attracted little comment and less criticism, it was a major development, reflecting a growing preference for sentiment over rationality. It is not at all clear how the experience of losing a relative yields a privileged insight into the nature of a disease, or any great wisdom into how to prevent or treat it. While official recognition of the families of victims reflects public acknowledgment of the particularly distressing effects of vCJD, their involvement in the wider aspects of the inquiry implicitly devalues scientific, clinical, and even political expertise.

These two key features—an appeal to worst-case scenarios and the inclusion of lay views—were paralleled in the Stewart inquiry into the safety of mobile phones (IEGMP, 2000), to quite a striking degree. In a comparative study of national responses to perceived health risks from mobile phones, researcher Adam Burgess notes that "[a]lmost by definition, what is a risk 'issue' is itself determined by the extent and character of government reaction," continuing: "There is also a more particular sense in which official risk responses potentially animate and cohere otherwise diffuse anxieties" (Burgess, 2002, p. 177).

According to this analysis, far from heading off potential accusations of complacency through a proactive strategy to "keep ahead of public anxiety" (Jowell, 1999, p. 1), the UK government's precautionary response through the establishment of the Independent Expert Group on Mobile Phones, led by Sir William Stewart, actually stimulated risk concerns, which increased subsequent to the inquiry. This is, according to Burgess, because "even balanced public information on negligible risks tends to increase anxiety, on the assumption that there must be something to worry about if the government is taking action" (2002, p. 179).

In a manner akin to the Phillips inquiry, Stewart and his panel acknowledged that "the balance of evidence does not suggest that mobile phone technologies put the health of the general population . . . at risk" (IEGMP, 2000, p. iii), but nevertheless the study called for a £7 million program of further research and for leaflets to be included in future purchases of mobile phones warning of the possible risks. The latter led one commentator to conclude that "in its rush to be open about communicating risk to the public, the government has simply forgotten that there was no risk to communicate" (Kaplinsky, 2000).

While not identifying any risk, other than that of using a phone when driving a vehicle, these leaflets suggest that the best way to reduce risk is to use the phone less. They also advise taking note of the specific absorption rate of phones, which measures their heating effect. This suggests that recording anything that was easy to measure became the key concern irrespective of the fact that it did not relate to the still-to-be-demonstrated "nonthermal" effects that campaigners worry about.

Again, the conclusions of the Stewart inquiry make remarkable concessions to the need to incorporate perceived public concerns and prejudice. Following the recommendations in the report, future research will now be required to take account of non-peer-reviewed and anecdotal evidence. Indeed, the inquiry itself went a considerable way to acknowledging and accommodating to such concerns by extending its remit beyond a review of the latest scientific knowledge on mobile electromagnetism to those concerns pertaining to the positioning of masts or base stations.

In a similar fashion, a recent Royal Society study on the safety of GM crops elevates these same two features—the exaggeration of risk beyond the available evidence and the now almost mandatory concession to include public concerns in such assessments (Royal Society, 2002). Despite finding no reason to doubt the safety of foods made from currently available GM ingredients, nor to believe that genetic modification makes food inherently less safe than conventional counterparts, the Royal Society gave prominence in its report to new hypothetical concerns, in an attempt to improve its own standing in the eyes of the public. This prompted a recent review of the study to comment that "it would appear that the Royal Society has not become more hesitant about the safety of GM crops and food—just more hesitant about saying so" (Gilland, 2002).

Despite concerns raised by some of the members of the working group that produced the report about the "extraordinarily selective" media coverage it elicited, this emphasis was triggered by the Royal Society's own press release, which was in turn influenced by the hesitancy of the report itself. It would appear that the scientists concerned now want to have it both ways, saying to fellow scientists, government, and industry that there is no reason to think GM products are unsafe, while assuring the public that safeguards should be strengthened. This incoherent approach is far more likely to backfire than reassure and recreate the trusting relationship they desire.

Examples of equivocation and obsession with including assumed public concerns by senior government officials and scientists abound. They are now the norm rather than the exception. Cases range from the Royal Society report on endocrine-disrupting chemicals (Royal Society, 2000), to the European Commission's restrictions on phthalate plasticizers (Durodié, 2000b), to the official inquiry into the Bristol Royal Infirmary children's heart surgery unit (Fitzpatrick, 2001). Variously, they cite "purported effects" or "public concern" as their instigators before exploring the limited evidence available as to any real problem and concluding with some kind of cautionary comment or call for public engagement.

The trend toward encouraging the public to decide on all scientific matters reached its logical denouement with the refusal of parents to allow their children to receive MMR (measles-mumps-rubella) vaccinations. Triggered by the response to a research paper based on a dozen selected cases that suggested a link between the vaccine and cases of childhood autism, the public understandably demanded to be able to opt for separate inoculations, which were not readily available through the UK National Health Service. Notably, the safety record of the separate vaccines was not interrogated.

Instead, hoisted by its own petard of criticizing scientists and the medical profession, as well as promoting personal choice in a health care market, the government was faced with the first significant outbreak of measles for many years in south London, where vaccination rates had fallen significantly below what could guarantee a herd immunity (Fitzpatrick, 2002). The government then had to set about educating parents as to the real risks and issues involved, often in an exaggerated manner, despite having done much to undermine public confidence in science in the first place.

But one of the real problems facing both government and scientists today is that the public tends to be bombarded with too much, rather than too little, information, particularly in matters pertaining to health. Accordingly, many have noted the rise of a new phenomenon—the worried well—who are literally worrying themselves sick over the multitude of agents and activities identified to them by well-meaning professionals as potential sources of danger (Durodié, 2003). And it is not at all evident that, having projected their own insecurities onto the public, the latter will trust reassurances coming from any proposed alternative system of regulation any more. The promotion of the virtues of risk awareness as a new moral framework for society would appear to have its limitations.

Values and Costs

Although science is necessary to inform democratic decision making within society, it is not in itself democratic. The contemporary preoccupation with the need for "public participation" within scientific decision making threatens to erode this distinction and demoralize scientists.

Rather than embracing uncertainty and change, as did previous generations, today we appear to reject them and highlight risks. What has really changed is not so much the scale of the problems that we face, but the outlook with which society perceives its difficulties, both real and imagined. These issues, though different, cannot really be described as greater than those facing previous generations, nor are they uniquely insurmountable. But our collective will and imagination to resist and overcome them appears to be much weaker, as evidenced by the advent of the precautionary principle.

The challenge to the old elites of society is possibly understandable, but the form it has taken—an attack on expertise *per se*—is inexcusable. Irrespective of the particular scientific inquiry concerned, this challenge has been expressed in similar terms. The BSE inquiry condemned the "culture of secrecy in Whitehall," while the Bristol inquiry under Professor Ian Kennedy attacked "club culture" within the medical profession. Outside the world of science, the Macpherson report into the murder of a black teenager, Stephen Lawrence, in a south London street challenged a "canteen culture" within Britain's police force.

At the same time, the specific prescriptions proposed for dealing with these varied problems have all proven to be remarkably similar—the need for greater openness and transparency through the inclusion of members of the public or public "values" into the decision-making process. But while consensus-seeking may go down well among woolly-minded bureaucrats in Whitehall and Brussels, it is a process largely unsuited to the needs of scientific inquiry.

Indeed, while civil servants, doctors, and scientists have been denigrated, what has been less discussed is the extent to which alternative sources of authority have accordingly been elevated. It is ironic that those who do not trust scientific expertise now invest their faith in a new breed of so-called experts—victims and ethicists—who are not required to submit their work for peer review or other ways of establishing the authenticity of their claims, and whose pronouncements are not open to any kind of experimental verification whatsoever.

Furthermore, members of ethics committees and special agencies are directly appointed by the UK government and are thus even less accountable to the public than politicians. Posturing as radical and democratic, this outlook invites a more authoritarian style of government over a more fatalistic, nervous society.

Clearly, there is tension between those who wish to include the public simply to keep them informed or supportive (Sainsbury, 2000) and those who genuinely hold that the public voice is a missing element for establishing scientific objectivity or accountability. This latter view appears to present a narrowly empirical model of science whereby truth, or an approximation of it, is to be reached through an averaging-out process of competing interested parties.

One significant difficulty for all concerned is how to include an increasingly disengaged public in such processes. The claims of various advocacy groups to representation of this wider audience have increasingly been questioned (Furedi, 1999).

At best, such bodies have a passive membership comprising a small percentage of any national population (Burgess, 2001). Whether members directly belong to such a lobby, or are hand-picked and carefully vetted outsiders, such an approach remains broadly unsatisfactory. Indeed, some have identified what appears to be a remarkable convergence of views between those officials presiding over such processes and the members of the public who participate in them (Appleton, 2001).

To get around the self-selected nature of advocacy groups, much emphasis has recently been placed on using quantitative research, such as polls and surveys, as well as qualitative research, including in-depth interviews, focus groups, and other stakeholder dialogue forums, to assess public opinion. The danger here is well documented. It includes projecting views and values through question-framing or selectively finding those selfsame views and values among the responses. Even identifying "what is not being said" (Wynne, 2001) requires prejudicial priorities among interviewers.

Hence, the danger is that, rather than recording the wishes of the majority, the inclusion of public views or "values" merely records a small subset of these, which are reflected back at researchers. Indeed, in the past, much of this research would have been called public opinion. Opinions are open to being challenged, interrogated, and altered. Labeling these as "values" effectively sets them apart from needing further inquiry.

Ironically, in many instances, corporations, governments, and the scientific establishment itself now appear increasingly willing to incorporate supposed public concerns into the decision-making process. The reasons for this may be varied, including a belief that this provides greater regulatory stability, despite the possibility that policy determined by popular prejudice is far more precarious. Another clear motive is an unwillingness to be held independently to account. This reflects an abdication of leadership and responsibility and a preference to deflect, diffuse, shift, or share the blame should things go wrong in the future (Hood, 2002).

Adhering to an increasingly cautionary and restrictive approach, under the banner of inclusion, reflects a diminished sense of ambition for society that not only limits scientific experimentation, but is also likely to preclude wider possibilities for social change.

It is precisely because the appearances of nature are deceptive that we need the methods of science, which commonly yield findings that contradict popular impressions and established traditions. Science is not about making us feel good. Many of its findings can be disconcerting, yet we owe much to those who took a stand against public perceptions and challenged prevailing prejudices. These principles are jeopardized by the philistinism of the contemporary political elite—a trend toward which many scientific authorities are, unfortunately, acquiescent.

Far from adding to the richness of scientific inquiry, lay views tend to focus on the immediate, rather than a more mediated or critical appreciation of available evidence. The ability to understand or transcend issues requires more diligence and discipline than inclusion and inspiration. To relegate the experienced and considered judgments of scientists to being just another point of view suggests that they merely represent a form of sectional interest. This forces an emphasis on quantity over quality in science that allows for manipulation through subjective impressions and vested interests.

Relabeling private views as public "values," and insisting that these should be included in the policy-making process, simply aggrandizes what remain personal opinions. This dilutes the science, denigrates the scientists, and both patronizes the public and panders to the conceit of those who claim to represent such "values" (Durodié, 2004). The elevation of opinion over professional expertise subordinates science to prejudice. Official recognition of these perceptions and beliefs then implicitly devalues the insights acquired through detailed experimentation and detached consideration. This undermines the confidence of scientists and marginalizes excellence.

Far from being egalitarian, this is an affront to a real democracy based on reason. Real exclusion begins when prejudice or opinion are taken to be a sound basis for decision making. Tragically, it appears that many individuals and institutions within the scientific establishment have abdicated their responsibility to judge and be criticized. Far from relieving them from pressure, this paralyzing diffidence will only further discredit and demoralize their profession.

Conclusions

Science clearly has an impact on society. But our contemporary obsession with, and misgivings about, this reality conceal a more important dynamic. That is, the impact of society on science.

A society focused through a positive aspiration and sense of direction encourages scientific development. One that highlights, and becomes paralyzed by, the inherent uncertainties in all decisions will not look kindly on those seeking to develop protocells—or anything else for that matter, other than technologies that restrict growth and development.

References

Adorno, T., & Horkheimer, M. (1989). *Dialectic of enlightenment*. London: Verso.

Appleton, J. (2001). The rise and rise of parents' groups. *Spiked*. Available online at: http://www.spiked-online.com/Printable/00000002D1F6.htm (accessed January 2007).

Beck, U. (1992). *Risk society: Towards a new modernity*. London: Sage Publications.

Better Regulation Commission (2006). *Risk, responsibility and regulation: Whose risk is it anyway?* London: Whitehall.

Boas, M. (1970). *The scientific renaissance, 1450–1630*. London: Fontana.

BSE Inquiry (2000). *The inquiry into BSE and variant CJD in the United Kingdom*. London: The Stationery Office.

Burgess, A. (2001). Flattering consumption: Creating a Europe of the consumer. *Journal of Consumer Culture, 1* (1), 93–117,

Burgess, A. (2002). Comparing national responses to perceived risks from mobile phone masts. *Health, Risk and Society, 4* (2), 175–188.

Carr, E. H. (1990). *What is history?* London: Penguin.

Coglianese, C. (1997). Assessing consensus: The promise and performance of negotiated rulemaking. *Duke Law Journal, 46* (6), 1255–1349.

Durodié, B. (2000a). Calculating the cost of caution. *Chemistry & Industry, 1* (5), 170.

Durodié, B. (2000b). Plastic panics: European risk regulation in the aftermath of BSE. In J. Morris (Ed.), *Rethinking risk and the precautionary principle* (pp. 140–166). London: Butterworth-Heinemann.

Durodié, B. (2003). The true cost of precautionary chemicals regulation. *Risk Analysis, 23* (2), 389–398.

Durodié, B. (2004). Limitations of public dialogue in science and the rise of new "experts." *Critical Review of International Social and Political Philosophy, 6* (4), 82–92.

European Environment Agency (2001). *Late lessons from early warnings: The precautionary principle 1896–2000*. Copenhagen: EEA.

Fitzpatrick, M. (2001). After Bristol: The humbling of the medical profession. *Spiked*. Available online at: http://www.spiked-online.com/Printable/00000002D1F4.htm (accessed January 2007).

Fitzpatrick, M. (2002). MMR: Injection of fear. *Spiked*. Available online at: http://www.spiked-online.com/Printable/00000002D39E.htm (accessed January 2007).

Fitzpatrick, M. (2004). *MMR and autism: What parents need to know*. London: Routledge.

Furedi, F. (1997). *Culture of fear: Risk-taking and the morality of low expectation*. London: Cassell.

Furedi, F. (1999). Consuming democracy: Activism, elitism and political apathy. Cambridge: European Science and Environment Forum. Available online at: http://www.geser.net/furedi.html (accessed January 2007).

Gilland, T. (2002). Putting fear before facts. *Spiked*. Available online at: http://www.spiked-online.com/Printable/00000002D40E.htm (accessed January 2007).

Gillott, J., & Kumar, M. (1995). *Science and the retreat from reason*. London: Merlin Press.

Hampson, N. (1968). *The enlightenment*. London: Penguin.

Hargreaves, I., & Ferguson, G. (2001). *Who's misunderstanding whom? Bridging the gulf of understanding between the public, the media and science.* Swindon, UK: Economic and Social Research Council.

Hobsbawm, E. (1988). *The age of capital: 1848–1875.* London: Cardinal Books.

Hood, C. (2002). The risk game and the blame game. *Government and Opposition, 37* (1), 15–37.

House of Lords (2000). *Science and society.* Select Committee on Science and Technology, Session 1999–2000, Third Report, HL Paper 38, London.

Hume, M. (2005). Introduction to S. Feldman & V. Marks, *Panic nation: Unpicking the myths we're told about food and health.* London: Blake Publishing.

Independent Expert Group on Mobile Phones (EIGMP) (2000). *Mobile phones and health.* Didcot: National Radiological Protection Board.

Jowell, T. (1999). *Minutes of evidence.* London: House of Commons Science and Technology Committee, The Stationery Office.

Kaplinsky, J. (2000). Mobile moans. *Spiked.* Available online at: http://www.spiked-online.com/Printable/0000000053FA.htm (accessed January 2007).

Kline, M. (1987). *Mathematics in western culture.* Oxford: Oxford University Press.

Ministry of Agriculture Fisheries and Food (MAFF) (2000). *BSE enforcement bulletin.* No. 43. London: The Stationery Office.

Manchester, W. (1992). *A world lit only by fire: The medieval mind and the renaissance.* Boston: Little, Brown and Company.

Marx, K. (1845). *The German ideology.* London: Lawrence & Wishart.

McNeill, W. H. (1976). *Plagues and people.* New York: Anchor Press.

Meek, J. (2002). "Elitist" royal society faces funding clash. *The Guardian,* February 4, 2002.

Morris, J. (Ed.) (2000). *Rethinking risk and the precautionary principle.* London: Butterworth-Heinemann.

O'Neill, B. (2001). Beefing up the debate. *Spiked.* Available online at: http://www.spiked-online.com/Printable/00000002D2A9.htm (accessed January 2007).

O'Riordan, T., & Cameron, J. (1995). *Interpreting the precautionary principle.* London: Earthscan Publications.

Parliamentary Office for Science and Technology (POST) (2001). *Open channels: Public dialogue in science and technology.* Parliamentary Office of Science and Technology, Report No. 153. London: The Stationery Office.

Pick, D. (1989). *Faces of degeneration: A European disorder, c1848–c1918.* Cambridge: Cambridge University Press.

Porritt, J. (2001). *Playing safe: Science and the environment.* London: Thames and Hudson.

Putnam, R. D. (2001). *Bowling alone: The collapse and revival of American community.* New York: Simon & Schuster.

Royal Commission on Environmental Pollution (RCEP) (1998). *21st report: Setting environmental standards*. London: Cm 4053.

Royal Society (2000). *Endocrine disrupting chemicals (EDCs)*. London: Royal Society, Document 06/00.

Royal Society (2002a). Do we trust today's scientists? *National Forum for Science*, 6 March 2002.

Royal Society (2002b). *Genetically modified plants for food use and human health—an update*. London: Royal Society, Document 4/02.

Sainsbury, D. (2000). Keeping the public on-side. *The Parliamentary Monitor (London)*, October 2000.

Turnbull, H. W. (Ed.) (1959). *The correspondence of Isaac Newton*. Vol. 1. Cambridge: Cambridge University Press.

Ungar, S. (2001). Moral panic versus the risk society: The implications of the changing site of social anxiety. *British Journal of Sociology*, 52 (2), 271–291.

Venters, G. A. (2001). New variant Creutzfeldt-Jakob disease: The epidemic that never was. *British Medical Journal*, 323 (8), 858–861.

Wynne, B. (1997). May the sheep safely graze? A reflexive view of the expert-lay knowledge divide. In S. Lash, B. Szerszynski, & B. Wynne (Eds.), *Risk, environment and modernity: Towards a new ecology* (pp. 44–83). London: Sage Publications.

Wynne, B. (2001). *Risk, democratic citizenship and public policy*. London: British Academy Conference, June 6–7, 2001.

Zizek, S. (1989). *The sublime object of ideology*. London: Verso.

II

Lessons from Recent History and Related Technologies

8

The Creation of Life in Cultural Context: From Spontaneous Generation to Synthetic Biology

Joachim Schummer

The artificial creation of life raises both strong fascination in scientists and strong concerns, if not abhorrence, in critics of science. What appears to be the crowning achievement of synthetic biology and the protocell research project is at the same time considered a major evil. That conflict, which perhaps epitomizes many of the cultural conflicts about science and technology in Western societies, calls for a deeper analysis. Standard ethical analyses, which would try to relate such conflicts to a difference in fundamental values, are difficult to apply here, because it is unclear what are the underlying values of such emotions as fascination and abhorrence. These emotions or affects, rather than just referring to what is morally right or wrong, seem to be rooted in our cultural heritage of desires and taboos of transgression.

My analysis in this chapter is primarily of a historical nature. By investigating ideas and emotions about the creation of life from the earliest times to the present, I aim to clarify their cultural origins. I argue that both the fascination and the abhorrence regarding the creation of life have a common religious basis. Moreover, unlike many commentators of nineteenth-century mad-scientist classics, from Mary Shelley to H. G. Wells, I argue that this basis has no ancient model in religious or mythological traditions but emerged only in the nineteenth century from an exchange between science and religion. Understanding the historical origin of these emotions will make social and ethical reflection on protocells better informed in several regards. It will help clarify the common cultural context out of which both emotions emerged, and thus the preconditions of what became a major cultural conflict about science and technology. My hope is that such an understanding might ultimately help overcome a fruitless conflict by redirecting the debate toward proper ethical issues of protocell research. For as long as these emotions dominate public debates, urgently required ethical deliberations about creating protocells are likely to be neglected.

Imagine a World Where Making Life Is Simple

Imagine a world where simple living organisms can easily be made from inanimate matter. Anyone can do it, provided they know how to combine the correct ingredients in the right way. Sometimes, when the ingredients happen to occur in the right combination and context, life even emerges spontaneously. Would any scientist care about synthesizing life? Would anybody be embarrassed or concerned about someone making life? Would anybody shout, "this is presumptuous! You are trying to play God!?"

Unfortunately, it has largely fallen into oblivion that our world was exactly like that up to the early nineteenth century. Spontaneous generation of life, or abiogenesis, as the phenomenon was called, was taken for granted since the earliest times. It was not part of some esoteric theory. Everybody had ample evidence from ordinary experience: Under favorable conditions, feces, dung, meat, straw, and so on are all perfect materials to generate different kinds of little organisms—even today, some anglers make use of that to obtain their baits.

Furthermore, the Bible as well as the Talmud, the Upanishads, and many other ancient texts and scriptures are full of stories of living organisms emerging from inanimate matter (Lippmann, 1933, ch. 2). For instance, in Genesis 1, all the plants and animals are not created like Adam and Eve; they emerge out of earth, water, and air on the Creator's fiat. In Exodus 8, the "magicians" make two of the plagues, lice and frogs, from dust and water, respectively. In almost any ancient culture, we find the notion that certain animals (mostly vermin, worms, insects, amphibians, snakes, and some birds and mammals like mice) and most plants owe their existence not to reproduction but to spontaneous generation under favorable conditions. If there was anything obscene in deliberately making creatures such as vermin, it was because nobody liked them.

Spontaneous generation was not merely a folk myth. It was the prevailing view among scholars since antiquity (see Lippmann, 1933; Farley, 1977). In the fourth century BC, Aristotle had studied the different ways that animals generate in greater detail than anybody before the eighteenth century. For those species for which he could find no causal relation to parents, such as the unicellular (!) *Testacea*, he considered the possibility of spontaneous generation out of nonliving matter enriched with "vital heat," that is, a material process that was in accordance with his general chemical views. Discussing the generation of *Testacea*, he wrote:

Animals and plants come into being in earth and in liquid because there is water in earth, and air in water, and in all air is vital heat so that in a sense all things are full of soul. Therefore living things form quickly whenever this air and vital heat are enclosed in anything.

When they are so enclosed, the corporeal liquids being heated, there arises as it were a frothy bubble. Whether what is forming is to be more or less honourable in kind depends on the embracing of the psychical principle; this again depends on the medium in which the generation takes place and the material which is included. (*On the generation of animals*, III, 11)

In contrast to Aristotle, late antique and early medieval authorities (such as Virgil, Ovid, Pliny, and Isidor of Sevilla), rather than performing their own investigations, collected the available folk knowledge and myths to build a growing standard set of views on how to make living beings. Such sets typically recommended the carcasses of cows for creating the useful bees, a flourishing art called *bougonia*, whereas those of horses and donkeys were only able to produce wasps and beetles, respectively. Late medieval Christian authorities, such as Albertus Magnus and Thomas Aquinas, basically repeated the received views, but emphasized the importance of astrological influence.[1] When some Renaissance authors tried to incorporate folk myths about geese and lambs growing on trees, criticism arose, but views on the spontaneous generation of simple animals and plants remained, with few exceptions, largely intact through the eighteenth century. Francis Bacon, in his utopia *New Atlantis* (1627), even devised an entire research program for the creation of entirely new animals and plants. He described how, starting from freshly made new simple organisms, one could breed higher species that perfectly meet human needs: "We make a number of kinds of serpents, worms, flies, fishes of putrefaction, whereof some are advanced (in effect) to be perfect creatures, like beasts or birds, and have sexes, and do propagate. Neither do we this by chance, but we know beforehand of what matter and commixture, what kind of those creatures will arise" (*New Atlantis*, para. 62).

It is important to note that there were no basic philosophical, scientific, ethical, or theological objections to spontaneous generation or artificial creation of life. Indeed, these were perfectly reconcilable with the biblical creation myth. Religious objections arose only after the invention of "creationism," which was provoked by various scientific developments in the nineteenth century. Before discussing these events later in this chapter, we need to have a brief look at another age-old topic: the creation of humanoids.

Imagine a World Where the Creation of Some Humanoids Is Ethically Acceptable

There are three rather unrelated traditions about the artificial creation of humanoids, which have nonetheless merged in many literary treatments of the topic since the nineteenth century: mechanical automata or androids, Kabalistic golems, and alchemical homunculi.

The first and the oldest tradition refers to mechanical devices that mimic the behavior of humans or animals. Greek mythology features at least two heroes of that art: Pygmalion, who sculptured an ivory figure of Aphrodite that was animated to become his wife Galatea, and Daedalus, the ingenious artisan who is said to have built various automata. The early Greek fascination with automata was likely more than mere folk myth, since the fifth-century BC poet Pindar, in his Seventh Olympic Ode, described animal-like automata placed in the streets of Rhodos for popular amusement. Ancient Greeks and Egyptians shared another fascination with "talking statues," in which hidden tubes transmitted the voice from a remote speaker, so as to animate the statues and give them authority for religious and prophetic messages. In antique Alexandria, where Greek and Egyptian cultures merged, the art of automata was perfected by the great engineers Ctesibius, Philo of Byzantium, and Hero of Alexandria, whose devices were copied and further developed by medieval Arab and, eventually, European engineers (Hill, 1996, ch. 11). Such toys made a great impression on our mechanical philosophers, including Descartes, Hobbes, La Mettrie, D'Alembert, and Kant,[2] particularly after the French engineer Jacques Vaucanson constructed an almost-perfect android in 1738. These philosophers all discussed whether perfect automata would be indistinguishable from natural animals, which most of them believed, and whether perfect androids would be indistinguishable from real humans, which some considered possible. However, none of them raised any ethical or religious concerns whatsoever about the mechanical manufacture of animals or humans.

The second tradition, Kabalistic golems, is more closely related to religion. Both the Greek Prometheus myth and the Jewish and Christian scriptures describe the divine creation of the first human out of clay or dust. Yet, only the Jewish tradition seems to have elaborated on the theme by deriving recipes for the artificial creation of humanoids, called golems (Idel, 1990; Newman, 2004, pp. 183–187). The oldest text is the brief and rather cryptic *Sefer Yezirah* ("Book of Creation"), probably of late antique origin, which inspired not only much of the Kabalah but also many medieval and early modern Rabbis to derive various recipes for making golems. Apart from forming a human figure out of clay or dust, they all made use of the magic of certain Hebrew words and letters, the command of which should bring the golem to life or to death. The motives for making golems greatly varied. A profane one was simply the want for cheap servants for housework. According to one account, the golem would slowly grow in size, turning from a servant into a threat to his creator, who would then destroy his creation by some Kabalistic magic (Scholem, 1969, pp. 200f). A second, more common motive was to prove the magic power of the Hebrew language, which was probably the original idea of the *Sefer Yezirah*. For religious people, the making of golems was also a way of worshipping and seeking closeness to God by repeating his creation, which was a highly revered motive (Idel, 1990, pp. xv–xvi). Yet the golem was usually described as speechless, stupid, and inferior to humans in order to point out the difference between human and divine creations. Only when people, after the thirteenth century, strived for the power of creation in order to compete with God was the making of golems severely criticized (pp. 98, 149). Thus, as with mechanical humanoids, there was no ethical objection specifically against the Kabalistic making (and killing) of golems.

The third tradition, alchemical homunculi, is related to alchemy, the forerunner of modern chemistry, and refers to the laboratory creation of homunculi. Compared to the widespread literary and artistic treatments of the topic since the nineteenth century, the alchemical sources in the form of explicit recipes are very rare; scholarly treatments of the topic are even rarer, despite its utmost importance in our context.[3] Indeed, much of the modern interest in homunculi seems to have arisen from misleading interpretations of the allegorical images and passages in alchemical texts that described chemical processes on the analogy of biological ones. However, there are at least three extant explicit recipes for homunculi from unknown authors that have attracted the attention of alchemists and others: two early medieval Arab texts (the pseudo-Platonic *Book of the Cow* and a passage from the Jâbir corpus) and a Renaissance treatise famously, but probably wrongly, attributed to Paracelsus (*De Rerum Naturae*).[4]

The alchemical homunculus tradition shows two important characteristics that distinguish it from the other traditions of creating life and humanoids. First, the essential ingredient was male semen that, following the Aristotelian theory of sexual reproduction, required for development the material matrix of menstrual blood, which the three authors sought to replace in the laboratory with various preparations, including the use of animal organs. Hence, the theoretical basis of creating homunculi was biological rather than mechanical or Kabalistic; it assumed that the male part was essential whereas the female part was replaceable by an alternative preparation. Second, the goals of creating homunculi were all related to ideas of perfecting nature or divine creation, because the artificial homunculi were considered to have improved qualities over natural humans. In the Arab texts, the homunculus, either in its entirety or in dismembered form, was believed to have medical, magical, or prophetic qualities. The pseudo-Paracelsian text points out that, because the homunculus was a product of art, he was acquainted with all the secret knowledge of art. Moreover, because his generation was not "polluted" by female contact, the author considered the homunculus a higher rational being.

Medieval and early modern concerns about making homunculi were largely of a theological nature.[5] For instance, critics argued that the creation was comparable to the usual practice of satanic demons, who, as they believed, stole human semen to breed giants and monsters. They argued that it was a temptation of God, who would be forced to create a new rational soul for the homunculi on demand. Furthermore, a newly created soul would lack the original sin stemming from Adam and thus undermine the predetermined religious order. Yet, the primary criticism of the making of homunculi has since early modern times always been the theological accusation of hubris, that is, of comparing one's creative power with that of the divine creator. Like the other objections, that is not an ethical issue but a problem deeply rooted in the intricacies of the Christian religion, which suggests that man was made in the likeness of an artisanlike creator god.

The few homunculi texts became famous, if only by rumor, because their authors all considered the creation of homunculi the crowning power of alchemy in surpassing the power of nature and even that of the divine creator. They thus considerably shaped the public image of alchemy, such that these presumptuous claims were also attributed to many other alchemists. The historical reason was that the claim of perfecting and surpassing nature was already highly debated before, in the field of metallic transmutation. Indeed, any simple chemical transformation was suspected to be a presumptuous change of the divine creation against God's will up to the eighteenth century (Newman, 1989; Obrist, 1996; Karpenko, 1998; Schummer, 2003a). Therefore, the homunculus could become an emblem of the hubris of alchemy altogether, which, as may be recalled, was the prototype of all laboratory sciences.

Compared to our views today, the assessments of artificial creations were almost inverted in medieval and early modern times. There were no ethical or theological objections against the creation of plants and animals, because that happened anyway all the time through spontaneous generation. Only if the creature was assumed to have a "rational soul," which some doubted even for non-European humans, was a battery of theological objections raised, from Satanism to tempting God to hubris. On the other hand, simple chemical transformations unknown from ordinary experience aroused strong suspicions of Satanism and hubris.[6] Surprisingly, there was an almost complete ethical vacuum, and the theological surrogates excluded just the realm of (nonhuman) artificial life.

Against that historical background, we may assume that today's public reactions to the project of creating life from nonliving materials are likely to refer to the two traditions that had previously raised concerns, that is, those of homunculi and alchemical substance transformation, rather than to those of automata and golems. However, the link to these traditions could become possible only through important developments in nineteenth-century biology and chemistry, which are briefly considered in the next two sections.

Religion Informs Science, Science Informs Religion

Christian creationism, as it is popular today particularly in the United States, relies on the idea that any living being owes its existence to the primordial divine creation. It might be hard for some to understand that this idea was largely formed only in the nineteenth century, and that it resulted from an exchange between science and religion. However, as long as people thought that spontaneous generation of life out of inanimate matter occurred every day, the notion that any living being owes its existence to the primordial divine creation was not very convincing.

There were alternative views to spontaneous generation since antiquity, however. The most prominent one was the Stoic doctrine of omnipresent sperms of quasi-material pneumatic nature, out of which living beings could grow under favorable conditions. Although the distinction between Aristotle's vital heat and pneumatic sperms was only a gradual and subtle one, it could make an important difference in the Christian doctrine. For instance, Augustine adopted the Stoic view and argued that all sperms were created during the primordial creation, such that any allegedly spontaneous generation was just an unfolding of the original seed of the Creator (Lippmann, 1933, pp. 23–24). However, despite some prominent followers, such as Leibniz and Buffon, that remained a minority view up to the nineteenth century.

Several seemingly unrelated scientific developments worked toward the formation of nineteenth-century creationism. First, late seventeenth-century mechanical philosophy, particularly in the hands of Boyle and Newton, was not only a scientific approach to the mechanical explanation of any phenomena, it also revived the old theological (and Islamic) idea of causal determinism, according to which any current phenomenon is the deterministic result of a chain of events that began with the primordial creation. In this natural theology, which became very popular in eighteenth-century philosophy and the dominant motive in natural history, anything, including any current living being, could be indirectly linked to divine creation and providence.

The second important scientific development was the increasing use and rigor of experiments in scientific studies, instead of the former collections of tales and second-hand observations. By careful study of the anatomy and reproduction of animals and plants, which was typically motivated by natural theology, the received standard set of spontaneously generating species incrementally shrunk. Each biological species for which the reproductive mechanism was explained could thus become a candidate for primordial creation. By the time of the great eighteenth-century taxonomies of Linneus and his followers, in which sexual reproduction played an important role, the possible candidates for spontaneous generation were reduced to microbes, most of which remained unobservable through microscopes for another century.[7] The most prominent candidates were then the so-called infusoria that appeared upon the infusion of dried plants, and that became the objects of rigorous experimentation in the mid-eighteenth century. The debate around 1860 over spontaneous generation between the chemist Louis Pasteur and the biologist Felix Pouchet illustrates the level of experimental sophistication (Geison, 1995, ch. 5; Schummer, 2003b). In order to prove the spontaneous generation of infusoria, Pouchet heated some grass at 300°C in an oxygen-free atmosphere and transferred it through mercury to a disinfected flask that contained nothing but freshly synthesized water and oxygen. In his famously celebrated counterproof, Pasteur demonstrated that dust could enter the flask via the mercury as a possible pathway for "germs" of Pouchet's infusoria. Pasteur left no doubt that his experimental rigor was based on religious motives. In his famous lecture at the Sorbonne in 1864, before the political and intellectual elite of France, he convinced his audience that the age-old idea of spontaneous generation would threaten the fundamentals of Christianity: "What a triumph, gentlemen, it would be for materialism if it could affirm that it rests on the established fact of matter organizing itself, taking on life of itself; matter which has in it all known forces! . . . What good then would it be to resort to the idea of primordial creation, before which mystery it is necessary to bow? Of what use then would be the idea of a Creator-God?" (Pasteur, 1922–1939, vol. II, pp. 328–346; English translation from Geison, 1995, p. 111).

To understand the radical religious shift toward creationism, from spontaneous generation being a banality to threatening the fundamentals of Christianity, we need to consider the third important scientific development: the theory of biological evolution, another offspring of natural theology. In Larmarck's theory of 1809, the elements that concern us here were already fully developed: He postulated an evolutionary "chain" going back from the most complex animal, the human species, to the simplest ones like worms and infusoria, which he thought arose from spontaneous generation.[8] Thus, with Lamarck one could already have argued that humans owe their existence to spontaneous generation. However, his evolutionary transformations were guided by a teleological principle of nature, which allowed for divine intervention, as both primordial creation and continuous guiding.[9] Moreover, Lamarck's ideas were soon discredited in the French Restoration, which linked spontaneous generation to materialism, atheism, and republicanism. It was another fifty years before Darwin published his *The Origin of Species* (1859) in which he carefully replaced Lamarck's teleology with causal determinism, which restored the principles of natural theology.[10] Although Darwin himself avoided publicly discussing spontaneous generation (Lippmann, 1933, pp. 106–107), the link to the generation of human beings was very obvious for contemporaries. Two years after the

French translation of his book appeared, Pasteur explicitly made this link in the same speech cited earlier: "Take a drop of sea water . . . that contains some nitrogenous material, some sea mucus, some 'fertile jelly' as it is called, and in the midst of this inanimate matter, the first beings of creation take birth spontaneously, then little by little are transformed and climb from rung to rung—for example, to insects in 10,000 years and no doubt to monkeys and man at the end of 100,000 years" (Pasteur 1922–1939, vol. II, pp. 328–346; English translation from Geison, 1995, p. 111).

As we have seen in the previous sections, humans were the only biological species about whose creation theologians had ever been concerned before, because only they had a "rational soul" that was immortal and administered by God. With Darwin's theory, the spontaneous generation of any simple organism could be seen as the first step toward the evolution of humans. The little infusoria thus challenged the core of divine creation, the making of Adam, and thereby the core of Christian salvation and moral theology, the immortal soul imbued with original sin. One answer to that challenge was nineteenth-century creationism, according to which any living being owes its existence to the primordial divine creation, for which scientists like Pasteur fought with the weapons of experimental science. While today's creationists see Darwin's theory as a threat to creationism, the historical order is rather inverse: Darwin's theory induced the formation of creationism because it linked spontaneous generation to the generation of humans.

Synthesis of Life Becomes a Challenge

The controversy over the spontaneous generation of life has never been settled (Farley, 1977). Rather, the candidates under debate incrementally moved from bacteria to viruses, to prions and simple self-sustaining molecular systems, increasingly blurring the distinction between life and nonlife. However, the controversy's religious offspring, nineteenth-century creationism, changed the Christian value system, even though it remained an extremist view within the variety of Christian doctrines. For millennia the creation of living beings, from simple organisms to simple humanoids, had been no matter of religious concern, as long as no "rational soul" or demon was involved. Now, even the most rudimentary approaches to the making of life, the chemical synthesis of organic compounds, could be accused of "playing God."

As mentioned earlier, alchemists had long been accused of changing the divine creation against God's will by simple chemical transformations of inorganic matter. That charge disappeared only during the eighteenth century based on systematic theories of chemical transformation, elements, and compounds. However, a corresponding charge came up again with the rise of nineteenth-century organic chemistry. As I have argued elsewhere, chemists became a target of severe criticism by hundreds of writers in many countries (Schummer, 2006a). These authors rediscovered the medieval literary figure of the "mad alchemist" and transformed it, by attaching some moral perversion, into the modern "mad scientist," which became the most powerful expression of the charge of hubris on contemporary chemists. To illustrate the problems that troubled nineteenth-century writers and the way they framed them, I quote a brief dialogue from Honoré de Balzac's novel *La recherche de l'absolu* (1834, ch. VI) between the chemist Balthazar Claes and his wife:

"I shall make metals," he cried; "I shall make diamonds, I shall be a co-worker with Nature!"

"Will you be the happier?" she asked in despair. "Accursed science! Accursed demon! You forget, Claes, that you commit the sin of pride, the sin of which Satan was guilty; you assume the attributes of God."

"Oh! Oh! God!"

"He denies Him!" she cried, wringing her hands. "Claes, God wields a power that you can never gain."

At this argument, which seemed to discredit his beloved Science, he looked at his wife and trembled.

"What power?" he asked.

"Primal force—motion," she replied. "This is what I learn from the books your mania has constrained me to read. Analyse [sic] fruits, flowers, Malaga wine; you will discover, undoubtedly, that their substances come, like those of your water-cress, from a medium that seems foreign to them. You can, if need be, find them in nature; but when you have them, can you combine them? Can you make the flowers, the fruits, the Malaga wine? Will you have grasped the inscrutable effects of the sun, of the atmosphere of Spain? Ah! Decomposing is not creating."

"If I discover the magistral force, I shall be able to create."

When Balzac wrote his novel, in which he aimed to describe "all the efforts of modern chemistry,"[11] chemists were able to synthesize only a handful of organic compounds from inorganic matter. Most chemists therefore still believed in "chemical vitalism," according to which organic compounds, unlike inorganic matter, were organized by a "vital force" that was largely out of the control of chemists. In the mid-nineteenth century, however, French and German chemists started what remains the biggest research project ever: In an effort to refute vitalism, they systematically synthesized anew, in their labs, all of the organic compounds that they had isolated before from animals and plants (Russell, 1987; Bensaude-Vincent, 1998, ch. 2; Bensaude-Vincent, 2002, pp. 29–32; Schummer, 2003a; Bensaude-Vincent, 2005, ch. 2). In the period from 1844 to 1870 alone, the number of known organic compounds thus rose from about 720 to 10,700 (Schummer, 1997). The project was meant to prove that the creative power of chemistry was comparable to that of "living nature." It did not stop there. As the synthetic toolbox expanded, chemists also produced new compounds, some of which served human needs better than natural products did. The number of compounds continued to grow exponentially, from about 100,000 in 1900 to 800,000 in 1950, to 18.5 million in 2000, as a gigantic proof of the creative power of chemistry.

In the first half of the twentieth century, when questions arose about the sense of synthesizing ever more compounds, chemists developed a metanarrative to provide historical meaning and future orientation, in a language reminiscent of alchemy. The early phase of refuting vitalism was now called a state of "learning from nature," which was followed by a state of "rivaling and surpassing nature" in producing better compounds for human needs. The next step then should be "mastering and designing nature," or, in the terms of the chemist Walden (1941, p. 49), "directing, in accordance to its conditions, the processes in the living organism and designing them for the benefit of humanity." Although the final step was frequently avoided or only indirectly hinted at, that metanarrative became the standard model for writing stories of progress in popular histories and popularizations of chemistry as well as in official reports of chemistry and, eventually, nanotechnology (Schummer, 2006b).

It is fair to say that the chemical synthesis of life has always been in the air since the mid-nineteenth century, from its religious critics and literati as well as aspiring chemists. Since Aldous Huxley's *Brave New World* (1932), at the latest, it has also been an established topic in the dystopian literature. For organic chemistry, on the other hand, the metanarrative made it the ultimate and natural goal, which should provide historical meaning to the entire discipline. There was no need for chemists to mention that goal explicitly, as long as their discipline was internally flourishing and externally regarded with suspicion.

During the Cold War era, however, chemistry lost much prestige and money to its rival physics, which was heavily involved in many military-related big-science projects, from nuclear energy and weapons research, to particle accelerators, radio-astronomy, and space exploration. Chemistry, on the other hand, had no such big-science project. Against that background, organic chemist Charles C. Price held his 1965 presidential address before the American Chemical Society. With an envious view on the latest "race-track accelerator," he called for "the setting of the synthesis of life as a national goal" (Price, 1965, p. 91).[12] "The political, social, biological, and economical consequences of such a breakthrough would dwarf those of either atomic energy or the space program. . . . The job can be done—it is merely a matter of time and money" (p. 91). Explaining where all these unprecedented consequences should come from, Price claimed that the synthesis of life would lead "to modified plants and algae for synthesis of food, fibers, and antibiotics, to improved growth or properties of plants and animals, or even to improved characteristics of man himself" (p. 91).

Price's synthesis of life never became a national goal, though. One reason was probably that the president of the American Chemical Society did not shy away from provoking ethical concerns, by explicitly relating the synthesis of life "to improved characteristics of man himself," to say nothing about religious concerns. I assume that this was a deliberate, albeit unlucky, provocation, intended to draw widespread public attention to his project. Another reason certainly was that he simply failed to provide a single argument as to what the synthesis of life should be good for. All of his arguments, which I have quoted here, refer to the modification rather than the *de novo* synthesis of life, the proper goal of the project. Lacking any utilitarian reason, the synthesis of life project was appealing only to Price's fellow chemists. It would have demonstrated the ultimate power of chemical synthesis, to the pride of chemistry and to the dismay of its religious critics.

Lack of Justification, Lack of Ethics

Currently, many diverse research programs are related to synthetic biology, protocell research, and, more recently, bionanotechnology. However, five of them should not be confused with the synthesis of life. The first is classical genetic engineering that modifies existing organisms on the genetic level by recombinant DNA technology, that is, by transferring DNA sequences from one species to another. The second is an offspring of genetic engineering, producing modified proteins by feeding the gene expression apparatus of organisms with synthetic DNA, RNA, or different amino acids. Neither project can be called synthesis of life, because they modify or use only existing organisms or parts thereof. Third, some engineers seek to employ DNA (or proteins) for digital data storage and processing, which has nothing to do with the synthesis of life. Fourth, there is an offspring of artificial intelligence, called *soft artificial life*, that develops algorithms for mathematically modeling existing or potential biological systems in order to understand their dynamics.[13] Fifth, the only approach that comes close to the synthesis of life, is laboratory simulation that tries to mimic chemical conditions on Earth several billion years ago to understand how the formation of organic compounds and simple prebiotic molecular systems could have happened.[14] While the first three projects typically seek justification (and support against ethical criticism) in their potentially beneficial products, the latter two projects do so by referring to improved understanding of the principles and origin of life.

Unlike these five projects, *de novo* synthesis of life consists of two other approaches. The first tries to replicate a particular naturally occurring organism (or its DNA) from scratch as, for instance, a group from SUNY did with the poliovirus (Cello, Paul, & Wimmer, 2002). The second approach seeks to assemble from a toolbox, which may or may not be derived from a variety of natural organisms, a new molecular system that complies with some definition of life, including the criteria of metabolism, self-replication, and natural selection (see Rasmussen et al., 2004).

Compared to the other five research programs, the ambitious goal of creating life *de novo* is conspicuously weak in substantial justification. Of course, reference is frequently made to both beneficial products and improved understanding, but the arguments are not always convincing. For instance, many arguments that refer to beneficial products rely on blurring the distinction between creating new organisms and modifying existing ones, as Price already did in 1965, and then draw their justification for *de novo* creation from the beneficial prospects of modifying existing organisms. Or they fail to provide convincing arguments as to why the much more difficult *de novo* synthesis would be a more promising and efficient strategy than modifying existing organisms. Sometimes, as in the much-discussed case of artificial cells that would create hydrogen to solve the world's energy problems, the arguments seem to draw on science fiction rather than the actual research in chemical catalysis for hydrogen production. Another frequent type of argument refers to possible spinoffs, according to which the *de novo* synthesis project would, as a side effect, enable some fundamental understanding or useful knowledge. However, as with all such arguments known from the Cold War era, it is questionable why one should hope for some unforeseeable spinoffs. If improved understanding or beneficial products are the goal, why not perform a research project that is tailored to that goal?

The relative weakness in justification for *de novo* synthesis of living organisms suggests that the historical motive is still important; that is, the leading motivation is to prove the "creative power of man," a symbolic act in the imagined rivalry with a metaphysical agency. Chemists have always called this rival "nature." Yet, as Robert Boyle already observed in 1682, that is a "semi-deity," "a kind of a goddess, with the title of nature," a substitute for the Christian god (Boyle, 1772, vol. 5, pp. 164, 191). Naturally that raises Christian concerns and the charge of hubris, of "playing God." It seems, therefore, that the scientific fascination with the creation of life and its religious abhorrence are but two sides of the same religious coin.

As I argued earlier, that religious coin was largely created in the nineteenth century, so neither the fascination nor the abhorrence have deep roots in Christian culture. Although constructed in response to eighteenth- and nineteenth-century scientific developments, the whole idea of "playing God" rests on the nineteenth-century ad hoc assumption that any natural living being owes its existence to the divine creation. That is not only bizarre from a scientific point of view but also theologically problematic, because it contradicts almost two thousand years of theology. It would be more useful to work on a sound theology than draw the emotions of fascination or abhorrence from the obscure concept of "playing God."

What is worse, however, is that we have been left with an almost complete ethical vacuum regarding the artificial creation of living organisms. That is not because "ethics is lagging behind technology," as a frequent excuse goes. On the contrary, for thousands of years people were convinced that they could create living beings, from simple organisms to humanoids, without ever bothering about any ethical issues. And as long as no rational soul was involved, there was no religious issue either. Of course, there are many modern ethical approaches to establishing respect for nonhuman living beings, from Jeremy Bentham's animal rights to Albert Schweitzer's "reverence for life." However, these approaches fail to address the artificial creation of life. The only viable approach thus far seems to be consequentialism, that is, assessing the creation of life from its prospective positive and negative consequences, and with precaution regarding its possible risks. Although that is not the topic of this chapter, I conclude with a brief ethical assessment.

Because the whole project of *de novo* synthesis of life thus far seems to lack substantial justification regarding both improved understanding and beneficial products, there is not much to say in its favor from a consequentialist point of view other than that the creators could pride themselves on their creation. On the negative side, there are predictable social costs, both in drawing research funding and attention away from more useful projects and in provoking a social conflict over a theologically obscure matter. Moreover, because creating does not presuppose or automatically produce a full understanding of the creations, protocells carry the risks of negative impacts on the natural environment. As self-replicating beings, they must interact with their environment; the closer their metabolism is to that of natural organisms, the more they will interfere with biological systems, on the organism and ecological levels as well as irreversibly on the evolutionary level. At the present state of biological knowledge, it is impossible to predict what the exact consequences will be, but only a fool would expect them all to be beneficial.

Conclusion

As today's scientists strive for the creation of life, they tend to dismiss the age-old history of creating life, because any former approach was based on scientific misconceptions. However, from a historical point of view, it does not matter much whether former views were right or wrong from our perspective, as long as the relevant contemporary scientists were convinced they were right. What matters instead is how these views were shaped and discussed in their cultural contexts. In such a history of the idea of creating life, which I have briefly sketched in this chapter, current efforts to create protocells are the latest effort that owes both its scientific motivation and mixed public reception to its prehistory.

I have argued that the crucial phase in that history was the nineteenth century, when the creation of life changed from a banality to a challenge, to both the Christian doctrine and the implicit goal of synthetic organic chemistry. In this period of vivid exchange between science and religion, the two emotions of fascination and abhorrence emerged as two sides of the same coin, created around the scientifically and theologically obscure idea of "playing God." Since then, public debates have been largely captured by these two irreconcilable emotions, such that appealing to one emotion on one side usually provokes the other emotion on the other side, and vice versa. Thus, whoever tries to promote protocell research by appealing to the fascination of creating life is likely to provoke public protest; and whoever raises the concern of "playing God" is likely to raise somebody's fascination for doing so.

Such public debates are not only counterproductive, they also have only a weak ethical basis. Fascination for and abhorrence of certain research are personal emotions that explain the approval or disapproval by an individual on psychological grounds; they do not, per se, provide ethical arguments for or against that research. As long as public debates are focused on these emotions, they continue to exclude ethical considerations about creating life. As I have argued throughout this chapter, there was never an ethical debate proper, not because creating life was considered impossible, but because nobody considered it an ethical issue. Instead, we have been left with an ethical vacuum about the creation of life that was filled only with quasi-moral surrogates. Since history tells us when and why these surrogates were created, we might be able to replace them with urgently required ethical deliberations.

Notes

1. "[I]n the case of animals generated from putrefaction, the formative power is the influence of the heavenly bodies" (Thomas Aquinas, *Summa Theologiae*, Pt. I, q. 71).

2. Descartes: *Discours de la méthode* (1637), V. 16; Hobbes, *Leviathan* (1651), Introduction; La Mettrie *L'homme machine* (1748); D'Alembert: "Androide" in his *Encyclopédie* (1758); Kant, *Kritik der Praktischen Vernunft* (1788), AA181.

3. In the following, I refer mainly to the excellent analysis of Newman (2004, ch. 4).

4. Paracelsus definitely wrote a treatise called *De homunculi*, which many modern authors, particularly Goethe scholars who comment on the homunculus in *Faust II*, have cited,

obviously without reading it, because it is largely a moral treatise against sodomy rather than a laboratory manual for creating homunculi.

5. The only nontheological objection seems to be that the idea of creating homunculi would denigrate women in human reproduction to the role of mere vessels (Newman, 2004, p. 193), but that objection addressed the underlying theory rather than the practice of creating homunculi.

6. The classical concern, which accompanied the entire history of alchemy up to the eighteenth century, was already formulated by the Latin church father Tertullian at the turn of the second century: If God wanted human beings to wear purple cloths, he would have created purple sheep; since he did not make purple sheep, the dyeing of wool is against God's will and therefore a sin, an alliance with Satan, as Tertullian emphasized (*De cultu feminarum*, I, 8); see also Schummer (2003a).

7. The resolution of microscopes was insufficient for the observation of most microorganisms because of both chromatic and spherical aberration. A first step with achromatic compound lenses in microscopes was made in the 1830s, but it took further decades to solve both problems by sophisticated lens systems, such that microscopes with satisfying resolution below 1 μm were commercially available probably only in the 1870s.

8. See J. B. Lamarck: *Philosophie zoologique* (1809), vol. 1, I. 6; for his views on spontaneous generation, see Lippmann, 1933, p. 75; Farley, 1977, pp. 41ff.

9. "And everywhere and always the will of the sublime Author of nature and of all that exists is invariably brought about" (Lamarck, 1809, vol. 1, I. 4).

10. "There is grandeur in this view of life, with its several powers, having been originally breathed by the Creator into a few forms or into one; and that, whilst this planet has gone cycling on according to the fixed law of gravity, from so simple a beginning endless forms most beautiful and most wonderful have been, and are being evolved" (*Origin of Species*, ch. 14, final sentence).

11. In a letter to Hippolyte Castille, Balzac explains, "Le héros de *La Recherche de L'Absolu* représente tous les efforts de la chimie moderne" (quoted from Ambrière, 1999, p. 401).

12. I am grateful to John Smith from Lehigh University for pointing me to this paper.

13. Of course, some adherents of "artificial life" would dispute my description, claiming that their algorithms are structurally equivalent to biological systems such that both are true forms of life but in different "media," and that writing an algorithm is creating life. I cannot comment here on these idealistic fallacies, which repeat former fallacies of AI, but refer instead to an earlier paper by Claus Emmeche (1992).

14. Strangely enough, reviews of "synthetic biology" frequently overlook that approach, although Stanley L. Miller and Harold C. Urey performed already in 1953 what is considered the classical experiment on the origin of life.

References

Ambrière, M. (1999). *Balzac et la Recherche de l'Absolu*. Paris: Presses Universitaires de France.

Aristotle (1912). On the generation of animals. In W. D. Ross (Ed.), *The Oxford translation of Aristotle* (vol. 5, A. Platt, Trans.). Oxford: Clarendon.

Bacon, F. (1627). The new Atlantis. In J. Spedding, R. L. Ellis, & D. D. Heath (Eds.), *The works of Francis Bacon* (vol. 5, pp. 347–414), London: Brown and Taggard.

Balzac, H. de (1834). *La recherche de l'absolu.* Paris: Vve Béchet (English trans. from *The Alkahest; or, The house of Claës,* trans. K. Prescott Wormeley. Boston: Roberts, 1887).

Bensaude-Vincent, B. (1998). *Eloge du mixte: Matériaux nouveaux et philosophie ancienne.* Paris: Hachette.

Bensaude-Vincent, B. (2002). Changing images of chemistry. In I. H Stamhuis et al. (Eds.), *The changing image of the sciences* (pp. 29–42). Berlin: Springer.

Bensaude-Vincent, B. (2005). *Faut-il avoir peur de la chimie?* Paris: Le Seuil.

Boyle, R. (1772). A free inquiry into the received notion of nature (1682). In T. Birch (Ed.), *The works of the honourable Robert Boyle* (vol. 5, pp. 164, 191). London.

Cello, J., Paul, A. V., & Wimmer, E. J. (2002). Chemical synthesis of poliovirus cDNA: Generation of infectious virus in the absence of natural template. *Science, 297,* 1016–1018.

Darwin, C. (1859). *The origin of species.* London: John Murray.

Emmeche, C. (1992). Life as an abstract phenomenon: Is artificial life possible? In F. J. Varela & P. Bourgine (Eds.), *Toward a practice of autonomous systems. Proceedings of the first European conference on artificial life* (pp. 466–474). Cambridge, MA: MIT Press.

Farley, J. (1977). *The spontaneous generation controversy from Descartes to Oparin.* Baltimore: John Hopkins University Press.

Geison, G. L. (1995). *The private science of Louis Pasteur.* Princeton: Princeton University Press.

Hill, D. (1996). *A history of engineering in classical and medieval times.* London: Routledge.

Idel, M. (1990). *Golem: Jewish magical and mystical traditions on the artificial anthropoid.* Albany, NY: SUNY Press.

Karpenko, V. (1998). Alchemy as donum dei. *Hyle: International Journal for Philosophy of Chemistry, 4,* 63–80.

Lamarck, J.-B. (1809). *Philosophie zoologique.* Paris: Dentu.

Lippmann, E. O. v. (1933). *Urzeugung und Lebenskraft: Zur Geschichte dieser Probleme von den ältesten Zeiten an bis zu den Anfängen des 20. Jahrhunderts.* Berlin: Springer.

Newman, W. R. (1989). Technology and alchemical debate in the late middle ages. *Isis, 80,* 423–445.

Newman, W. R. (2004). *Promethian ambitions: Alchemy and the quest to perfect nature.* Chicago: University of Chicago Press.

Obrist, B. (1996). Art et nature dans l'alchimie médiévale. *Revue D'Histoire des Sciences, 49,* 215–286.

Pasteur, L. (1922–1939). *Œuvre de Pasteur,* 7 vols. Paris: Masson et Cie.

Price, C. C. (1965). The new era in science. *Chemical and Engineering News*, Sept. 27, 90–91.

Rasmussen, S., Chen, L., Deamer, D., Krakauer, D., Packard, N. H., Stadler, P. F., & Bedau, M. A. (2004). Transitions from nonliving to living matter. *Science, 303*, 963–965.

Russell, C. A. (1987). The Changing Role of Synthesis in Organic Chemistry. *Ambix, 34*, 169–180.

Scholem, G. (1969). *On the Kabbalah and its symbolism.* New York: Schocken.

Schummer, J. (1997). Scientometric studies on chemistry I: The exponential growth of chemical substances, 1800–1995. *Scientometrics, 39*, 107–123.

Schummer, J. (2003a). The notion of nature in chemistry. *Studies in History and Philosophy of Science, 34*, 705–736.

Schummer, J. (2003b). Chemical versus biological explanation: Interdisciplinarity and reductionism in the 19th-century life sciences. *Annals of the New York Academy of Science, 988*, 269–281.

Schummer, J. (2006a). Historical roots of the "mad scientist:" Chemists in 19th-century literature. *Ambix, 53*, 99–127.

Schummer, J. (2006b). Providing metaphysical sense and orientation: Nature-chemistry relationships in the popular historiography of chemistry. In I. Malaquias, E. Homburg, & M. E. Callapez (Eds.), *Chemistry, technology and society: Proceedings of 5th International Conference on the History of Chemistry, Estoril & Lisbon, Portugal, 6–10 September 2005* (pp. 166–175). Aveiro: Sociedade Portuguesa de Química.

Tertullian (1844). *De cultu feminarum.* In J.-P. Migne (Ed.), *Patrologia Cursus Completus. Patrologia Latina* (Vol. 1). Paris.

Walden, P. (1941). *Geschichte der Organischen Chemie seit 1880.* Berlin: Springer.

9

Second Life: Some Ethical Issues in Synthetic Biology and the Recapitulation of Evolution

Laurie Zoloth

What Are We Watching?

After seeing the scientist's PowerPoint demonstration, the moral philosopher walks up to the podium to talk to the molecular biologist and to watch, once more, the little movie he has shown during the talk. It is a short looping clip of small lipid globules that line up, the circles linking into chains, and the chains moving back and forth across the screen. They are moving, it would appear, in response to the magnesium concentration in the solution they are in. In another loop, they cluster and link and move across temperature gradients. If they were cellular organisms, one would describe this as responsive behavior and view the linkage of cell to cell to form chains as primitive and evolutionary organization. However, these are not living cells—not quite. They are entirely artificial, made as surely as any machine, but of logical and naturally occurring components that, in their assemblage, create something entirely new. What they are (beings in the act of being) and what they are doing (*ex nihilo* generation and supple adaptation; see Bedau, 1996) has long been the subject of philosophical discussion.

This project—the building of an artificial cell—is part of a larger goal for both the philosopher and the biologist. As the project of synthetic evolution continues, what are the issues in ethics and philosophy that it evokes? What are we really asking when we ask which norms and values are engaged when one tries to recapitulate life itself? Is it a matter of ethical permissibility? Transgressive research? Are we "seeing" lifelike patterns because we shape data into familiar stories, "biologizing" physical events? Is life a matter of descriptive criteria, or is it a matter of activities? This chapter contends that asking about life's origins not only uncovers intriguing questions about chance, meaning, and order, but also allows a rather startling discourse about the nature of knowledge in a world with physical realities and enduring cultural narratives.

Professor Jack Szostak has long been interested in how life evolved from the molecules available in the early Earth environment. A new understanding of the role of RNA caught his interest early in his career. Previously thought to be capable only of transmitting information, RNA, according to discoveries in the early 1980s, was capable of catalyzing enzymes as well. Consider the following depiction of Szostak's research:

A Nobel Prize–winning discovery in the 1980s by Tom Cech and Sidney Altman propelled Szostak down a new research path. The pair independently demonstrated that RNA, the sister molecule of DNA, can catalyze certain chemical reactions inside cells, a job previously thought to be the exclusive domain of proteins. Until then, RNA was thought to have just one function: storing the genetic information cells need to build proteins. This new revelation about RNA's dual role suggested to some scientists, including Szostak, that RNA likely existed long before DNA or proteins because it might be able to catalyze its own reproduction. Their discovery made it easier to think about the origin of life, Szostak says. "They inspired me to try to think of ways to make RNAs in the lab that could catalyze their own replication." By 1991, Szostak had shifted the entire focus of his lab to evolving new functional RNAs and other molecules in a test tube. As the basis for his work, Szostak developed a technique called *in vitro* selection to study the evolution of biological molecules. This method screens vast numbers of molecules for a predetermined function, such as the ability to catalyze a specific chemical reaction or bind a target molecule. Those that don't fit the desired profile are filtered out and the process is repeated over and over again until researchers find the molecule that does a particular job. Using *in vitro* selection as a way to apply the forces of natural selection in a laboratory setting, Szostak and his colleagues evolved RNAs that bind to ATP, a common biological substrate, from a massive library of 1,000 trillion random RNA sequences. (HHMI, 2006)

Once he made RNAs that could bind to ATP, Szostak was curious about how he could create catalytic RNA, that is, make artificial ribozymes akin to the actual ribozymes that act as catalytic substrates in a living cell. He was not only about to generate such molecular structures, but to generate them with far more complexity than is actually found in nature—which led him to speculate as to whether the defining specificity in the actual world was the end of a longer evolutionary process.

This is critical because scientists have largely come to agree and accept the basics of Darwinian evolutionary theory of adaptive mutation to ecological niches, which, while a satisfying explanation for the speciation and specification of biological life, still leaves open the question of first cause in evolution: How was the step from inorganic chemicals in a coincident neighborhood to self-replicating, self-organizing single cells made? Since this is the core question of origins—of the nature of life itself—why is it so difficult to explain how this shift occurs? Szostak is interested in that borderline question. Could one set in place the core chemical requirements for biological life by assembling the elements presumed to be available in the primordial Earth, and, by creating conditions for evolutionary processes, set them in motion, create a model system, and watch it evolve? What would be the implications of such a project: an artificially created "RNA world" set in motion by single-stranded RNA?

Finding the most interesting and efficient nucleic acid sequences was itself an evolutionary project, in which selective pressures and amplification allowed the discovery of the most functional coding sequences, which Szostak called aptamers—nucleic acids able to bind a wide range of molecules that are a part of every cell.[1] The next step was to place this biologically capable, information-bearing RNA strand inside of something like a cell—a small sack of lipid-based molecules—and see what would happen, given carbon molecules, warmth, and water.[2] Vesicles were induced to form by manipulation of the basic environment. As Szostak noted: "Other researchers had observed that if fatty acid micelles, which are stable at basic conditions, are exposed to more acidic conditions, they spontaneously assemble into vesicles. This reaction has a long lag period, and some sort of nucleation surface is required to trigger the process. We reasoned that if the right kind of mineral surface was present, this lag phase would be eliminated" (HHMI, 2006).

As his group considered this problem, they turned toward testing a variety of common (and thus probable) materials, and came across the idea of adding small quantities of the clay montmorillonite to the fatty acid micelles. This turned out to greatly accelerate the formation of vesicles, and their ability to stabilize. They also discovered that many other substances with negatively charged surfaces similarly catalyzed formation of vesicles. When the researchers loaded montmorillonite particles with a fluorescently labeled RNA and added those particles to micelles, they detected the RNA-loaded particles in the resulting vesicles. Going a step further, Szostak and his colleagues showed that when they encapsulated labeled RNA alone inside vesicles, it did not leak out (Hanczyc, Fujikawa, & Szostak, 2003). Thus, they had created something that looked very much like a cell.

The chance discovery of clay as a catalytic agent was key in introducing a tantalizing possibility: Since montmorillonite is ubiquitous, was it just a coincidence that the action in the lab model so closely mimicked the observations in biology?

If "clay and other mineral surfaces accelerate vesicle assembly, but assuming that the clay ends up inside at least some of the time, this provides a pathway by which RNA could get into vesicles," said Szostak. However, he said, "even primitive, non-living, cell-like structures need a mechanism to grow and divide." Thus, the scientists explored the behavior of vesicles to which micelles had been added—finding that acidic conditions induced the micelles to become unstable and somehow incorporate themselves into a growing vesicle. "After we showed that efficient growth was possible, the next problem was how to complete the cycle by persuading these vesicles to divide," said Szostak. The scientists discovered that if they extruded larger dye-containing vesicles through smaller pores, the result was a proliferation of smaller vesicles, which still contained dye. "Exactly how this proliferation happens is not clear, and there are different models for the processes," said Szostak. "The important thing is that it all works." (HHMI, 2006)

The spontaneously organized vesicles displayed interesting properties: Physical parameters, not specific genetic instructions, seemed to drive selective competition. Vesicles with differing chemical properties (e.g., sucrose or lack thereof, variant volumes of RNA) were more or less successful at growing larger, and then more complex, "creating" membranes across chemical gradients. Selection could be accelerated by forcing the vesicles through a gradient net that would break them into smaller vesicles, with clay and genetic material in each. Next:

"We proposed that the genetic material could drive the growth of cells just by virtue of being there," he said. "As the RNA exerts an osmotic pressure on the inside of these little membrane vesicles, this internal pressure puts a tension on the membrane, which tries to expand. We proposed that it could do so through the spontaneous transfer of material from other vesicles nearby that have less internal pressure because they have less genetic material inside." In order to test their theory, the researchers first constructed simple model "protocells," in which they filled fatty-acid vesicles with either a sucrose solution or the same solvent without sucrose. The sucrose solution created a greater osmotic pressure inside the vesicles than the solvent alone.

There are several important differences between the early versions of an engineered cell and the actual ones found in nature. However, it may well be the case that this level of sophistication was not found in primordial cells. In any case, the process of growth could be clearly noted. . . . When the scientists mixed the two vesicles, they observed that the ones with sucrose—in which there was greater membrane tension—did, indeed, grow by drawing membrane material from those without sucrose. "Once we had some understanding that this process worked, we moved on to more interesting versions, in which we loaded the vesicles with genetic molecules," said Szostak. The researchers conducted the same competition tests using vesicles loaded with the basic molecular building blocks of genetic material, called nucleotides. Next, they used RNA segments, and finally a large, natural RNA molecule. In all cases, they saw that the vesicles swollen with genetic material grew, while those with no genetic material shrank. "It's a nice simplification of the whole process," says Szostak. Different replicator-vesicle packages compete with each other to become more numerous, so Darwinian evolution can occur with relatively simple molecular systems. Once these simple cells start competing, Szostak believes, there is a snowball effect. "You start to get additional functions evolving, and that's going to lead to changes in the membrane composition. The whole system is going to be under pressure to get a lot more complicated pretty quickly." (HHMI, 2006)

Over a period of seven years, as each step in the process unfolded and the artificial "RNA world" became more interesting, what the increasingly sophisticated vesicles actually were became a more nuanced question. Clearly, they were ordered chemical systems or processes that followed observable physical rules. In reflecting on the progress of this work, one can turn to Erwin Schrödinger, a physicist confronted with new tools in molecular biology and information-bearing DNA. He noted:

What is the characteristic feature of life? When is a piece of matter said to be alive? When it goes on "doing something," moving, exchanging material with its environment, and so forth, and that for a much longer period than we would expect an inanimate piece of matter to "keep going" under similar circumstances. When a system that is not alive is isolated or placed in a uniform environment, all motion usually comes to a standstill very soon as a result of various kinds of friction . . . after that, the whole system fades away into a dead, inert lump of matter. . . . It is by avoiding the rapid decay into the inert state of "equilibrium" that an organism appears so enigmatic . . . everything that is going on in Nature means an increase in the entropy of the part of the world where it is going on. . . . (Schrödinger, 1944, p. 70)

Szostak's intellectual project continues to evolve. Each year, the lipid vesicles and their organelles, which I call *event entities*—for they are both things and processes—continue to "act" more complexly. These event entities become less process and more thing, at which point we may call them, after the seventeenth-century automata, *synbiotica*. This raises the question of when they could reasonably be used as signifiers for "life." Is it when they become more organized, and thus compete more vividly, acting more and more lifelike? Is it when the events themselves seem to follow historically understood evolutionary paths?

What do we actually mean by calling such event entities markers of "life"? And what sort of category are such newly synthesized things creating along with their "behaviors"? These questions were theorized long before the science was possible. The problem of complexity and entropy is tantalizingly counterintuitive. To Schrödinger, theorizing about the nature of life before the discovery of RNA or DNA (he guessed that order might be created by a mysterious "aperiodic solids" somewhere on the chromosomes), the capacity for the ordered state he would call "life" emerging from the disorder of random atoms in space (matter) must be a matter of some new type of physical law:

To a physicist the state of affairs ("the organisms' astonishing gift of centering a 'stream of order' on itself and thus escaping the decay into atomic chaos—of 'drinking orderliness' from a suitable environment") is not only plausible but most exciting, because it is unprecedented. (Schrödinger, 1944, p. 70)

But certain aspects of the actual work are not completely without precedent, of course, for as soon as the work was published it became front-page news. This was true for a variety of reasons, from the precedents in text and in history; the shaping of lifelike beings using assembled parts has been a scientific fascination for centuries. Using clay as a catalyst deepens the emotive force and symbolic character of the work. This is because of the powerful hold that clay as a signifier of creation itself has on the narrative imagination in many religious traditions. Within the Islamic world, the research drew immediate attention: "Humans are a mix of dust and divine. For Muslims, clay holds a special significance – it is not quite dust and not quite water, it can be shaped, fired, permanent yet fragile, an indispensable part of civilization, a central metaphor for life from non-living form" (Umar, 2003, p. 3).

For Jewish ethicists, as well as the textual and theological borderlands adjacent to life, clay and creation were well-understood metaphors. Two creation narratives deal with clay and the passage from thing to creature. The first is in Genesis 2:6–7, in the first narrative of creation: "And a mist went up from the earth, and watered the whole face of the ground. And the Lord God formed man of the dust of the ground, and breathed into his nostrils the breath of life; and man became a living soul." The metaphor of the body as made of clay—fragile and pliable—appears in Jeremiah 17 ("Behold, as the clay is in the potter's hand, so are you in my hand, O house of Israel"), Isaiah, and Job 5 ("Your hands have made me and fashioned me; yet you turn around to destroy me . . . Remember, I beseech you, that you have made me of clay; and will you bring me back to dust? . . . You have granted me life and favor, and your providence has preserved my spirit").

The second creation narrative, a story repeated from the Gemora to the present day in Jewish fiction, deals with the frank ability to create life from material, not *ex nihilo*, but from a parts list of possible available things. "Things" is understood here both as material things, like clay or mandrakes, and as specific words, like names or incantations. "Rava said: If the righteous wished, they could create a world, for it is written, 'Your inequities have been a barrier between you and your God.' For Rava created a man and sent him to R. Zeira. The rabbi spoke to him but he did not answer. Then he said: 'You are from the pietists: return to dust'" (Zoloth, 2001, p. 4; Babylonian Talmud, Tractate Sanhedrin).

The trope is repeated in medieval and contemporary Jewish fiction, and is recalled by many ethicists as a cautionary tale about the moral instability of such an act of creation (in virtually every article in *Jewish Bioethics* written about cloning or robotics; see, e.g., Wolpe, 1997; Dorff, 2004). There is a persistent narrative that a humanlike being could be created, part by part, by the use of clay and incantation of God's secret name. This creature, called a golem, stands as the liminal case between artificial and real life. (In secular terms, the sign of the catalytic clay becomes the sign of electricity, and the golem becomes the Frankenstein myth.) In contemporary Jewish literature, it is recalled by Cynthia Ozick, who writes a series of golem novels, and Michael Chabon (2000), who wrote a Pulitzer prize–winning novel retelling the story of the golem, made from clay by the cleverest of scholars, and turned into a humanoid creature with the use of sacred letters. The parallels to the use of RNA and clay are inescapably tantalizing, so much so that any resulting creative synthetica become signs for a series of values and fears about power, creativity, scholarship, and the shifting boundaries between self and non-self.

For all Abramic traditions, the creation narrative in which clay is a key element in making human beings is a repeated trope. It offered a satisfying substantiation to the sign of creation that is signified by any of the research. Yet it is also morally destabilizing, for the closer artifice draws toward lifelikeness, the more we feel we owe the thing that we have brought into being. The ethical challenges emerge when the duty toward the other that is understood as alive is mistakenly undermined by the sense that the not-quite being can be destroyed, lest it become uncontrollable.

In other texts, philosophers have noted the uncertain quality of descriptors of life. For Michael Wheeler (1996), *animats* or artificial autonomous agents can be categorized as life. Yet animats are at a stage of existence prior to such a definition; they are largely events or processes, rather then entities. As the editors of this volume note, the vesicle systems made in the lab are a sort of thing—one can track them, weigh them, photograph them acting in time. Hence, perhaps an entirely new category is called for in describing things whose lifelikeness consists almost entirely of the barest gesture toward being, the first gesture of their change in state and nonentropic organization; hence the search for the proper name: event entities, eventities, or synbiotica.

"What are we watching?" the moral philosopher asks as we watch the event entities repeat their mesmerizing loops. "I do not know," the molecular biologist says, "but it is very interesting."

Let me conclude this section with a cautionary note. Many things in the chemical world move about in response to other things: iron filings in response to magnetic fragments, leaves in response to wind, lint in response to static electricity, soap bubbles in response to friction. Yet their movements evoke in us only a sense of unseen power, not of purpose, or instinct. What is powerful about the cascade of events taken by the Szostak synthetica is their changing and derivative evolution into increasingly complex things, with relative advantages and rapid exploitation.

So what is it that makes a single cell, even one lacking key parts, like a virus, "alive" and not a thing? It is worth noting that at different periods in history, the borders between life and nonlife were drawn slightly differently, and the world was more animate than it is at present.

Yet the project of synthetic life, the creation of "molecular machines" and the use of chemical engineering and biological construction can be seen as a twenty-first-century manifestation of a process of construction that began in the modern sense in the eighteenth century. It is here that we see another powerful text—that of the history of automata. In 1774, two such automatic replications of humans were first exhibited by Jaquet-Droz and his son, in Nuechatal (the replicas still perform there monthly). They are perfect replicas of human children—one is drawing and one writing, gesturing, and blinking. It was said that people came from miles away to see them, and these artificial beings have enchanted, frightened, and perplexed their viewers ever since. The writing automaton writes "I think therefore I am." Clearly, what was at stake was consciousness and its limits.

In 1766, the two inventors added to the exhibit a "musical lady," who played music, and this was followed by the worldwide exhibition of a "chess-playing Turk," adding the allure of the "East" and the power of one of humans' most complex achievements to the synthetic life. The more human the automata seemed, the more intense the reaction was. This reaction of anxiety, marvel, and fear was consistently expressed in articles about the automata. According to Gaby Wood, underlying the anxiety is a sense "that we can be replaced all too easily, and we are uncertain of what it is that makes us human" (Wood, 2002, p. xvii), a fear articulated by Freud in his discussion of the phenomenon.

It has been clear since the eighteenth century that the project of making or reconstituting artificial life is longstanding. Further, early social theorists considered such melodramatic efforts, drawing links to Greek mythology, in particularly to the Prometheus myth. In William Godwin's *Lives of the Necromancers*, he refers to Socrates, Daedalus, and Aeschylus, all of whom considered the ethical problems of making their statues "alive," which "if untethered would give you the slip like a runaway slave" (Wood, 2002, p. xviii). His daughter, Mary Shelley, drew the metaphor in its classic form, in her revision of the story *Frankenstein: The Modern Prometheus*. It is Shelley who focuses her account of artificial life—in which the organism, like Szostak's synthetica, grows and adapts to its environment—on the moral problem of duty toward the thing that is made. Like the earliest narratives, the line between life and nonlife comes with a price, a moral responsibility that creates an absolute claim on the creator.

The Larger Project of Synthetic Biology

Szostak is not alone in the quest to create synthetic organisms *de novo*. This project is part of a larger discipline—a recent merging of chemistry, engineering, and molecular and computational biology into a field called synthetic biology. I will consider some of the many ethical issues engaged in this field and its emerging community of basic research scientists.

The field uses the following self-definition: "Synthetic biology is a) the design and construction of new biological parts, devices and systems, and b) the redesign of existing natural biological systems for useful purposes" (Wolbring, 2006). Note that, unlike earlier efforts with genetic intervention, the goal is not only the transformation of existing cells, but the creation of entirely new ones using the basic chemicals or models that exist in nature as a sort of parts shop, and in some cases, as a set of plans, rather like the Sears home building blueprints of the 1920s. It is a profound engineering project using the models and hacker abilities of computer science. The idea is to understand and replicate the basic, minimal machinery of cells. Szostak's project is a "bottom-up" version, using the processes in nature to recapitulate alternative paths to synthetic life. Others use construction from sequences completely unseen in nature; still others use a stripped-down model, in which cells are deconstructed gene by gene to create a minimal working cell. Synthetic biologists form a loose, interactive community that stresses open discourse, creativity, and transparency in the development of "tools." Some offer plans for a "top-down" engineering approach, called *BioBricks*, which describes the idea of having inter-changeable, standard biological sequences for a plug-and-play synthetic biological production. On the joint Web site that collects opinions on the self-organization of the field, and attempts to offer one definition of the project, some members of the community add:

We are a group of individuals from various institutions who are committed to engineering biology in an open and ethical manner. We are currently working to help specify and populate a set of standard parts that have well-defined performance characteristics and can be used (and re-used) to build biological systems; to develop and incorporate design methods and tools into an integrated engineering environment; reverse engineer and re-design pre-existing biological parts and devices in order to expand the set of functions that we can access and program; reverse engineer and re-design a "simple" natural bacterium. (synthetic biology project Web site, 2007)

This language is rich with assumptions; such moral and linguistic claims imply a particular framing of the issue and a particular stance toward "naturalness" and "usefulness" (not that such framing endows the field with a set of a priori ethical principles). Indeed, research on these topics is warranted and is the ongoing subject of other work.[3] The task of this chapter is more limited in scope: to consider how one project—the design of an artificial, self-organizing, evolving cellular environ-ment—may cross the border between things that are nonliving and things that are living, that is, that eat and breathe, and grow, and replicate. Szostak is not alone in this intellectual project. Here are some other examples:

David W. Deamer, professor emeritus of chemistry and biochemistry at the University of California, Santa Cruz, and a cadre of pioneers expanded the quest three decades ago, launch-ing an attempt to build a "protocell." According to Deamer, such an entity must meet 10 requirements for life including having membrane enclosures (1) that can capture energy (2), maintain ion gradients (3), encapsulate macromolecules (4), and divide (5). Macromolecules must be able to grow by polymerization (6), evolve in a way that speeds growth (7), and store information (8). Add to that information store the ability to mutate (9) and to direct growth of catalytic polymers, and you have (10).

Albert Liebchaber of Rockefeller University engineered a DNA plasmid to express proteins and put them into membranous sacs. They could produce proteins for a few hours but would eventually peter out when the raw materials ran low inside the compartment. They needed to keep the supply coming. So, he and Vincent Noireaux, now an assistant professor at the University of Minnesota, designed them to produce a channel-forming protein, alpha hemolysin. Suddenly, finished proteins tagged with Green Fluorescent Protein inserted themselves into the artificial membrane allowing nucleotides and other molecules to enter. These "cells" survive for up to four days, but it's only a small victory. In the quest to build life, defining success is hard, Liebchaber says. Is it success simply to create a cell that functions? Or must it also reproduce? "I think in our case at least, the first step has been achieved." Next, he wants to make them divide, something that's only been done thus far through physical manipulation. (Lucentini, 2006, p. 30)

The idea of developing a synthetic biological organism that can divide into copies of itself is more than a goal of purely investigative research. One clear plan for the work is in its application and use. For engineers, such restructuring is a part of how their field is intended. Using biological parts is merely a difference in kind, not intent. In this manifestation, naturally occurring solutions to problems may be discarded in favor of ones with better (more intelligent) design.

Indeed, in attempts to create artificial life, researchers at Los Alamos National Laboratories in the United States have thrown out many of the conventions found in nature. They have turned the protocell model inside out, designing a micelle with information coding and metabolic machinery on its exterior. Extant thus far mostly on paper, these micelles use peptide nucleic acids (PNAs), DNA mimics with a pseudopeptide backbone conjugated to a light-sensitive molecule. When exposed to light, the photosensitive chemical discharges an electron-triggering chemical reaction to convert nearby nutrients into new fatty acids and PNA, based on the PNA template. These incorporate into the micelle, which grows until it spontaneously pinches in half and divides.

How We Define Life Leads to Our Duties Toward It

Although it is philosophically satisfying and interesting to speculate about how and under what circumstances the line from inanimate to animate object, from "evententity" to entity, is crossed, the ethical issues emerge only when we consider our duties toward the things we create. We are led to ask: What, if any, moral statuses do the new things we make have? What duties, if any, do we have toward them? If one begins a process of evolution, that is, forces evolution, how far can this process be allowed to go? What are the constraints on this sort of artificially created RNA world in terms of safety, subject protection, and so on?

We have some duty toward created things—even crafts and manufactured goods have meaning and worth, for what sort of moral creature is someone who wantonly and without justification destroys things around her? Surely, "eventities" that may well be alive also have meaning. The moral duty might not be much, of course. We might discard the clay bubbles, unmake them in order to understand them, or use them (not in any way we wish, but that prohibition emerges from issues of policy or safety, not of moral status).

Is Uneasiness a Moral Category?

For many, the ethical issue at stake is not the problem of the actual creation of new life forms and our possible duties toward them. Rather, it is the problem of disruption, or the uneasy liminality in the borders of nature that this research implies, where all life that is exists, and is merely to be discovered. It is precisely the eighteenth-century Swiss automata discussed earlier that led Sigmund Freud to define a category of reaction to all such phenomena. This definition serves us well in considering why we find the Szostak experiments so complexly disturbing. Freud calls this *the uncanny*: the feeling that arises whenever there is intellectual uncertainty about the borderline between nonliving and living.

If the natural order of the world of things is disturbed, and the basic lines between nonliving and living are blurred, then something about the fixity of our perceived world is awry. First among these objections is the argument that humans possess an essential nature, and live within an essential natural order that cannot be altered without harm. For C. S. Lewis, this is expressed as a concern that the very acts of rational science—dissection, analysis, and quantification—are a violation of the sacred integrity that lies behind all of nature:

Now I take it that when we understand a thing analytically, and then dominate and use it for our own convenience, we reduce it to the level of "nature," we suspend our judgments of value about it, ignore its final cause (if any), and treat it in terms of quantity. This repression of elements in what would otherwise be our total reaction to it is sometimes very noticeable and even painful: Something has to be overcome before we can cut up a dead man or a live animal in a dissecting room. (Lewis, 1998, p. 274)

For Lewis, the understanding of the body as replaceable is disturbing: "The real objection," he says, "is that if man chooses to treat himself as raw material, raw material he will be, not raw material to be manipulated by himself as he fondly imagined, but by mere appetite" (p. 274). (Lewis imagines that new transformative technology will be manipulated by "controllers" who will eventually transform man into mere matter.)

Callahan echoes his concern, both in the sense that limits need to be placed on what is decent to do to nature, and that such action is part of a larger danger—that power in the hands of medicine to heal is really power in the hands of the elite or the state to manipulate and control. He argues:

The word "no" perfectly sums up what I mean by a limit—a boundary point beyond which one should not go. . . . There are at least two reasons why a science of technological limits is needed. First, limits need to be set to the boundless hopes and expectations, constantly escalating, which technology has engendered. Advanced technology has promised transcendence of the human condition. That is a false promise, incapable of fulfillment. . . . Second . . . limits [are] necessary in order that the social pathologies resulting from technologies can be controlled . . . while [technology and science] can and [do] care, save, and free, [they] can also become the vehicle for the introduction of new repressions in society. . . . (Callahan, 1973, p. 5–7)

However, the judgment that we can be guided by our moral intuitions has a danger as well. Joseph Fletcher raises a disagreement:

The belief that God is at work directly or indirectly in all natural phenomena is a form of animism or simple pantheism. If we took it really seriously, all science, including medicine, would die away because we would be afraid to "dissect God" or tamper with His activity. . . . Every widening and deepening of our knowledge of reality and of our control of its forces are the ingredients of both freedom and responsibility. (Fletcher, 1970, p. 59)

The transgressive nature of the research is clearly on display when critics claim that such an act violates our human limits or recapitulates creation itself, or that a second round of creation may well end in a cataclysm.

To Make Is to Know: Notes on an Old Problem about Knowledge

The classic debates of the 1970s are not the only problems engendered in the history of ethical responses to the technological gesture at the heart of synthetic biology—the special kind of knowledge that such making implies is also at stake. For Aristotle and the Hellenists, useful knowledge, "practical wisdom," was *phronesis*, which implied actually doing an act, making, in order to know. The act of making, not those of perception or contemplation alone, determined how wisdom, rationality, and power were achieved.

Furthermore, the use of technology within the body of the patient is a different matter than the use of technology to essentially enhance the body of the practitioner. For all earlier technology, the sense perception of the doctor was the object of transformation or enhancement. Stethoscopes and otoscopes make sounds of the body more audible; X-rays, CT scans, and MRIs make the inner vistas of the body visible; EEG and EKG make the electrical currents that animate the central and peripheral nervous system quantifiable; microscopes make invasive bacteria visible at the microscopic and, increasingly, molecular level. These earlier technologies extended the reach of what Bacon increasingly trusted, and the Greeks did not – the perception, observation, and deliberate perturbation of the phenomena of the world.

Bacon's method presupposes a double empirical and rational starting point. True knowledge is acquired if we proceed from lower certainty to higher liberty and from lower liberty to higher certainty. The rule of certainty and liberty in Bacon converges. . . . For Bacon, making is knowing and knowing is making (cf. Bacon IV 1901, 109–10). Following the maxim "command nature . . . by obeying her" (Sessions, 1999, 136; cf. Gaukroger, 2001, 139 ff.), the exclusion of superstition, imposture, error, and confusion are obligatory. Bacon introduces variations into "the maker's knowledge tradition" when the discovery of the forms of a given nature provide him with the task of developing his method for acquiring factual and proven knowledge. (Stanford Encyclopedia of Philosophy, 2003, p. 5)

Thus, the world is known through understanding of its parts, and from that, theorizing (knowing) by induction, to principles, axioms or laws of nature, physics, and chemistry. In contemporary science, knowing is done largely by "unmaking," that is, by deconstructing the component parts in ways scientists of Bacon's era were unable to imagine. Many of these "unmaking" techniques, such as splicing alternative DNA or manipulating cellular structures, allow a sense of inherent interchangeability, as if the real world of biological being and indeed, each of our bodies were merely a set of Lego parts, awaiting clever recombination.

Time and the Other

Szostak's work is important for moral philosophy in another way: For phenomenologists, defining the category of being, and hence nonbeing, is important. Heidegger's project is to use the actuality of the physical world to define the limits of being. Thus, of course, being is real only if related to temporality, and to the death of the being; for Heidegger, nonbeing can be finally known only through death. The distinction between life and death creates a clear categorization.

For Levinas, thinking of what "being" actually means, the event of death presents no solution to the borders of a being, for indeed, the death of any being allows a space to be filled by the living immediately—the being's existence is complexly maintained (Levinas, 1998, p. 3). What, then, can be said about being and otherness? Levinas notes that there is a nonbeing in the profound otherness of the established world; living others are different from rocks and shoelaces, for example. This otherwise-than-being is then clearly further queried by things, phenomena, or perhaps actual beings (with some narrative possibility) that have a point of nonexistence, and would not exist save for their deliberate creation. As these phenomena change over time, they share time as well as terrain with us. Thus, calling into being a new sort of thing—a protocell, a mimicry, a synbiotica—is a complex ontological event.

What Is a Thing? The Perils of Deconstruction

Heidegger asks "what is a thing?" and in so reflecting, understands a thing as an object separate from the self. But what of a made thing, an object that becomes the self? Is a thing that is made into a creature, or that evolves along a path that is set in place but not designed, a thing at all? Is it, rather, a creature with a lifespan? Making actual cells, and making actual beings, thus in mimesis of the real being of the actual world extends the Baconian act in radical ways. Here, the experimental perturbation is the making of a protocell, which may or may not be the basis of actual evolution, only in a faster, more controllable form. This cannot help but excite concern about the nearly infinite possibilities for technological shaping of larger beings, and ultimately, of course, of the self.

Cells, and perhaps new creatures that grow and replicate, do what no static thing can do, and this replication using chemical codes for information storage and transmission distinguishes creatures from mere things; it is what makes cell division and bubble division very different acts.

Here, then, is the question: Do the objects we see in the slides possess enough of the chemistry of living beings to qualify as making the transition from nonliving to living beings? If so, it is precisely this act that is so utterly uncharted. What sort of an act is it to restart the clock of creation, the Lego-by-Lego-block being as it were, with the standard parts for the being list actually for sale on the Internet, or for free exchange (a longstanding goal of the above-mentioned BioBricks project)?

In fact, this idea of making simulacra using standard chemical formulas is not new. The idea of synthetic life was not limited to making life as a kind of machine, for like synthetic biologists, earlier scientists became fascinated with the idea that "life matter" was a series of forces, or entities that could be recombined, if only they could figure out the recipe. Making gold was only one goal of the alchemist. Making life from the manipulation of parts and context was another.

Not all attempts to create artificial life were mechanical. During the Renaissance, a number of astrologers and alchemists developed recipes of more magical creations. Cornelius Agrippa believed, along with his fellow sorcerers, that humans could be grown from mandrake roots.... Paracelsus published instructions for the manufacture of homunculi.... Human semen could be put in an airtight jar and buried in horse manure for forty days..., magnetized, preserved at the temperature of a mare's womb and fed human blood for forty weeks. (Wood, 2002, p. 16)

The eighteenth-century automata returned to this concept. But in this instance, the signs of the new discipline of science were used to explain the goal:

[The] ambitions of the necromancers were revived in the well respected name of science. Life advances in scientific instruments, and a fondness for magical tricks, meant that automata were thought of as glorious feats of engineering or philosophical toys. (Wood, 2002, p. 16)

As philosopher Umberto Eco noted (Wood, 2002), it took the emerging power of the industrial revolution to fully use the new power of machines that took over labor, and thus were "substituted" for actual people, using mechanics instead of magic and potions. The uncanniness of the assembly line thus stems from two features. First, human craftsmanship, tool-making, which is the very sign of humanness, can be entirely recapitulated by a machine. Second, the objects themselves, and the acts of labor, can be broken down to a series of parts that can be endlessly recreated, moved about and used in unforeseen ways, and combined and recombined into a variety of shapes and forms. In essence, notes Wood, the ultimate victory of the new technology was not making machines into humans, but rather, in factories and assembly lines, making humans into something more like machines. It remained only for genetics to then appropriate the semiotic principle—parts of a gene are called the *cassette*, which can be "swapped out" and changed, "molecular machines" of one organism can be imported into another, mouse models can be "made" and then sold in catalogs that list their specifications.

It is this context in which Szostak's project is located. Part of the problem lies in the way modernity itself is understood, and how our narrative is shaped by the very technology that is the most lifelike and artificial, the creation of the entirely false world of filmmaking, in which bits of action can be cut together to make a story. For Gaby Wood, the invention of photography, and of moving photography, enabled films to create an "artificial life," seemingly completely accurate in a way that painting was not, yet to alter the element of time by splicing pieces of film to create the illusion of a coherent narrative (Wood, 2002, ch. 3).

Let me develop this concept. Much of the substantive act of genetic modification is based on a similar idea and relies heavily on a similar technological advance in visual narrative, whose typology then both elucidates and distorts the object of the gaze. Genetic modification takes slices of narrative (in that the genetic code is developed over time and in specific contexts and narrative histories) and "cuts and pastes" them into new creatures. This creates a fictitious narrative—a made-up evolution, as it were—within the organism, which is then displayed in a PowerPoint movie, with certain features (molecules or cells) labeled (e.g., with GFP) as if in a specific costume to be seen and understood by the audience in a particular role. Szostak's "eventities" are proven to be evolving or changing because we can see them moving in the movie described at the beginning of this chapter. His work, like much of modern molecular biology, is a sort of "tape rewind" in which the great narrative of evolution is wound backwards, and in this case, played forward again in slow motion.

This is important, in more than the ironic cultural-studies sense, because it speaks to the concerns often expressed about the disruptive effect of this research (which is understood by the synthetic biology community in the postdoctoral fellow's scientific joke: "44 billion years of evolution! Time for a change"). Synthetic biology extends the idea of any new science, which is really a disruptive project, the known undone by the discovered. Molecular biology, nanotechnology, and their synthesis have deconstructed the core narrative of all of life. However, they extend the project even further because, unlike descriptive biology or even basic research on how to repair flawed or nonfunctional DNA that causes illness, they rewind the "movie" back to the point of origin, allowing for entirely new plot endings. Understanding this deepens our problem, for short of entirely disallowing the project, at what possible point could society proscribe it? How far should the recapitulation go? It is in the name of these cultural anxieties that normative urges arise.

Whose Life Is It Anyway?

Is creation "ours"? Do we stand outside it, or within it, and if the latter, are the acts we struggle to learn and do really just instrumentation as a part of our inbred genetic character? In other words, is making things in this way simply more of the same sort of agricultural production, the most enduring form of technica? Is it science that we have come to accept, simply more precursors to the technology that we depend on? Or are synthetic biology and its attempts to make new cells substantially new moral gestures, unlike, for example, animal husbandry, at the nanoscale?

In reflecting on this issue, and in moving from descriptive analysis to actual policy, source texts provide our usual recourse. Synthetic biology in its replicative and recapitular modes draws on classic tropes and on established scientific history. But which source texts matter? In whose interests are they interpreted, received, and understood? If we remember that practitioners of artificial life in earlier times drew from their knowledge of secret magic tricks, then secrecy and the forbiddenness of certain sorts of knowledge, and problems with the transmission of this knowledge, have a long ancestry in this context. If we find that engineering literature is the source for synthetic biology, then we need to focus on the conditions of use for such an experiment, its utility, and whether applications need even be expected for such speculative research—a project closer to cosmology or basic chemistry than biology. If we conceive of the project within the context of biosecurity, then the key questions are who should know the data, who should learn the techniques, and under what constraints should the new knowledge be disseminated. If we imagine this work to be a variant on genetic engineering, then the questions we would bring to the table are ones of regulation. Who should regulate these sorts of trials, which fall outside traditional categories of genetic manipulation, human or animal subjects, or dangerous materials standards?

The tension between and curiosity about the borderline between made and begotten, artifice and "natural," has fascinated people since antiquity. This tension reached a turning point in the eighteenth century, when machinery was used in elaborate magic acts that were thought to be reconstructions of living beings, and when "making a copy of the machinery of biology" meant literally understanding the human being as a machine, an assemblage of mechanical parts (Wood, 2002). This idea has been a part of science ever since, the metaphors shaped by the most current mechanical device. Hence, in the eighteenth century, the body was "bellow, pumps and pulleys," in the nineteenth, the work of the body could be replicated and efficiently mass-produced, and in the early twenty-first century the body is understood as a series of information codes that drive the functions of the molecular machine. The ongoing idea that machines are like people and vice versa is replicated in several ways in the quest for self-assembling artificial life, or artificially assembled real life.

First, the impulse to create the machine as a device to better understand the processes of the real world shares an unstable border across the limen with the project of deconstructing the human to machinelike pieces toward the same end—as a metaphor and model for the understanding of the human body. Second, the automatons of the eighteenth and nineteenth century were a mix of magical toy, scientific device, and marketable object. Philosophers commonly argued that understanding a creature's body as a "thing or a machine" made it understandable, manageable, and ultimately describable. That process is reversed when constructing a "thing" or a set of machined parts (as in BioBricks) into a life.

Yet, even in the seventeenth century this created an immediate categorization problem: Is the created thing a being, a model, or a commodity? Commenting on the new artificial automata, and indeed any of the new machines of this century, Diderot writes of the philosophic problem. He is frankly disturbed by what machines were: "[W]hat difference is there between a sensitive and living watch [Ed: here he means a human] and a watch made of gold, iron, silver or copper?" Virtanen, writing of the same problem, notes that determining what artificial life actually is remained a difficult task; describing the problem of watching the oddly realistic, clever automata he writes: "Would you not believe they had a soul like your own, or at least the soul of an animal?" (Wood, 2002, p. 16).

Such a history raises the question of who actually "owns" the things or life forms that are "made." If a thing has "even an animal soul," how do we decide its moral status? Can an unnaturally instigated phenomenon that becomes a part of the natural world be owned, patented, and sold? How do we imagine the phenomena are controlled? What does the making of artificial life say about our desires and propensities? Driving the ethical debates in earlier periods, when automata excited moral concern, was the idea that the artifice might be better, more capable or more real than the actualities of life. This is not only a moral panic in the case of synthetic biology. In fact, making a more successful organism, one shaped more clearly for human use, is one stated and tantalizing goal of synthetic biology. Engineering science could make a "better" or more authentic yeast cell or *E. coli* bacteria: standard, replicable, and engineered for mass production.

Moving Toward Normativity: Challenges for an Ethics of Synthetic Biology

From its very inception, synthetic biology has faced serious ethical questions. Some of these are based on the longstanding concerns noted previously, the questions about "naturalness," scientific hubris, error, and safety. Synthetic biology, however, also came of age at a particular time, after the rise of global terrorism and the emergence of new diseases with lethal epidemic potential. Hence, in addition to the concerns about recombinant DNA was the concern that the very openness and "coolness" of the research, and its portability and scope, might allow for use by bioterrorists. Synthetic biology deals with minimalism as a core concept. Thus, small sections of DNA can be imagined (or replicated from naturally occurring alleles) and synthesized by a lab, the instructions arriving through the Internet, the product arriving through FedEx. Then chemical units can be linked to create longer units— whole chromosomes. Thus, the possibility of a malignant construction, or one designed for destructive purposes, has been long noted in the field (see, e.g., Maurer & Scotchmer, 2006). Is this a problem of similar concern with the creation of an evolving organism?

The response to this question seems to depend on the level of risk deemed reasonable. The core of the method, after all, is stochastic, exploitative in the Darwinian sense, and thus fully unpredictable in its details. Synthetic biology is related to but unlike genetic manipulation, and if it is to proceed, the ethics must evolve right along with the science. We cannot yet know the complexity of the issues, and it is a thin response indeed to merely note that science must be "cautious," for what, precisely, does such an admonition actually mean?

No Direction Home: A Complete Unknown

The ethical considerations that attend to the creation of artificial life in any form, and in the form of the phenomena that then seem to be a step away from synthetic simple organisms (synthetica), largely surround the uncertainty of outcomes that accompanies all basic research. Yet the recapitulation of life's origins raises special issues. Can such a process, which is intended to be based on evolution's principles of mutation, random selection, and transformation, be said to be regulated at all? What is the meaning of a call to regulate such experimentation? The elusive nature of such questions raises special considerations for ethics. Unlike many questions in bioethics, in which burden, risk, harms, and benefits can be rationally foreseen and considered, the recapitulation of life's trajectory is entirely speculative.

At this point in such discussions, bioethicists tend to turn toward concrete normative suggestions for public policy. In the particular case described in this chapter, the basic science is new, speculative, and entirely inapplicable in its present form. Surely, all such work is constrained by the obvious and primary need for safety and containment of any *de novo* organism. But ought there to be more of an "ethics of synthetic biology"?

I suggest that the main goal of ethics is a stance of deeply informed and thorough *attention*, rather than calls for moratoria or cessation of activities. Such attention will require a careful reading of history, some degree of ironic recollection of the history of automata, as was briefly related here, and some degree of intellectual courage. The synthetic biology project might fail (as so many of the projects ultimately did fail) and become merely an interesting and clever toy (as many early efforts were). It might turn in directions that we have yet to anticipate or understand fully. Much of synthetic biology falls outside of classic review structures, for it does not use animal or human subjects in the ordinary sense. Creating ethical guidelines for this field will require humility and patience. These are qualities of all reflective discourse, perhaps borrowed from earlier centuries as surely as the science. The ethical quandaries at the boundaries of artifice and inquiry will require, most of all, an ongoing and interactive conversation between philosophers and scientists, who are both learning what it is that we see. It is this last consideration, the question of the use of the technology and its effect on human societies, that turns our ethical gaze from ontological concerns toward normative ones. Science, to be normatively ethical, must most of all be witnessed. In large part, the tasks of this chapter have been to elucidate the case and its history, and to raise core questions in moral philosophy and ethics yet to be fully understood by a watching world. The public gaze, the academic process, and the commercial users who wait even at the margins of Szostak's theoretical project are a part of such witness—it is both a fascinating and necessary task.

Notes

1. Of their research, Szostak's group says: "One fundamental question that we are attempting to address through RNA aptamer selections is the relationship between information content

and biochemical function. It seems intuitively obvious that more information should be required to specify or encode a structure that does a better job at performing some function, such as binding a target molecule" (HHMI, 2006).

2. It is a truism of extremophile biology that in every single Earth environment where these three things are present, life forms are found. See NASA (2005).

3. These issues are obvious to any bioethicist, humanities scholar, historical or social theorist who reads the preceding definition, and include the following senses: that nature is a commodity whose existence can be improved for human use, that the biological world can be understood as a sort of machine, with parts that bear scant relationship to a particular whole, that the givenness of order as we find it is mutable and temporal, and that our intervention is sanguine, for such a world can be taken apart and reconstructed much as any machined element of society. A positive, pragmatic idea that the power of basic research can be harnessed is also reflected. This footnote is not intended as a critique of these ideas, but rather to note that the authors and the reader understand these ideas are behind the declarative sentences. The ontological and epistemic phenomena of this work are not within the scope of this chapter.

References

Bedau, M. (1996). The nature of life. In M. Boden (Ed.), *The philosophy of artificial life* (pp. 332–360). Oxford: Oxford University Press.

Callahan, D. (1973). *Science, limits and prohibitions.* Hastings Center Report 3, November.

Chabon, M. (2000). *The amazing adventures of Kavalier and Clay.* New York: Picador Press.

Dorff, E. (2004). *Matters of life and death: A Jewish approach to modern medical ethics.* Philadelphia: Jewish Publication Society.

Fletcher, J. (1970). Technological devices in medical care. In K. Vaux (Ed.), *Who shall live.* Minneapolis: Fortress Press.

Hanczyc, M. M., Fujikawa, S. M., & Szostak, J. W. (2003). Experimental models of primitive cellular compartments: Encapsulation, growth and division. *Science, 302,* 618–622.

Howard Hughes Medical Institute (HHMI) Web site (2006). Investigator page, online at: http://www.hhmi.org/research/investigators/szostak_bio.html (accessed March 2007).

Levinas, E. (1998). *Otherwise than being.* Pittsburgh: Duquesne University Press.

Lewis, C. S. (1998). The abolition of man. In S. Lammers & A. Vehey (Eds.), *On moral medicine* (pp. 270–276). Grand Rapids, MI: Eerdmans Publishing Co.

Lucentini, J. (2006). Is this life? *The Scientist, 20* (1), 30–35. Available online at: http://www.the-scientist.com/article/display/18854 (accessed February 2008).

Maurer, S., & Scotchmer, S. (2006). Open source software: The new intellectual property paradigm. In T. Hendershott (Ed.), *Economics and information systems,* Vol. 1. New York: Elsevier Press.

NASA (2005). ESA to search for life, but not as we know it. Available online at: http://astrobiology.arc.nasa.gov/news/expandnews.cfm?id=9329 (accessed March 2007).

Schrödinger, E. (1944; reprinted 1992). *What is life?* Cambridge: Cambridge University Press.

Stanford Encyclopedia of Philosophy (2003). *Francis Bacon.* Available online at: http://plato.stanford.edu/entries/francis-bacon/#5 (accessed June 2007).

Synthetic biology project Web site (2007). Available online at: http://syntheticbiology.org/Who_we_are.html (accessed June 2007).

Umar, M. (2003). *Human in the image of God.* Atlanta, GA: Society for Scriptural Reasoning, American Academy of Religion.

Wheeler, M. (1996). *From robots to Rothko.* In M. Boden (Ed.), *The philosophy of artificial life* (pp. 209–236). Oxford: Oxford University Press.

Wolbring, G. (2006). Synthetic Biology 2.0. *Innovation Watch.* Available online at: http://www.innovationwatch.com/choiceisyours/choiceisyours.2006.05.30.htm (accessed August 2007).

Wolpe, P. R. (1997). If I am only my genes, what am I? Genetic essentialism and a Jewish response. *Kennedy Institute of Ethics Journal, 7* (3), 213–230.

Wood, G. (2002). *Edison's Eve: A magical history of the quest for mechanical life.* New York: Random House.

Zoloth, L. (2001). Born again: Faith and yearning in the cloning controversy. In P. Lauritzen (Ed.), *Cloning and the future of human embryo research* (pp. 132–144). Oxford: Oxford University Press.

10

Protocell Patents: Property Between Modularity and Emergence

Alain Pottage

The terms of the forthcoming engagement between patent law and protocell inventions are still a matter of speculation. This is true even if the frame of reference is expanded to take in the field of synthetic biology in general. Although there have been a number of patent applications relating to DNA synthesis and to the metabolic components that are the most likely elements of gene switches and programmable interfaces, the throughput of inventions is still too limited to have had any formative effects, and the U.S. Patent Office has not yet established a specific digest of prior art for patents in synthetic biology (see Kumar & Rai, 2007). In the spirit of speculation, what questions might be useful to pursue? The obvious one is that of institutional inertia and adaptability. What resistances—doctrinal, bureaucratic, or economic—are protocell inventions likely to encounter as they make their way through the patent system?

Although these issues are entirely pertinent (see the final section of this chapter), they often bring with them the assumption that the premises and mechanics of the patent system are settled, closing off the questions that are most likely to reward speculative inquiry. Protocell research emerges at an interesting point in the history of modern patent law, when the terms in which the patent system has been explained and justified for more than two centuries are being thoroughly reconsidered. And although this process is obviously stimulated by the radical novelty (for law) of biological and informatic technologies, it has much to do with a broader set of conditions that have disturbed the political and institutional settlement on which modern patent law was founded. The evolution of communication technologies, investment practices, technological form, and scientific collaborations has called into question nineteenth- and early twentieth-century representations of invention, and has given renewed force to the argument that the limited monopolies granted by patent rights are not always the most desirable or effective ways of generating and diffusing innovation. It may be that the more productive question to be asked about protocell inventions is not *what will be patentable?*, but *how will patents be mobilized, and how will they combine with alternative vectors of innovation?*

To begin with an example that has obvious relevance to the future course of protocell research, the adverse effects of patent practices on scientific collaboration (see Eisenberg, 1987; Heller & Eisenberg, 1998) have prompted a search for vectors of innovation that open, rather than restrict, the cognitive energies generated by diverse interests and perspectives. The most developed model of collaborative science is proposed by the CAMBIA-BIOS project (www.cambia.org), which has drafted a form of license applying the principles of the open-source software movement to agricultural biotechnology. Rather like open-source software licensing, the BIOS license seeks to turn the patent system against itself by stipulating conditions that keep the core technologies open while at the same time feeding subsequent improvements or enhancements back into the common pool (www.bios.net). Whereas the basic mission of the modern patent system is to create private rights in the public interest, the open-source movement recruits private motivations to create public goods. The meaning of the term *public* changes significantly. Consider the ambitions of the Patent Lens database developed by CAMBIA:

The goal of the Patent Lens is to use the power of informatics and community to harmonize and make transparent the world of patents, so that thoughtful individuals, institutions, and agencies can guide thoughtful and humane reform of the innovation system and to spur efficient and socially relevant innovation. This is an essential platform if we are to make use of the patent system itself to expand and protect a technology commons. (Jefferson, 2006, p. 31)

This consists of something more than an intensification of the traditional idea of disclosure (namely, the idea that the grant of a patent is conditional on full disclosure of the invention and its principles of construction to the public, or to a group of artisans standing in for the public). Although the objective is partly to allow participants to understand patent texts without going through the restrictive codes maintained by a self-interested "patent clergy," the scheme also seeks to "provide the public with tools to recognize and overturn [egregious] patents where they undermine progress" (Jefferson, 2006, p. 31). This implies a more critical and active mode of public engagement, in which the patent system itself—not just each patented invention—is a *res publica*, a matter of public concern and reflection. In theory, transparency frees the patent system from the grip of technical expertise, thereby opening its norms and operations to a democratic process of scrutiny and ongoing revision.

These developments matter to the constitution and future development of protocell research. The conventional assumption is that the patent system precedes invention. According to doctrinal and economic theories that took shape from the late eighteenth century onwards, an effective patent system induces inventors to disclose their inventions and secures the investments needed to turn a raw technology into marketable manufactures and applications. With technologies such as synthetic biology and nanotechnology, the patent system precedes invention in a quite different sense. Programs of research in synthetic biology and nanotechnology have from their very inception been alert to public or "civil society" commentary on their social and ethical implications, and this sensitivity has produced collaborations between biological science and social science that would once have seemed improbable. In the case of protocell research, central themes in public commentary are biosafety and bioterrorism, but thematizations of the new technology also take in the lessons learned from public reactions to the marketing (and patenting) of the products of agricultural biotechnology.

The question of intellectual property is central to the process of establishing an ethical warrant for research into protocells and synthetic biology more generally. A socially aware science is one that takes into account the potentially constrictive effects of patents; a *Scientific American* editorial on the prospects for synthetic biology observes that "the problem is not patents as such but bad patents" (2006). What makes a bad patent? The question cannot be answered from within the narrow framework of patent doctrine. Instead, one has to look to the broader instrumentality of patents: How are patents mobilized, and with what consequences for scientific collaboration, market competition, and public participation? At this point, ethical thematizations of science coincide with a broader academic reconsideration of the political and institutional settlement of modern patent law.

For some time now, intellectual property scholars have been developing a sustained critique of the role of patents in the field of biomedicine. A particularly influential formulation of the critique is expressed in the notion that research in the field is facing a "tragedy of the anticommons:" "Anticommons property can best be understood as the mirror image of commons property. A resource is prone to overuse in a tragedy of the commons when too many owners each have a privilege to use a given resource and no one has a right to exclude another. By contrast, a resource is prone to underuse in a 'tragedy of the anticommons' when multiple owners each have a right to exclude others from a scarce resource and no one has an effective privilege of use" (Heller & Eisenberg, 1998, p. 698).

The full potential of contributions to a heterogeneous or distributed research field such as biomedicine cannot be realized if research tools are made available only as proprietary products, wrapped in restrictive licensing terms and transferred subject to prohibitive royalty terms. In setting these terms, each contributor has a tendency to overvalue his or her own contribution to research paths, and current patenting practices concretize this differentiation of interests because "intellectual property rights in upstream biomedical research belong to a large, diverse group of owners in the public and private sectors with divergent institutional agendas" (Heller & Eisenberg, 1998, p. 700). Empirically, the theory of the anticommons might be questionable (compare Walsh, Cho, & Cohen, 2005, with Murray & Stern, 2005), but it expresses a broadly shared concern about the erosion of public science, both as an ethos and as a domain of open knowledge. And, although "public science" may be a myth fostered by the particular structure of research funding in the Cold War period (see Mirowski & Sent, 2002), and the ideal of scientific openness might need to be qualified by reference to regimes of scientific credit (see Biagioli, 1998), the concerns of intellectual property scholars have generated new theoretical tools, such as the concept of "peer production" (Benkler, 2006) and novel practical or institutional initiatives such as the CAMBIA-BIOS open-source biology project (see also chapter 11 of this volume).

From the perspective of intellectual property, protocell research is already captured in this potentially complex intersection of property and commons, private and public, restrictive and dynamic. The complexity of these tensions may generate certain strategic problems for protocell researchers, both as consumers and as producers of research: More precisely, what combination of proprietary and open-source licensing regimes is likely to advance academic and commercial objectives?; or, how might licensing regimes be combined in such a way as to reconcile tensions between those objectives? From the perspective of intellectual property, on the other hand, protocell research and synthetic biology in general become a kind of testing ground for exploring different hypotheses about the constitution of the patent system; for intellectual property lawyers, synthetic biology offers an occasion to "rethink the boundary lines between intellectual property and the public domain" (Rai & Boyle, 2007, p. 389). This convergence between ethical self-reference in both science and intellectual property is likely to be a continuing feature, at least as long as the research agenda in synthetic biology and protocells is set by academic scientists. Initially, the terms of the convergence can be explored by looking at the field of synthetic biology in general, and specifically the representation of bioengineering as the program of constructing organisms that "behave as expected, as we want them to behave" (Drew Endy, cited in Stone, 2006, p. 567).

Modularity

The widely diffused representation of synthetic biology as "engineering" has engaged the imagination of patent lawyers:

The scale and ambition of synthetic biology efforts go well beyond traditional recombinant DNA technology. Rather than simply transferring a pre-existing gene from one species to another, synthetic biologists aim to make biology a true engineering discipline. In the same way that electrical engineers rely on standard circuit components, or computer programmers rely on reusing modular blocks of code, synthetic biologists wish to create an array of standard modular gene "switches" or "parts" that can be readily synthesized and mixed together in different combinations. (Kumar & Rai, 2007, p. 1745)

The sense of synthetic biology as "a true engineering discipline" rests on a comparison between synthetic biology and biotechnology. Although recombinant DNA technologies sometimes look like engineering projects (see Knorr-Cetina, 1999), they fall short of rational construction because they have not yet managed to conceive and instrumentalize the complexity of life. A recent European Union report observed that biotechnologies have so far been based on "tinkering rather than rational engineering" (European Commission, 2005, p. 11), and went on to identify one of the reasons: Biotechnology addresses life as an organization of parts, each of which is "a unique—messy—result of millions of years of evolution and frequently is subject to a large amount of cross-regulation from functionally rather distant elements of cellular function" (European Commission, 2005, p. 24).

Whereas recombinant DNA technology is the "equivalent of writing a paper using a photocopier, scissors and a stick of glue," with the advent of DNA synthesis technologies "biologists have received their metaphorical typewriter" (Marguet, Balagadde, Tan, & You, 2007). Engineering therefore presupposes a means of complexity reduction that allows "messy" living materials to be fully instrumentalized. In synthetic biology programs, complexity is reduced through a process of modularization, by constructing modules (DNA segments) that are engineered to perform according to specified tolerances and can be linked together by means of universal connectors. According to the home page of the BioBricks initiative, these modules are "standard DNA parts that encode basic biological functions. Using BioBrick standard biological parts, a synthetic biologist or biological engineer can already, to some extent, program living organisms in the same way a computer scientist can program a computer" (BioBricks Foundation, 2007). Given modularity or, more precisely, a nested hierarchy based on the organization of parts into devices, and devices into systems (see Endy, 2005), bioengineers can work at one of these levels while externalizing the complexities involved in the other levels. Life becomes entirely open to rational design and construction; more precisely, it becomes entirely amenable to invention.

How does the theme of modularity play out in patent law? In a somewhat abstract sense, property lawyers of all species are quite at home with the notion of modularity. According to one view of things, property institutions are themselves techniques of modularization: Each private property right is a module engineered (juridically) to reduce the complexity of social interaction. Potentially disruptive (and complex) conflicts over access to scarce resources are simplified by modularizing the entitlement in the form of a private property right and allowing these modules to interact according to standardized interfaces (transactional forms) that can be activated only by the holders of rights (on this theme, see Langlois, 1999). So in one sense, the patent system itself is a modularizing institution. But in relation to patent law, the theme of modularization has two dimensions, each of which opens into a different reading of the premises of modern intellectual property. Taken as a cipher for rational design and construction, the notion of modularity resonates very closely with the basic precepts of nineteenth-century patent law. At the same time, however, the theme of modularity opens into another representation of the patent system, one that questions the models of motivation, knowledge production, and public interest that underwrite the modern framework of intellectual property. In this alternative representation, modularity opens novel ways of imaging the organization of collaborative production, and thereby becomes the premise of alternative vectors of innovation.

The nineteenth-century definition of *invention* anticipated the contemporary notion of modularity. It described invention in terms of engineering, or rational construction. To take a particularly apt example, mechanical inventions were seen as artificial "organisms," or as combinations of parts that were entirely ordered and animated by an inventive idea.[1] According to the classic definition, invention was an art of composition, combination, or interrelation: "over the existence of matter itself, [man] has no control. He can neither create nor destroy a single atom of it; he can only change its form, by placing its particles in new relations" (Curtis, 1849, p. xxv). This definition proposed both a theory of technological creativity and a theory of the distinction between invention and discovery, or between proprietary knowledge and the public domain. Inventors added something to natural forces or properties by combining them to produce a new means or effect, in much the same way as a sculptor added something to clay or marble by shaping it into a new form. Just as the authorship of the sculptor was concentrated on the artistic design expressed in the form of the sculpture, so the inventorship of the mechanic was concentrated on the idea of combining given materials to produce a new and useful means or effect. The materials, forces, or properties themselves remained in the public domain, open to incorporation in other inventive combinations, and accessible to scientists interested only in adding to knowledge about those forces and properties as such. So invention was an art of origination: "engineering" rather than "tinkering." And if this capacity for origination is what synthetic biology brings to biotechnology, it might almost be said that after various adventures with complexity, emergence and epigenesis, bioengineering is finally maturing into a classic nineteenth-century technology.

But there is a crucial difference between the nineteenth-century notion of engineering and the modularity of synthetic biology. In the project of creating standard biological parts, engineering happens twice over: Biological parts are engineered to be engineered. For example, the purpose of Drew Endy's project of constructing a synthetic equivalent of the T7 bacteriophage is not only to gain an understanding of how the virus works by taking it apart and reconstructing it according to transparent principles or linkages, but also to build a "device" with competences that will be predictably reproduced outside its native laboratory (see Chan, Kosuri, & Endy, 2005). Each of the parts contributed to the BioBricks project is a composite of elements—essentially DNA segments—that has to be constructed, tested, and controlled with a view to its incorporation in a variety of devices and systems. And this is the most difficult aspect of the enterprise. Most of the energies of the Berkeley/Amyris project of using synthetic biology to manufacture artemisinin, a precursor to antimalaria drugs, are devoted to the rectification of unintended interactions between "parts" (Henkel & Maurer, 2007, p. 1). So the construction of parts according to standard tolerances has to both anticipate their incorporation in "higher-level" engineering projects and participate in the even more open-ended project of developing the kind of standardized molecular syntax that would enable parts to communicate and interconnect. What matters here is the notion of participation. In engineering parts for undefined projects, according to tolerances with still undefined principles of calibration, contributors to the BioBricks project are working within a particular ethos of science. Engineering is both a technical and an ethical enterprise. In the process of engineering modules to be accessible to other engineers, synthetic biologists are quite literally building a normative program into the components that they contribute. In this sense, at least, the program of synthetic biology exemplifies the logic of peer production, and it does so precisely because it deploys modularity in ways that run counter to the classical notion that was the basic premise of inventorship and ownership.

According to the theory of peer production, which describes the operation of a range of collaborative projects, from Wikipedia and online gaming to biological science, commons-based production is more likely to take off when projects are modular: "[W]hen a project of any size is broken up into little pieces, each of which can be performed by an individual in a short amount of time, the motivation to get any given individual to contribute need only be very small" (Benkler, 2002, p. 378). Here, modularity describes a mode of production that engages and motivates creativity not by attributing originality or origination to a single idea or inventor, but by facilitating contributions from a number of interdependent contributors. These contributions—parts—have to be compiled effectively, which calls for "low-cost integration," meaning "both quality control over the modules and a mechanism for integrating the contributions into a finished product" (Benkler, 2002, p. 379). In a sense, this version of modularity is perfectly represented in the BioBricks project, and more precisely in the institution of MIT's Registry of Standard Biological Parts (http://parts.mit.edu). First, the parts collected in the Registry are donated by scientists whose contributions are shaped by their own specific, self-determined levels of motivation. Second, because the function or value of each part will emerge only when synthetic biology reaches the stage where the hierarchical process of parts, devices, and systems becomes effective, it is simply more efficient to modularize the overall project rather than attribute it to the work of a single inventive entity. "Low-cost integration" is achieved in various ways, notably the reincorporation of users' experiences and observations into the common pool of parts, in much the same way as software sharing enhances and debugs the original code, and community control is invoked as a safeguard against the challenges posed by biohackers (see the Declaration of the Second Meeting on Synthetic Biology, 2006). Significantly, the work of developing a common syntax for the interconnection of parts might itself have the effect of embedding the ethos of peer production. To the extent that the task of regulating protein signals is the most crucial and difficult part of synthetic biology programs, if or when it succeeds it will create a kind of industry standard for the connectivity of synthetic biology parts.

Property

In sketching out these two dimensions of modularity, it is important to note that they are not so much alternatives as alternations. As the example of the Venter Institute's prospective *Mycoplasma laboratorium* patent suggests, the use of modularity as a premise for intellectual property does not exclude alternative modes of modularized collaboration. On the other hand, the development of modularized peer production within the rubric of the BioBricks project seems to be entirely compatible with the proprietary strategies of the DNA synthesis corporations founded by some of the leading players in open-source synthetic biology.

When the text of the Venter Institute's prospective *Mycoplasma laboratorium* patent was published in May 2007, critical commentary focused on two issues. For many commentators, the patent application was significant because it signaled the imminence of the dangers commonly associated with synthetic biology: ecological corruption, bioterrorism, and technological hubris (see generally ETC Group, 2007a, 2007b). Others focused more sharply on the instrumentalities of the patent, arguing that "Venter's enterprises are positioning themselves to be the Microsoft of synthetic biology, putting foundational technologies under monopoly ownership and control" (ETC Group, 2007b), or that the application was "tasteless" because nothing in the text explained how to make the claimed protocell (cited in Kaiser, 2007). Venter's mission is explicitly patent-based: "We're certainly patenting all the methods we're making, because we've had to create all these processes ourselves. Obviously, if we made an organism that produced fuel, that could be the first billion- or trillion-dollar organism. We would definitely patent that whole process" (Venter, cited in Sheridan, 2007).

The *M. laboratorium* patent, which claims the first true protocell invention, is the latest phase in the project of finding and exploiting the (potential) capacity of marine microorganisms to produce new biofuels or to metabolize industrial carbon dioxides. If it works, Venter's protocell will function as the minimally complex (and hence minimally disruptive) host into which a variety of protein-encoding genes could be inserted, initially to test for promising functions, and ultimately to produce an alternative to carbon-based fuels.

The text of the *M. laboratorium* patent application draws on the theme of modularity, observing that in "the new field of synthetic biology [there] is an emerging view of cells as assemblages of parts that can be put together to produce an organism with a desired phenotype" (USPTO, 2007, para. 4). This version of modularity restates the program of "combinatorial genomics," which was once described by Venter as "one of my better ideas if it works. In fact, it's one of my better ideas if it doesn't work" (cited in Shreeve, 2004, p. 8). The first step toward Venter's "top-down" protocell (see the introduction to this volume) was taken in November 2003, when a team led by Venter assembled a replica of the complete genome of the ΦX174 virus. A set of commercially produced oligonucleotide strands was composed into a genome, by first using computerized sequencing technologies to match the "ends" of sequence fragments, and then splicing this genome into a host bacterium. In the press release announcing the sequencing of the ΦX174 genome, the project was described as the first step toward producing "cassettes of particular genes or pathways that could be inserted into host organisms to conduct many types of functions" (IBEA, 2003). The publication of the *M. laboratorium* patent heralded the realization of this ambition. By editing the genome of *M. genitalium* down from 482 protein-coding genes to 381 apparently "essential" protein-coding genes, the project seems to reach the degree zero of life, and the first effective host organism. Commenting on the *M. laboratorium* patent application, Venter observed that "we'd certainly like the freedom to operate on all synthetic organisms" (cited in Kaiser, 2007, p. 1557). In light of this observation, the application might be interpreted as the vehicle of a strategy aimed at gaining control of all minimal genomic protocells.

The principle of the *M. laboratorium* invention is almost transparently implicit in the notion of a "minimal bacterial genome," but it is worth spelling things out.

Unlike a whole genome invention (on this form of invention, see generally O'Malley, Bostanci, & Calvert, 2005, in particular the discussion of the *M. genitalium* patent), the Venter Institute's application seeks to define a novel kind of genomic composite. The provisional patent claims present the invention as a "set of protein-coding genes," and this set is constructed in such a way as to define the invention not as a variant of the *M. genitalium* bacterium but as an entirely novel kind of genomic composite. It is quite routine in patent practice to claim sequence variations that extend the scope of genomic patents beyond a strict sequence ID: "With the aim of optimizing the legal protection of genetic inventions, what is claimed is not only the literal sequence specified in the patent application, but also a set of sequences related to it. The current state of knowledge in the field does not enable rules to be established for specifying the boundaries for a set of sequences that can contribute to a given function" (Dufresne & Duval, 2004, p. 231). Although it employs similar drafting maneuvers (particularly in the specification of functional equivalence), the strategy of the *M. laboratorium* patent is distinctive because the "function" that it claims is not a particular biological process but the "function" of living as such.

The description of the invention divides the 482 protein-encoding genes of the *M. laboratorium* genome into two categories comprising, respectively, those genes that are dispensable or nonessential, and those that perform essential biological functions. The "set of protein-coding genes" that composes the invention is defined in two alternative ways: positively, as a gene set including a certain proportion of the 381 essential genes, and negatively, as a gene set lacking a certain proportion of the 101 deleted genes (the negative definition reflects the fact that the 381 "essential" genes were identified in a process of elimination, by identifying nonessential genes). If this set of 381 genes really is the minimum set required for life, not just numerically but substantively, in the sense that researchers starting with any other bacterial genome would end up with a set containing more or less each of the genes listed in the category of essential genes, then the patent, by carefully calibrating the values of "more" and "less," effectively claims the ultimate genomic protocell. In its initial form, the patent application does not actually enable the splicing of this minimal genome in a ghost cell, nor does it determine whether the true minimum number of protein-coding genes is 381 or 386 (plus 43 structural RNA genes). The description of the invention states only that the "data suggest that a genome constructed to encode the 386 protein-coding and 43 structural RNA genes could sustain a viable synthetic cell" (USPTO, 2007, para. 58). Nevertheless, the provisional claims seek to encompass the gene set as embodied in various forms or media (a chromosome, a free-living organism grown in a bacterial culture, a computer-readable medium, and, most significantly, a "free-living organism" cultured so as to produce hydrogen or ethanol). Assuming that the question of enablement is addressed successfully at a later stage, the resulting patent might effectively monopolize a foundational technology.

There is more to this strategy than "freedom to operate" in relation to all forms of protocellular chassis. Turning to the other side of the *M. laboratorium* project, namely, Craig Venter's Global Ocean Sampling Expedition—the *Sorcerer II* project—it becomes apparent that a proprietary claim to all forms of minimal genomes would affect the proprietary status of the marine sequences collected by the expedition. In August 2003 the *Sorcerer II* embarked on a much-publicized voyage to collect marine genomic diversity into what Venter called "the mother of all gene databases" (Shreeve, 2004, p. 8). Most of the samples of microbial DNA collected by the *Sorcerer II* were taken from international waters, or were collected in territorial waters under permits granted by national authorities. There were, however, a number of controversial episodes. In particular, the *Sorcerer II*'s visit to Ecuadorean waters prompted complaints from Ecuadorean environmental organizations and international pressure groups that Venter had failed to obtain all the necessary permissions for prospecting, and that samples taken from territorial waters had been improperly obtained. Counsel for the Ocean Sampling project wrote to affirm that "no patents or other intellectual property rights will be sought by IBEA on these genomic sequence data" (ETC Group, 2004, p. 26). This dispute was symptomatic of the legal and political confusion that surrounds proprietary expectations relating to genetic resources (see, generally, Pottage, 2006). Some of this confusion arises from the iteration on a global, and more complex, scale of the question confronting protocell research projects, namely, how to resolve the tension between open (public) and closed (private) vectors of innovation. For present purposes, however, what matters is that, though the data derived from the mission of the *Sorcerer II* were immediately made public through the GenBank database (see Venter et al., 2004), the monopolization of minimal genome technologies might give the patent holders effective control of a good proportion of these ostensibly public sequences.

From one perspective the *M. laboratorium* patent application looks like an instrument of bad modularity, or an example of precisely the kind of patenting strategy that the open-source biology movement seeks to mitigate or overcome. But things might not be so straightforward. Craig Venter has indicated that any foundational synthetic biology technologies patented by his corporate enterprises will be licensed on terms that do not unduly impede academic research (Kaiser, 2007, p. 1557). This begs the question of how academic research is supposed to be distinguished from commercial applications, but it does suggest the sense in which diverse vectors of innovation can intersect across or within the patent system. Indeed, the ostensibly dynamic modularity of open-source technology and peer production collaborations intersects in interesting ways with the classical modularity of property institutions.

This is very clear in the case of synthetic biology. A good proportion of the patenting activity relating to synthetic biology is concentrated in the fields of the DNA synthesis technologies that will be needed to turn the virtual parts indexed in the Registry of Standard Biological Parts into actual material components (see Kumar & Rai, 2007, p. 1748, who note that the proprietary positions of large-scale gene synthesis corporations "are likely to be enhanced, not diminished, by the widespread availability of the information necessary for making parts, devices, and systems"). Strikingly, many of those who are at the forefront of the movement for open-source biology and a synthetic biology commons also play leading roles in these gene synthesis corporations (see ETC Group, 2006). One can expect the same patterns to reproduce themselves in the structuring of "bottom-up" protocell research, so the question is what combination of intellectual property and peer production is likely to emerge, and to what effect?

Emergence

The terms of this question take us back to the question of inertia and adaptability that was set aside at the start of this chapter: *What will be patentable?* In part, that question matters because both the traditional use of intellectual property instruments to provide incentives or security for investment and the contrasting use of those instruments in open-source licensing will work only if the early-stage products of bottom-up protocell research are patentable. It matters also because the patent system will play a role—along with contract law—in determining when, for example, a particular research part or product has become the occasion or ingredient of a new invention. It is obvious why patentability matters to the traditional use of patent rights, but projects of open-source biology present specific difficulties. To what extent can software be taken as a model for synthetic biology? Biological parts are not like source code, in the sense that enhancements to biological parts are not immediately capitalized in the "text" of the code, and furthermore, the patent system is less immediate and more expensive than the software system (on this latter point, see Rai & Boyle, 2007). More generally, if "the institutions that made embedded Linux work were, in some sense, an accident" (Henkel & Maurer, 2007, p. 3), to what extent could one replicate a similar scheme by design? Is dynamic modularity necessarily an emergent or evolutionary attribute? Can it be (juridically) engineered? Might the quality of emergence that synthetic biologists seek to engineer out of life return by means of legal institutions?

Beginning within patent doctrine, one could ask a number of questions about the likely patentability of protocell inventions. For example: At what point will innovations in protocell research be deemed to have made the transition from disinterested science to patentable technology?[2] Will protocell inventions be claimed as methods or molecules, and how would that distinction be made? Would these inventions have to be enabled by the deposit of viable specimens or starting materials, as is the case with many biotechnological inventions, or would protocells be the first living inventions to satisfy the written description and enablement requirements by means of text alone? (On the relation between written description and enablement, see Merges, 2007).

With emergence in mind, another question seems even more pertinent. The project of creating protocells from the bottom up might be significantly different from the program of synthetic biology as it is embodied in the BioBricks initiative and the protocell project described in the text of the Venter Institute's prospective minimal bacterial genome patent (discussed below). Whereas the latter projects seek—each in their own way—to master the emergent processes that characterize living process, the virtue of bottom-up protocells is their capacity for "self-repair, open-ended learning and spontaneous adaptability" (see the introduction to this volume). Ironically, the most "artificial" approach to life technologies reproduces its most characteristic feature. How would this attempt to recruit emergence be accommodated in existing criteria of patentability? In particular, how would one define the scope of such a protocell patent? For example, whereas genomic inventions are defined by reference to a sequence ID, or a sequence "set" in the case of the Venter Institute's "top-down" protocell, how would one claim a protocell invention of changing configuration? What degrees of functional variation could an inventor claim to have enabled?

Although these questions—or questions of this sort—will undoubtedly be litigated, it is not clear that patent doctrine will have much difficulty in recognizing and accepting even emergent protocell inventions. Doctrine has taken a broadly inclusive approach to postindustrial technologies. In *Diamond v. Chakrabarty* (1980; 447 U.S. 303) the Supreme Court addressed the most basic criterion of patentability: Did an artificial bacterium fall within the statutory subject matter limitations—"any new and useful process, machine, manufacture, or composition of matter"? Rejecting the argument that the bacterium could not qualify as a "composition of matter," the court held that the meaning of the statutory formula was governed by the prefatory term *any*: "In choosing such expansive terms as 'manufacture' and 'composition of matter,' modified by the comprehensive 'any,' Congress plainly contemplated that the patent laws would be given wide scope" (p. 308). The result was to foreground utility as the dominant criterion of patentability (see the decision in *State Street Bank v. Signature Financial Group*, 149 F.3d 1368, Federal Circuit, 1998). Critically, one might say that this approach compromises the basic mission of the patent system,[3] but what matters here is the point that patentability in the narrow doctrinal sense may be less significant than in other layers of the patent system. Instead, the terms of protocell patents are likely to be determined by developments in the patent office.

The patent bureaucracy is an obvious place to begin. In 2004 the U.S. Patent Office (USPTO) created a separate digest of prior art for inventions in the field of nanotechnology. Although researchers in the field had for some time been arguing that this was necessary, the USPTO's response before 2004 was that the volume of patents and applications had not reached required "critical mass." What lessons might there be here for patents in synthetic biology and protocell research? In one interpretation, the creation of a separate digest of prior art for synthetic biology and protocell research would facilitate the process of applying for a patent, and might resolve doctrinal uncertainties in favor of inventors. Collecting prior art in a specialized index would allow examiners to assess the novelty of an invention more effectively. The problem with emergent technologies is that prior art tends to be distributed across a range of existing indexes, making it too easy to grant patents for similar inventions and difficult to decide interference disputes between two competing claimants to an invention. A specialized index brings with it a specialized group of examiners (subject to the obvious question about recruitment), each of whom would understand the technology in its own terms and would be able to give meaningful content to criteria of patentability that turn on the expectations of a "person of ordinary skill in the art." The point is not that patents would necessarily be easier to obtain; indeed, to the extent that a dedicated index makes for stricter and better informed examination, it might produce patents with greater resistance to litigation. Rather, the point is that the recognition of a new field of technology by the USPTO fosters the adaptation of doctrinal language to the specific features of synthetic biology and protocell inventions, thereby easing their incorporation into the system.

To end by adding abstraction to speculation, how does emergence return to programs of synthetic biology? The theory of peer production assumes a certain quality of coherence in collaborative projects, so the project of writing a novel, for example, is not a good candidate for peer production. Collaborative fiction enterprises "suffer from the fact that modularity and granularity lead to disjunction relative to our expectations of novels" (Benkler, 2002, p. 379, n. 18). A peer-produced novel would lack the linear development and narrative economy of other novels. But what if scientific research were like collaborative fiction? More precisely, what if the meaning and value of each advance in research were essentially conditional on the emergence of future advances? What if innovation itself were emergent and recursive, so that successive inventions recharacterized their predecessors? Of course, the bioengineer's model of a hierarchy of standardized parts, devices, and systems assumes that research is not like collaborative fiction. As the concept of BioBricks implies, parts can be engineered to have determinate functional competences and facets, and to fit together in coherent permutations. But precisely because engineering is not just a technical activity—it simultaneously builds norms and practices of collaboration into standardized parts—the modularity and granularity of each module is also conditioned by the other social techniques that are involved in the constitution of modes of collaboration, notably intellectual property law. In other words, the "size" of a part is not just a matter of engineering: it also depends on the way in which patent law, or the licensing practices that give effect to patent rights, delimit contributions to the functionality of each part. Whether they are made with exclusionary intent or as part of an open-source project, licenses have to address the problem of when an enhancement of the licensed technology should count as an independent modification or improvement. When, if at all, should a licensee be entitled to patent an invention that depended on access to the licensed technology?

This question is problematic, both normatively and doctrinally. Normatively, the issue is where to draw the line between property and its alternatives. To what extent is peer production compatible with the development of applications that depend on high levels of investment? If it turns out that researchers in the fields of synthetic biology and protocell research lack the motivation to contribute to a common stock of parts, would it be preferable to shift to proprietary vectors of innovation, and if so, which particular kind of vector? Or, granted that the motivation to engage in peer production has limits, how should one balance the two species of incentive (property and open source)? The existing differentiation between the construction of standardized parts (peer production) and DNA synthesis technologies (property) strikes the balance in one way.

Things become more complicated when the difference is made by a license that has to map out in advance the legal mutations of a single technological resource. The BIOS license makes the difference between openness and exclusion by distinguishing between the basic technology and two kinds of enhancement: Licensees are entitled to seek intellectual property protection for independent products but not for mere improvements (see www.bios.org). What is the difference between an improvement and an independent product? When or how does one species of module metamorphose into another? This is precisely where the classical doctrinal question of how to calibrate the increments of an innovation process (inventions) arises. And it is also the point at which temporal recursivity of innovation becomes problematic. Quite simply, when is an enhancement already contained, potentially, in licensed research materials, and when is it an independent, supervening product of those materials? Here, the hierarchies of synthetic biology are complicated by a legal technique that makes the modularity and granularity of parts dependent on the institutional negotiations through which recursive trajectories of innovation are patterned. So, from the perspective of intellectual property, the program of bottom-up protocell research may offer a more illuminating metaphor for modularity than programs of rational construction in synthetic biology.

Notes

1. "[A machine] is an artificial organism, governed by a permanent rule of action, receiving crude mechanical force from the motive power, and multiplying, or transforming, or transmitting it, according to the mode established by that rule. This rule of action, imposed by the inventor on the material substances of which the machine consists, is what the courts have called the 'principle of the machine'" (Robinson, 1890, p. 257).

2. This is not a difficult transition to make: See the federal circuit decision in *Re Fisher*, 421 F.3d 1365 (Federal Circuit, 2005), and especially the dissent by Judge Rader.

3. "The United States Court of Appeals for the Federal Circuit, entrusted by Congress to manage the patent system, has deliberately remolded that system to protect investment as such, rather than discontinuous technical achievements that elevate the level of competition. The patent system has accordingly degenerated to protecting incremental slivers of know-how applied to industry, including the very business methods that were formerly the building blocks of the free-enterprise economy" (Maskus & Reichman, 2005, p. 21).

References

Benkler, Y. (2002). Coase's penguin, or, Linux and the nature of the firm. *Yale Law Journal, 112*, 396–446.

Benkler, Y. (2006). *The wealth of networks*. New Haven, CT: Yale University Press.

Biagioli, M. (1998). The instability of authorship: Credit and responsibility in contemporary biomedicine. *The Journal of the Federation of American Societies for Experimental Biology, 12*, 3–16.

BioBricks Foundation (2007). Available online at: http://www.biobricks.org/ (accessed September 2007).

Chan, L.Y., Kosuri, S., & Endy, D. (2005). Refactoring bacteriophage T7. *Molecular Systems Biology, 10*, 1038.

Curtis, G. T. (1849). *The law of patents for useful inventions*. Boston: Little, Brown.

Declaration of the Second International Meeting on Synthetic Biology (2006). Berkeley, CA, May 29, 2006. Available online at: http://openwetware.org/wiki/Synthetic_Biology/SB2Declaration (accessed September 2007).

Dufresne, G., & Duval, M. (2004). Genetic sequences: How are they patented? *Nature Biotechnology, 22* (2), 231–232.

Eisenberg, R. S. (1987). Proprietary rights and the norms of science in biotechnology research. *Yale Law Journal, 97* (2), 177–231.

Endy, D. (2005). Foundations for engineering biology. *Nature, 483,* 449–453.

ETC Group (2004). *Communiqué 84: Playing God in the Galapagos.* March/April 2004, available online at: www.etcgroup.org (accessed August 2007).

ETC Group (2006). *Terminator: The sequel.* Available online at: http://www.etcgroup.org/en/materials/publications.html?pub_id=635 (accessed August 2008).

ETC Group (2007a). Extreme genetic engineering—An introduction to synthetic biology. January 2007, available online at: www.etcgroup.org/upload/publication/602/01/synbioreportweb.pdf (accessed August 2007).

ETC Group (2007b). Backgrounder: J. Craig Venter Institute's patent application on the world's first human-made species. June 2007, available online at: www.etcgroup.org (accessed August 2007).

European Commission (2005). *Synthetic biology—Applying engineering to biology. Report of a NEST high-level expert group (EUR 21796).* Luxembourg: Office for Official Publications of the European Communities.

Heller, M., & Eisenberg, R. (1998). Can patents deter innovation? The anticommons in biomedical research. *Science, 280,* 698–701.

Henkel, J., & Maurer, S. M. (2007). The economics of synthetic biology. *Molecular Systems Biology, 3,* 1–4.

Institute for Biological Energy Alternatives (IBEA) (2003). *Press release: IBEA researchers make significant advance in methodology toward goal of a synthetic genome.* Available online at: http://www.jcvi.org/cms/press/press-releases/full-text/browse/8/article/ibea-researchers-make-significant-advance-in-methodology-toward-goal-of-a-synthetic-genome/?tx_ttnews%5BbackPid%5D=67&cHash=d4fae30116 (accessed March 2008).

Jefferson, R. (2006). Science as social enterprise: The CAMBIA BIOS initiative. *Innovations, 1* (4), 13–44.

Kaiser, J. (2007). Attempt to patent artificial organism draws a protest. *Science, 316,* 1557.

Knorr-Cetina, K. (1999). *Epistemic cultures: How the sciences make knowledge.* Cambridge, MA: Harvard University Press.

Kumar, S., & Rai, A. K. (2007). Synthetic biology: The intellectual property puzzle. *Texas Law Review, 85,* 1745–1768.

Langlois, R. N. (1999). Modularity in technology, organization and society. Available online at: http://ssrn.com/abstract=204089 (accessed August 2007).

Marguet, P., Balagadde, F., Tan, C., & You, L. (2007). Biology by design: Reduction and synthesis of cellular components and behaviour. *Journal of the Royal Society Interface, 2007,* 1–17.

Maskus, J. H., & Reichman, K. E. (2005). The globalisation of private knowledge goods and the privatisation of global public goods. In J. H. Maskus & K. E. Reichmann, *International public goods and transfer of technology under a globalized intellectual property regime* (pp. 3–45). Cambridge: Cambridge University Press.

Merges, R. (2007). Software and patent scope: A report from the middle innings. *Texas Law Review, 85,* 1627.

Mirowski, P., & Sent, E.-M. (2002). *Science bought and sold: Essays in the economics of science*. Chicago: University of Chicago Press.

Murray, F., & Stern, S. (2005). Do formal intellectual property rights hinder the free flow of scientific knowledge? An empirical test of the anticommons hypothesis. NBER Working Paper, available online at: www.nber.org/papers/w11465 (accessed August 2007).

O'Malley, M., Bostanci, A., & Calvert, J. (2005). Whole genome patenting. *Nature Reviews Genetics*, 6 (6), 502–506.

Pottage, A. (2006). Too much ownership: Bio-prospecting in the age of synthetic biology. *Biosocieties*, 1, 137–158.

Rai, A., & Boyle, J. (2007). Synthetic biology: Caught between property rights, the public domain, and the commons. *Public Library of Science: Biology*, 5 (3), 389–393.

Robinson, W. C. (1890). *The law of patents for useful inventions, volume 1*. Boston: Little, Brown.

Scientific American (Eds.) (2006). How to kill synthetic biology. *Scientific American*, May 22, 2006 issue, available online at: http://www.sciam.com/article.cfm?articleID=000D6682-91AE-146C-91AE83414B7F0000&sc=I100322 (accessed September 2007).

Sheridan, B. (2007). Craig Venter's next quest. *Newsweek International*, available online at: www.msnbc.msn.com (accessed August 2007).

Shreeve, J. (2004). Craig Venter's epic voyage to redefine the origin of the species. *Wired Magazine*, 12, available online at: www.wired.com/wired/archive/12.08/venter.html (accessed August 2007).

Stone, M. (2006). Life redesigned to suit the engineering crowd. *Microbe*, available online at: http://www.asm.org/microbe/index.asp?bid=47155 (accessed March 2008).

USPTO (2007). Patent application 20070122826: Minimal bacterial genome. Published May 31 2007.

Venter, J. C., Remington, K., Heidelberg, J. F., Halpern, A. L., Rusch, D., Eisen, J. A., et al. (2004). Environmental genome shotgun sequencing of the Sargasso Sea. *Science*, 304, 66–74.

Walsh, J. P., Cho, C., & Cohen, W. M. (2005). View from the bench: Patents and material transfers. *Science*, 309, 2002.

11

Protocells, Precaution, and Open-Source Biology

Andrew Hessel

Biological technologies are poised to shape this century in the same way that electronics and information systems reshaped the last. Like computers, biological technologies are becoming more powerful every day, yet less expensive and easier to use, foreshadowing a tipping point when their use will become mainstream. As this occurs, it will greatly expand the size and scope of the present biotechnology industry, and deliver powerful economic, medical, and agricultural benefits—but also unprecedented new risks.

Life has been hard to engineer. It was only with the appearance of recombinant DNA technologies in the early 1970s that the direct manipulation of DNA and, by extension, metabolism became possible. But the full potential—and hazard—of genetic engineering has not been fully realized, mainly because the early techniques were so limiting. Newer technologies automate and digitize genetic work, allowing any DNA to be designed and assembled from scratch with special synthesizers. This lowers the economic and technical barriers to genetic engineering, and effectively turns DNA into a programming language for biological systems. Almost overnight, a new field has emerged—*synthetic biology*—with capabilities that are growing exponentially. With over 100 billion bases of DNA already deposited in public databases, and more flooding in daily, biological programmers have ample data on which to base genetic designs. Potentially *anything* biological can be made, from single enzymes, to protocell components, to fully synthetic cells—or, just as easily, synthetic biological weapons.

The acceleration of biological research will dramatically change the business of biotechnology, affecting, among other factors, the landscape of the global industry, how intellectual property (IP) is managed, and how regulatory oversight is achieved. How to manage biological research and development (R&D) remains an open question, complicated by the fact that biological knowledge can be used to save lives, or to more efficiently take life. If R&D outpaces the ability for societies to understand, monitor, and regulate developments, it could result in the proliferation of gray-market biotech. Alternately, if R&D is overly restricted, it could prevent necessary or beneficial good from reaching markets, also harming people. Most worrisome is the possibility that R&D could spark an endless arms race, with perils on par with nuclear weapons. Moving forward, precaution is warranted. Policies and strategies that maximally lead to constructive applications and international cooperation, while working to minimize biological hazards, are necessary if we are to shape, rather than just react to, the biological future.

Open-source biology could help bring meaningful progress on these issues. Open source emerged in the 1990s as an alternative route to software development. Reducible to "share and share alike," open strategies led to developer communities willing to pool and focus resources on shared or common goals. In the process, it created a new—and, some argue, better—route to software development, doing so in the shadow of the established software industry. If transferable to biotechnology, the advantages of open-source software—useful products, low costs, maximum access, and high security—could expand biological capability while helping to maintain oversight and security.

Although the proprietary ownership of biology is desirable for business development, in reality biology is difficult to own or regulate. Biological materials are easy to procure and biological knowledge is widely disseminated. For this reason, open-source strategies may be a good fit. But the mechanics of open source remain poorly defined, even for computer software. With only a handful of successes, the business models remain poorly understood and poorly tested. There are no guarantees that open source can work or will work for biological product development, which is more complex and faces greater social challenges than software development. If it does work, though, open-source biology could become an important counterbalance to proprietary biotechnology, helping to ensure fair prices and fair markets.

This chapter explores how open source is currently being explored in biological research, and speculates on greater roles for open biology in the near future. It examines how open biological standards are being used in engineered biology, how openness might play a role in global bioeconomics and biosecurity, and how open-source biology communities could organize to deliver cost-effective R&D, drawing where it can on past experience with open-source software. It also highlights some of the challenges, like national security, that could prevent the use of open-source biology. These issues are already being encountered by the nascent synthetic biology community, in which both proprietary and open forces are active. How the issues are conceptualized and resolved by synthetic biologists is likely to be relevant to other groups working at the forefront of biological technology, including researchers working to bring protocells into everyday use.

Appearance of Engineered Biology

DNA, with few exceptions, is the substance of heredity for living things. As biological source code, it has the potential to direct the synthesis of virtually everything biochemical. The genome sequence of an organism is roughly equivalent to a computer operating system, except for format. DNA is written in nucleic acid bases instead of electronic zeros and ones. Both formats, however, are digital, and, with the technologies of sequencing and synthesis, interchangeable (Cox, 2001).

The systematic reverse engineering of genetic code, protein synthesis, and metabolism is rapidly revealing the biochemical bases of life and disease. Now biology is moving beyond classification and deconstruction and entering a constructive phase. Founded on the ability to synthesize long-chain DNA, synthetic biology opens a door to the design and manipulation of almost any biological system, including protocells. With synthetics, DNA manipulation becomes trivial, driving researchers to think more about *what* to make with biology, not about how to make it in the lab. Constructive biology marks the maturation of biological science into a true engineering discipline, and its willingness to play catch up. Almost everything in the modern world has been engineered to some degree—except for life. But this is changing quickly. Students are gravitating to synthetic technologies as earlier generations did to computers, and delivering impressive results. Universities are paying attention to this trend, and are rolling out new departments and programs focused on this work.

Engineered biology promises to distinguish itself from historical genetic "engineering" efforts, which are by comparison crude hacks. Engineered organisms will be increasingly understood as machines—their design, function, and evolution completely knowable, unlike the organisms of the natural world. Ideally, they will exist only because a conscious decision was made to build them, and persist because their utility, cost, or benefit has made them worthy of avoiding the biological scrap heap—selection mediated by conscious decisions, or what might be considered *memetic* evolution. But engineered organisms will also bring great responsibility, since somebody will ultimately be accountable and liable for their creation and their actions.

A foundation for engineered biology is emerging. Following in the footsteps of other engineering fields, such as very large-scale integration (VLSI) electronics and software engineering, *abstraction hierarchies*, and preliminary *biological standards* have been created for synthetic biology (Endy, 2005). Abstraction hierarchies compartmentalize systems into layers and components, and permit parallel development streams—necessary if groups are to work collaboratively and efficiently on complex projects. Biological standards work to provide engineers with detailed specifications about parts and systems. Together, they have led to BioBricks, modular genetic components developed by the Massachusetts Institute of Technology (MIT) fashioned after LEGO™ toy blocks.

With BioBricks, DNA sequences are organized into *parts* that have defined regulatory or structural functions when expressed in a suitable cell, such as *Escherichia coli* or other bacterium. For example, one part might specify the function "in the presence of the sugar glucose, turn on," while another might encode "if on, make Green Fluorescent Protein (GFP)." Various parts can be assembled to build *devices*, the next level up in the abstraction hierarchy. For example, the two aforementioned parts would combine to create a sugar biosensor device that would produce a visible GFP signal in the presence of glucose.

To "run" this program, the DNA for the device is first assembled, either by physically joining together the appropriate BioBrick plasmids (the unique design of BioBricks allows the iterative assembly of any number of parts, and also robotic assemblies), or by synthesizing the computed DNA sequence of the complete device; the end result is identical. Once made, the DNA is uploaded into cells through transformation. Expected outputs are measured quantitatively, generating functional data about the device and components for downstream designers. In a similar fashion, one or more devices can be assembled to produce even more complex *systems*, the third abstraction layer; and so on.

Taking a lesson from the open source community, MIT researchers have chosen to make BioBricks an open standard for engineering biological systems, opting to put both parts and data in a public collection, the *Registry of Standard Biological Parts* (http://parts.mit.edu). The expectation is that this approach will encourage more researchers to use the parts for their own work, and to deposit new parts back into the collection. In theory, the more biological parts that become available, the more devices and systems that can be designed and built. In turn, the more devices and systems that are built, the more data that can be generated about the performance of the systems and parts, increasing predictability. A key hope is for this approach to facilitate the design of biological systems *in silico*, eventually allowing metabolism to be modeled and tested in software in the way buildings, automobiles, and aircraft are now developed. The early results, like biological photographic film (Levskaya et al., 2005), have been encouraging.

Although open standards are not absolutely necessary to further engineered biology, experience with electronic standards has taught that only open standards—like the IP and HTTP protocols used by the Web—can support *both* proprietary and open development, maximizing flexibility and utilization. In contrast, proprietary standards serve mainly to reinforce monopolies. As engineered biology expands, perhaps to include modular protocell development, other technical standards are likely to emerge. But technical standards alone are not sufficient. Legal standards that facilitate the exchange of parts and use of multiple parts in combination are also necessary, as are standards for biological security that allow authentication, real-time monitoring, or forensic analysis. A major focus of the BioBricks foundation (http://biobricks.org) is to create the technical and legal standards required for synthetic biologists to cooperate.

The Emerging Bioeconomy

With synthetic capabilities already growing at a good clip, the continued expansion of biotechnologies is virtually assured, barring drastic interventions. Biotechnologies seem ideal to meet the global challenges of providing food, medicine, and ecologically friendly structural materials in a cost-effective and sustainable fashion. Already, the global biotechnology landscape is beginning to shift as more countries come on board. International interest has grown over the last decade, with countries such as Singapore, India, Korea, and China investing heavily in biotech. Their progress has been rapid. Some now compete effectively with the United States and Europe in areas like stem cells and gene therapy, a trend that could continue given the strong culture of entrepreneurialism in these regions. As with computers, the competition for market leadership and market share could become fierce, also driving innovation.

The potential for economic abuse and political interference is unsurpassed in biotechnologies, given the size of the global markets and the power these technologies hold over people. Life is a priceless commodity, distorting market forces, even in the absence of monopolies. In addition to being the source of many medicines and therapies, living things also provide many of the essentials of daily life. The outputs of a bioeconomy will thus include products that people, quite literally, may not be able to live without. This reason alone should be a powerful motivator for people to demand open accountability and, to the greatest extent possible, open markets. But biotech is still well below the radar for most people, a science distant from their everyday lives.

Born going head-to-head with global pharmaceutical companies, the biotech industry got tough quick. It used patents aggressively to secure investment capital and to fend off competitors, creating a tangled web of intellectual property (IP). This "thicket" may already result in a significant drag on new products reaching consumers (Heller & Eisenberg, 1998). Patents will continue to be important for commercial biotech development, but reforms may be necessary as the industry booms. The utility of patents correlates with the pace of development, the rigor with which patent claims are observed and enforced, and the ease of copying the product. Medicines in particular have thorny IP issues. If powerful and beneficial biotechnologies are created that could save or enhance lives, but cannot reach the marketplace because of IP, black or gray markets will arise to compensate. Similarly, if monopolistic pricing makes essential products unattainable, revolts are likely. Brazil, in trying to provide HIV medicines to patients, enacted new laws to strong-arm drug manufacturers to keep drugs affordable (Gottlieb, 2001). Manufacturers complained, but found little support. Meanwhile, individual workarounds could be as simple as a Google search, which today returns dozens of online pharmacies that will ship products worldwide.

DNA code, closer to software, seems to warrant more flexible IP. With its inherently faster evolutionary pace, computer software has largely been protected by copyright, although software patents are also starting to be used more frequently, creating new problems for software developers (Wren, 2006). Unfortunately, synthetic biology, at the intersection of computing and biology, could suffer from flawed biotechnology and software laws (Rai & Boyle, 2007). The ability to invoke copyright for digital DNA may not be clear, making the use of patents necessary. But patents may not be feasible if a modular approach to biology becomes widely adopted, since negotiating usage rights for every component in a construct might be impossible. Rai and Boyle (2007) note that *copyleft licenses*—licenses used in open source to enforce openness and stimulate the creation of ever-expanding commons—have worked brilliantly for software developers. In theory, they could also work very well for synthetic biology (Hessel, 2006).

But the use of open biological IP faces large challenges. First, its legitimate use requires that it not violate existing patent claims, or that patent-holders exempt open source developers from litigation. Second, it might simply not work to support the development of commercial biotech products. Software is an informational good, with relatively low development costs. Biological products are expensive to develop—especially for pharmaceuticals, which can cost upwards of hundreds of millions of dollars. IP strategies that do not provide exclusivity or monopoly may not attract enough investment capital to support commercial development. This could change, however, if the cost of developing drugs can be reduced, or if new sources of unattached funding—like private philanthropists—can be tapped.

It might not be IP, blocking some products from sale, that bottlenecks biological development. It could be regulatory approval, able to block *all* sales, that proves the stumbling block. Unlike computer software, which can be sold without review, biological products are heavily controlled in most countries. In the United States, this work is done by the Food and Drug Administration (FDA), the Environmental Protection Agency (EPA), the Department of Agriculture, and the Drug Enforcement Agency (DEA). Their degree of scrutiny is expensive, slow, and tends to be overcautious. As innovation accelerates, the pressure will grow for regulatory agencies to keep pace with the times. New classes of products, from genetically engineered pets to gene therapies, are beginning to enter the system. There is little evidence that these agencies can scale to accommodate faster review, particularly as biological products become more complex, nor can they relax and maintain any authority. The FDA in particular has struggled with scientific impartiality, corruption, and drug safety issues, despite ongoing reforms.

If biological products begin to flood into the system, especially if they are made in foreign countries, it may become impossible for regulatory agencies to maintain tight control. They could be reduced to following behind manufacturing and distribution trends, working more like the DEA. Upset by delays, consumers are already taking the FDA's policy of blocking individual access to unapproved medicines to court (Hede, 2006). Reformers argue that seriously ill patients should be given access to drugs following Phase I (safety) trials, letting patients determine efficacy for themselves. But even this may be too conservative for those desperate for medical options. Health consumers already demonstrate a willingness to seek and acquire medicines online, caring more about their cost than their legality, or to travel to more permissive regions. Gray markets and medical tourism could flourish.

New Biological Threats

Thirty years ago, when recombinant DNA technology first appeared on the scientific scene, the risks of genetic engineering were unknown. There was great concern that the techniques would lead to dangerous or unexpected results, perhaps threatening the safety of laboratory workers, or polluting the environment. In a bold, precautionary move, a voluntary moratorium on genetic experimentation was enacted to explore the concerns (Berg et al., 1974). The restrictions were eventually lifted after the now-famous Asilomar conference of 1975, which brought together world scientific leaders, government officials, lawyers, and the press to discuss how best to proceed. The meeting led to the creation of biological containment and handling protocols that are still in use (Berg et al., 1975; Singer & Berg, 1976). Now, thirty years after Asilomar, safety concerns have increasingly given way to the concern that these technologies will be used as terrorist tools, or to create weapons of mass destruction. Another Asilomar-type meeting has been suggested for synthetic biology, to reign in the technology before it can race out of control (Ferber, 2004).

The development of biological weapons is outlawed by the 1972 Biological and Toxin Weapons Convention (http://www.opbw.org), with signatories agreeing not to develop or maintain stocks of biological agents or toxins. The convention makes no provisions for inspections, however, and widespread violations are thought to have occurred (Chyba, 2001). In reality, few forces other than international goodwill have been working to prevent proliferation of these weapons, and most nations have voluntarily restricted work with biological agents to small-scale defensive studies. Despite this lack of formal policing, only a few instances of biological terrorism have been documented. The exact reasons for this are not known, since biological agents are not hard to acquire from natural sources.

Acquisition is getting easier, too. Synthetic biology significantly alters the potential risks associated with biological weapons development and proliferation. First, it allows, by direct DNA synthesis, access to many of the viruses, microbes, and toxins capable of being weaponized. This access was demonstrated in 2002 by the manufacture of functional poliovirus from mail-order DNA fragments (Cello, Paul, & Wimmer, 2002). Synthetics has since been used to resurrect the 1918 influenza strain that killed millions (Tumpey et al., 2005), and a 5-million-year-old progenitor of the human endogenous retroelement HERV-K (Dewannieux et al., 2006). Viruses may be particularly attractive to people intent on creating social or economic disruption, because of their ability to self-replicate and spread widely. The SARS coronavirus emerged from a single source in China to reach over 30 countries in less than a year, infecting 8,000 people and killing about 800, before disappearing in 2004 (Guan et al., 2004). Synthetic recreations of natural agents like SARS could wreak international havoc, yet be misidentified as natural outbreaks. Soon, the ability to create such attacks could fall within the reach of novices.

Still more worrisome, synthetic technologies could also be used to create even more dangerous *designer pathogens*. Such advanced agents could be engineered to be fast-acting or spread stealthily, to act broadly, or affect only certain crops, animals, people, or cell types. Agents engineered to resist known antimicrobial or antiviral agents, or to circumvent or target the immune system, could also be exceedingly dangerous. The successful development of agents such as synthetic prions (Legname et al., 2004), enhanced mousepox (Jackson et al., 2001), and silencing RNAs that can precisely turn any gene off (Clark & Ding, 2006) serves to reinforce the possibility of advanced agents and their likelihood to appear in some form. The worst-case scenarios are horrifying, easily on par with nuclear attack.

Designer pathogens could also give rise to biological "hacks" or biological "spam," more prankish than deadly, created for bragging rights or financial gain. Possible examples include agents able to produce an unsightly but otherwise harmless blemish, treatable only by a specially formulated and expensive ointment, or neuroactive compounds that can subconsciously shift moods or intentions. In time, pernicious threats could prove more disruptive to society than overt attacks. Along similar lines, synthetic biology increases the possibility that biotechnologies will be employed to make illicit drugs, today produced mainly with organic chemistries. The necessary DNA code could be made available online. Expressed in everyday plants or bacterial cultures, or via gene delivery systems inserted into an individual's own body, biointoxicants or stimulants could become relatively easy to make and distribute and almost impossible to detect, further complicating the work of regulatory agencies.

Global culture that has learned not just to expect but to demand rapid change is increasing biological risk. The pace of biological development has been glacial in comparison to computer technologies, perhaps a factor that has until now kept us safe. Ready or not, biological innovations are going to come faster and faster. And, unlike the virtual world of computers and computer networks, biological software is released into the environment of our own genosphere. Early computer viruses such as the Morris worm (Erbschloe, 2005) brought chaos to networks, because no electronic barriers were in place to check inappropriate proliferation. With virtually no biological defenses in the real world (Pejcic, De Marco, & Parkinson, 2006), there is a similar risk that our early experimentation with biotechnologies could cause widespread harm, intentional or not. Only, unlike computers, the natural environment cannot be easily rebooted or reformatted. We may have to live a long time with our biological mistakes.

Balancing Biological Risks and Rewards

As the hazards of advanced biotechnologies become better understood, a new balance between risks and rewards should emerge. Already, consensus threats are being realized as "opportunities" for groups to develop new biodefense products and technologies, including vaccines and diagnostic kits. But the lessons from information technologies presage the challenges ahead. Consumer desire for new products, including those that may be frivolous, results in markets that businesses are eager to fill. Genetic technologies will be no exception. Already, personal markets have emerged for genetic testing, gene banking, and even gene-engineered pets. Moreover, people are demonstrating a willingness to shop the world for the health products they want, and for the best prices. With online prescriptions and medical tourism changing the way consumers access therapeutic and cosmetic treatments, patriarchal controls are unlikely to be effective as more biotechnologies become available. China already has a licensed gene therapy, Genidicine, that has been used to treat thousands of cancer patients (Peng, 2005; Wilson, 2005). Other treatments, and other countries, cannot be far behind.

Reaching consensus on how to safeguard society from emerging biological threats is proving very difficult. One reason is simply a lack of information about what constitutes a true threat. Take emerging diseases as an example. Molecular surveys of the natural world have only just begun. Familiar pathogens such as Ebola and Marsburg come and go without shedding new light on their natural reservoirs, while new studies of the DNA present in air, water, and soil are revealing millions of viral and bacterial species previously unknown (Venter et al., 2004; Goldberg et al., 2006). Similarly, we have only started to understand how the human organism works in health and disease at the molecular genetic level. Even the fundamental structure of our genome is proving more complex than previously thought (Redon et al., 2006).

Another reason is that a profound leadership vacuum exists with respect to bio-logical technologies. Governments are reluctant to slow the pace of biological development in the absence of immediate risk, particularly since strong restrictions by one nation may only provide advantages to others while compromising economic growth, social welfare, and even national security in the process. Biologists are also struggling to find consensus on how to manage emerging risks. Although the syn-thetic biology community has taken steps toward guidelines and best practices, the process is incomplete (Maurer, Lucas, & Terrell, 2006). Scientific self-governance remains strongly favored over external legislation, which could restrict the ability to conduct research. Precautionary avoidance no longer appears on anyone's agenda. Not that any call for moratorium would be taken seriously, as the times have changed significantly since Asilomar. Back then, no multibillion-dollar biotechnol-ogy industry existed, nor were so many biological scientists and their institutions involved with industry. And experience with information systems suggests that the appearance of an organized, technically elite hacker culture is not just possible, but inevitable.

If the forward progress of biotechnology cannot be slowed or halted, the issues of concern will have to be dealt with on the fly. Speaking out against restrictive measures, the National Research Council recently advised "to the maximum extent possible, promote free and open exchange in the life sciences" to create a "web of protection" against biological risks (National Research Council, 2006). In other words, everyone—be they corporate, academic, or simply concerned members of the public—will have to work together. From a security perspective, this position would have been surprising even twenty years ago, when nonproliferation ideas were firmly rooted in the nuclear age, and globalization was less of a factor. It took the Internet to practically demonstrate that security in an open environment was even sensible, by first rendering rigid controls impractical, and then allowing a dis-tributed approach to emerge. Now it is understood that "bottom-up" approaches, bound by common goals—such as keeping networks up and running—are robust and fault tolerant, able to work even if some of the components fail. But the idea that letting go can achieve greater stability is still very counterintuitive.

Taking this approach one step further, open-source software puts source code for critical software—like the Linux operating system—in the public domain, available to hackers and developers alike. Yet even this radical step has not led to widespread failure. In fact, it has had the opposite effect, pushing programmers to make system software that is both highly secure and reliable, and fostering deep security expertise to deal with problems as they arise from time to time. However, at least when it comes to developing code, software developers have a distinct advantage over biologists. Should a vulnerability be found in Linux, creating a patch or distributing a revision is trivial, whereas updating our own genomes or immune systems to fend off threats is next to impossible, at least with current technologies.

The public acknowledgment of vulnerabilities may be an important first step to rectifying them. In his seminal essay on open source, *The Cathedral and the Bazaar*, Eric S. Raymond notes that Linux security arises in part because "given enough eyeballs, all bugs are shallow" (1998). Beyond just finding errors, recent studies show that tapping large communities can also generate novel solutions (Lakhani, Jeppesen, Lohse, & Panetta, 2007). Records from InnoCentive, a company that collects and distributes R&D problems from industry, show that "broadcasting" unsolved problems to "solvers" can rapidly uncover answers. Solvers—who did not have to be experts or hold degrees—often had solution information already in hand, or brought fresh perspectives that made the difference. Similar approaches to harvesting innovation have been successfully used by the Defense Advanced Research Project Agency (DARPA) to create autonomously driven vehicles, and by the X-Prize Foundation (http://xprize.org), a group specializing in incentive prizes, to make suborbital spacecrafts. Prizes for highly sought-after biological goals are also expected to be successful. Recently, the X-Prize Foundation announced a $10 million award for fast (10 genomes/day) and cheap ($10,000 or less) human DNA sequencing.

Although these organizations offer cash for solutions, this is not absolutely necessary as a motivator. Many solvers enjoyed the problem-solving experience itself, considering it fun and stimulating (Lakhani, Jeppesen, Lohse, & Panetta, 2007). Similarly, many contributors to open-source projects gain little in the way of recognition or status in their communities, but simply like writing good code. These results suggest that finding solutions to problems created by advancing biotechnologies could be as straightforward as raising awareness of them, and having a mechanism in place to collect and prioritize reasonable solutions, no matter where they originate.

Toward Open Biological R&D

In the last few years, open-source biology has moved beyond speculation into early practical experimentation. A major area of progress has been an increase in open access to biological information. Open access journals, spearheaded by the Public Library of Science (http://www.plos.org), are leveraging inexpensive online publishing to ensure unfettered access to articles. Wikipedia (http://wikipedia.org), an online open-source encyclopedia, is also becoming an excellent source of biological knowledge, delivering what amounts to continuously updated review articles on major topics.

Biological education is also shifting toward more open access. MIT in particular has aggressively pursued bioeducation initiatives, offering free access to advanced biology courses through their OpenCourseWare initiative (Margulies, 2004), and supporting the synthetic biology research community with their OpenWetWare wiki (http://openwetware.org). In addition to these data resources, MIT also coordinates the international Genetically Engineered Machine (iGEM) competition, an annual contest in which undergraduate and high school students create BioBrick-based devices. Growing about 300 percent each year, the 2006 competition attracted 380 participants from 37 universities, resulting in projects that ranged from biosensors for arsenic contamination to engineered bacteria that smell like bananas (Jan, 2006). Mirroring open-source software initiatives, iGEM teams are strongly encouraged to provide detailed experience and measurement data on the BioBricks they use or create, and deposit new parts with the Registry of Standard Biological Parts, a resource available to the entire scientific community.

Open software projects may begin with personal motivation, but they use networks effectively to attract more developers, share code and other resources, fundraise, advertise, distribute products, and collect client feedback. The network becomes central to the creation and evolution of the software product. Even proprietary companies now leverage aspects of the network—for example, sharing customer feedback on their sites. Scientists, on the other hand, though adept at navigating peer networks, have been slow to tap the social networks accessible via the Web, perhaps because they place great emphasis on individual discovery and personal reputation. Although increasingly open to blogging, sharing protocols or raw data, and putting lectures on YouTube, most biologists still operate within laboratory-sized collectives, with limited perspectives. Professional barriers remain high, and massively collaborative efforts like the Human Genome Project (Lander et al., 2001) are very rare.

Wider collaborations may be necessary as engineered biology matures. As biological designs increase in complexity, input from diverse intellectual sources may be required, including input from regulatory groups. If biological engineers can learn to organize as open-source software developers have, major research initiatives could attract the interest of large numbers of contributors from many countries. Studies could be devised in which wet-bench work is divided among those equipped to do the work, with each scientist donating fractions of their time and research funds, and pooling and sharing the results. The entire community could work to analyze and discuss the data. This way, large-scale biology could be done relatively easily, even if the scientists were geographically dispersed.

Community-based research projects would benefit from an intrinsic and dynamic review process. Conceivably, any project could be proposed, but only projects that were able to attract support would advance—a natural way to create peer review and to avoid political or scientific biases. The human and financial resources flowing to a project would directly correlate with the collective interest, perceived need, and perceived chance of success. And with no upper limit on the size of research teams, projects with strong appeal could grow to a size impossible to support with more top-down approaches. Transparent management would also help ensure that research efforts were not duplicated or squandered needlessly.

And if massively collaborative research is feasible, why not massively collaborative development? Expansion of open research communities to include product development or business experience could go far to advance promising research toward commercialization. The result could be open-access, virtualized biotechnology companies able to operate at a fraction of the cost of wholly proprietary groups because of donated efforts, yet fully able to bring new products to the market. These virtual companies could prove ideal for developing low-margin or high-liability products, such as vaccines and antibiotics, highly personalized medicines, or gene therapies that could lead to cures, that would be unattractive to for-profit pharmaceutical companies and would implode markets.

An open path to R&D could significantly increase the reliability of biological engineering at all levels of research and development. Beyond just open standards, open-access biology would encourage the publication of all available research data, positive or negative, not just the subset of illustrative results. Negative data may be particularly revealing, since it could offer clues as to why so many biological experiments fail, necessary information if biological engineering is to become more predictable. Furthermore, if an open development path can be created, public access to development and clinical trial data would increase significantly, facilitating effective safety reviews and data meta-analysis, whereas today such data remains sequestered within the FDA and in various company databases.

In some ways, open-source biology research efforts could prove easier to support financially than open-source software projects. The professors and students likely to be participating in open-source biology efforts in most cases would already have their living expenses covered by salaries, stipends, and grants. In addition, academic scientists would stand to gain in reputation and experience from their participation in open-source projects. And finding real dollars to support the cost of collaborative biological research might also be straightforward, as academic groups and institutions are adept at seeking public support.

But open-source biology is no panacea. It could complicate the legal and regulatory environment that works to support biological science, and facilitate the expansion of biotechnologies beyond industry and academic walls. It could provide advanced technologies to nations otherwise ill-equipped to safely manage and monitor their use. Open-source biology could also face strong resistance from the proprietary biotechnology industry if it begins to threaten key markets. The dominant biotechnology players have used acquisition and patent litigation to hold competitors at bay. But open-source biotechnology could be resistant to these defenses, and be seen as a credible threat. Already, open source is beginning to encroach on commercial arenas. Open access journals are making strong advances into territory that was once held exclusively by for-profit scientific publications. Another example is TransBacter, a gene transfer system for plants that circumvents patents covering *Agrobacterium*-based transformation, eliminating restrictive license fees in agricultural biotechnology (Broothaerts et al., 2005). Soon, open-source projects could even threaten pharmaceutical markets, the lifeblood of the industry, prompting more aggressive measures.

Given a chance to survive, open-source biology, by tapping the power of communities, could play a significant role in making biological technology more accessible, efficient, and synchronized with people's needs. By ensuring access, it can strip away artificial professional, economic, and geographic barriers, allowing greater participation in R&D and in regulatory oversight. Although openness could potentially facilitate some experiments of concern, it is far more likely to seed a diverse community of problem-solvers willing and able to help and, by keeping scientific innovations and communications open, to diffuse the risk of a global arms race.

In the software world, open source occupies a small economic niche among industry giants, but wields considerable influence. It has forced proprietary software developers to work more efficiently and innovatively, which has led to higher productivity and profit. If open-source biology can do the same in biotech, it could become a significant and positive influence on the industry, helping to bring about reforms that lead to the sustainable development of better products, while ensuring fair access for all players to the international marketplace. But for open-source biology to really work, people will need to do more than just share data and discoveries. They will also need to consider what type of biological world they want to live in. As biological technologies advance, they will increasingly become a tool for the expression of human ideas—and these vary widely in cultures, communities, and individuals. More than the technology itself, our common intentions and desire to let creativity flourish will give shape to our biological future.

References

Berg, P., Baltimore, D., Boyer, H. W., Cohen, S. N., Davis, R. W., Hogness, D. S., et al. (1974). Letter: Potential biohazards of recombinant DNA molecules. *Science, 185* (148), 303.

Berg, P., Baltimore, D., Brenner, S., Roblin, R., & Singer, M. (1975). Summary statement of the Asilomar conference on recombinant DNA molecules. *Proceedings of the National Academy of Sciences of the United States of America, 72* (6), 1981–1984.

Broothaerts, W., Mitchell, H. J., Weir, B., Kaines, S., Smith, L., Yang, W., et al. (2005). Gene transfer to plants by diverse species of bacteria. *Nature, 433* (7026), 629–633.

Cello, J., Paul, A. V., & Wimmer, E. (2002). Chemical synthesis of poliovirus cDNA: Generation of infectious virus in the absence of natural template. *Science, 297* (5583), 1016–1018.

Chyba, C. F. (2001). Biological security in a changed world. *Science, 293* (5539), 2349.

Clark, J., & Ding, S. (2006). Generation of RNAi libraries for high-throughput screens. *Journal of Biomedical Biotechnology, 2006* (4), 45716.

Cox, J. P. L. (2001). Long-term data storage in DNA. *Trends in Biotechnology, 19* (7), 247–250.

Dewannieux, M., Harper, F., Richaud, A., Letzelter, C., Ribet, D., Pierron, G., et al. (2006). Identification of an infectious progenitor for the multiple-copy HERV-K human endogenous retroelements. *Genome Research, 16,* 1548–1556.

Endy, D. (2005). Foundations for engineering biology. *Nature, 438* (7067), 449–453.

Erbschloe, M. (2005). *Trojans, worms, and spyware: A computer security professional's guide to malicious code.* Burlington, MA: Elsevier.

Ferber, D. (2004). Synthetic biology. Time for a synthetic biology Asilomar? *Science, 303* (5655), 159.

Goldberg, S. M., Johnson, J., Busam, D., Feldblyum, T., Ferriera, S., Friedman, R., et al. (2006). A Sanger/pyrosequencing hybrid approach for the generation of high-quality draft assemblies of marine microbial genomes. *Proceedings of the National Academy of Sciences of the United States of America, 103* (30), 11240–11245.

Gottlieb, S. (2001). US concedes on cheaper drug production in Brazil. *BMJ, 323* (7303), 12.

Guan, Y., Peiris, J., Zheng, B., Poon, L., Chan, K., Zeng, F., et al. (2004). Molecular epidemiology of the novel coronavirus that causes severe acute respiratory syndrome. *Lancet, 363* (9403), 99–104.

Hede, K. (2006). Patient group seeks overhaul of FDA clinical trial system in court. *Journal of the National Cancer Institute, 98* (18), 1268–1270.

Heller, M. A., & Eisenberg, R. S. (1998). Can patents deter innovation? The anticommons in biomedical research. *Science, 280* (5364), 698–701.

Hessel, A. (2006). Open source biology. In C. DiBona, D. Cooper & M. Stone (Eds.), *Open sources 2.0: The continuing evolution* (pp. 281–296). Sebastopol, CA: O'Reilly Media, Inc.

Jackson, R. J., Ramsay, A. J., Christensen, C. D., Beaton, S., Hall, D. F., & Ramsaw, I. A. (2001). Expression of mouse interleukin-4 by a recombinant ectromelia virus suppresses cytolytic lymphocyte responses and overcomes genetic resistance to mousepox. *Journal of Virology, 75* (3), 1205–1210.

Jan, T. (2006). Genetic jamboree draws innovators. *Boston Globe*, November 5.

Lakhani, K., Jeppesen, L. B., Lohse, P. A., & Panetta, J. A. (2007). The value of openness in scientific problem solving. Harvard Business School Working Paper number 07-050. Available online at: http://www.hbs.edu/research/pdf/07-050.pdf (accessed August 2008).

Lander, E. S., Linton, L. M., Birren, B., Nusbaum, C., Zody, M. C., Baldwin, J., et al. (2001). Initial sequencing and analysis of the human genome. *Nature, 409* (6822), 860–921.

Legname, G., Baskakov, I. V., Nguyen, H. B., Riesner, D., Cohen, F. E., DeArmond, S. J., & Prusiner, S. B. (2004). Synthetic mammalian prions. *Science, 305* (5684), 673–676.

Levskaya, A., Chevalier, A. A., Tabor, J., Simpson, Z. B., Lavery, L. A., Levy, M., et al. (2005). Synthetic biology: Engineering Escherichia coli to see light. *Nature, 438* (7067), 441–442.

Margulies, A. H. (2004). A new model for open sharing: Massachusetts Institute of Technology's opencourseware initiative makes a difference. *Public Library of Science, Biology, 2* (8), E200.

Maurer, S. M., Lucas, K. V., & Terrell, S. (2006). From understanding to action: Community-based options for improving safety and security in synthetic biology (Draft 1.0).

National Research Council (2006). *Globalization, biosecurity, and the future of life sciences.* Washington, DC: The National Academies Press.

Pejcic, B., De Marco, R., & Parkinson, G. (2006). The role of biosensors in the detection of emerging infectious diseases. *Analyst, 131* (10), 1079–1090.

Peng, Z. (2005). Current status of gendicine in China: Recombinant human Ad-p53 agent for treatment of cancers. *Human Gene Therapy, 16* (9), 1016–1027.

Rai, A., & Boyle, J. (2007). Synthetic biology: Caught between property rights, the public domain, and the commons. *Public Library of Science, Biology, 5*. Available online at: http://biology.plosjournals.org/perlserv/?request=get-document&doi=10.1371%2Fjournal.pbio.0050058 (accessed August 2008).

Raymond, E. S. (1998). The cathedral and the bazaar. *First Monday, 3* (3). Available online at: http://firstmonday.org/htbin/cgiwrap/bin/ojs/index.php/fm/article/view/578/499 (accessed November 2008).

Redon, R., Ishikawa, S., Fitch, K. R., Feuk, L., Perry, G. H., Andrews, T. D., et al. (2006). Global variation in copy number in the human genome. *Nature, 444* (7118), 444–454.

Singer, M., & Berg, P. (1976). Recombinant DNA: NIH guidelines. *Science, 193* (4249), 186–188.

Tumpey, T. M., Basler, C. F., Aguilar, P. V., Zeng, H., Solórzano, A., Swayne, D. E., et al. (2005). Characterization of the reconstructed 1918 Spanish influenza pandemic virus. *Science, 310* (5745), 77–80.

Venter, J. C., Remington, K., Heidelberg, J. F., Halpern, A. L., Rusch, D., Eisen, J. A., et al. (2004). Environmental genome shotgun sequencing of the Sargasso Sea. *Science, 304* (5667), 66–74.

Wilson, J. M. (2005). Gendicine: The first commercial gene therapy product. *Human Gene Therapy, 16* (9), 1014–1015.

Wren, J. D. (2006). Theory and reality for software patents: Good in concept, not so good in practice. *Bioinformatics, 22* (13), 1543–1545.

12

The Ambivalence of Protocells: Challenges for Self-Reflexive Ethics

Brigitte Hantsche-Tangen

A new wave[1] of ethical questions in the field of biotechnology is closely connected to the project of decoding and changing human genomes. Transferring methods of manipulation and cloning from plants and animals into the human realm strongly challenges many ethical sensibilities and public acceptance. Skepticism toward science has arisen in response not only to these innovations, but to Western societies' experiences with new technology in the twentieth century—notably in warfare[2] and technological disasters[3]—and more recently to the impact of technology on the environment. This skepticism has done away with the "old" belief that scientific progress automatically implies human progress (Stehr, 2004; Grimm, 2005, p. 3). Furthermore, globalization and international competition press different cultural traditions and civilizations to interact within a "money-driven regime" (Pestre, 2005, p. 33) that evokes experiences of alienation and the need for new social constructions. In contrast, the world economy feeds off scientific and technical innovation, and has thus become a driving force in realizing desires and wishes for the future, a self-organizing force for change driven by curiosity and innovation (Nowotny, 2005b). Scientific progress, the promise of sociocultural achievements, and the application of new technologies are all situated in these grounds.

Mark Bedau (2003, 2005) describes artificial cells, or protocells, as self-replicating, self-organizing entities evoking "culture shock," following the publicity at the release of Michael Crichton's *Prey* (2002). Although protocells are currently more of a vision than a working product, there is already talk of the possible risks, including the potential to fundamentally change the natural environment.

Coping with the long-term effects of modern technologies was the starting point for Jonas (1984), who proposed an ethics of responsibility for future generations, which includes sustainable relations with nature. Given the discovery that climatic change is affected by human action (Stehr & von Storch, 1999; Viehöfer, 1997) and the controversy over the long-term effects of genetically modified plants, Jonas's proposal reveals its urgent relevance.

The dissolution and blurring of mankind's relationships with both the polis and nature are key aspects of the social environment in which protocells are being developed. These aspects are also relevant to questions about the social and ethical implications of protocell research.

Scientific discovery and technological innovation challenge the familiar self-conceptions of man. There are many of these challenging technologies, and they affect some of the core values of modern Western societies like identity, individuality, privacy (Nowotny, 2005a), democracy, and control (Giddens, 1996).

Looking for responses to perceived risk requires orientation. Institutionalized national committees and commissions (e.g., the German National Ethics Council), working together with other national and international ethics commissions and institutions (e.g., the European Group on Ethics in Science and New Technologies, or EGE, International Bioethics Committee, and Intergovernmental Bioethics Committee), indicate a demand for discussion in the public and for new regulations in the hot fields of cloning, stem cell research and genomic alteration (Honnefelder, Taupitz, & Winter, 1999, p. 11). Moreover, the need for an integration of ethical questions in research was addressed by the Sixth Framework Program of the European Union.

If ethics can be understood as a special "text" in the symbolic universe, giving response to a perceived trouble (Böhme, 1997, p. 7), ethical discussion might be able to provide this orientation and thus might be a principal reason for the surge in demand for ethical thought.

Questions arise, such as "What kind of troubles and what kinds of challenges do protocells raise?" Terms like *Frankencells* (Mooney, 2002) and literary creations like self-organizing swarms (Crichton, 2002; Schätzing, 2004) draw analogies to recognized figures of European literature,[4] expressing tragic conflicts between scientific curiosity, knowledge gain, and creation of "creatures." Other conflict lines address knowledge gain and seduction to the "evil" (Faust), the fear of losing control over unchaining powers, and fears of not being able to integrate the created artifact into society (Frankenstein). It seems that one challenge of protocell research is that it activates disturbing feelings associated with the creation of, and confrontation with, a novel autonomous entity. Protocells are intended to replicate, self-organize, and evolve in response to their environment. Considering them as living entities raises questions about whether they are friends or enemies, how far they can be integrated and accepted into our world, and how they will influence our world.

Science fiction can be valued as proactive imagination of possible risks. Regarding technological development, Ahrendt (1958) claimed that a quality of new technology in the twentieth century is the "unchaining of natural processes"[5] in a cosmic dimension. I will argue that protocells have the potential to be a mighty tool sharing high ambivalence as a feature with other advanced technologies. In a more rational context of risk assessment, the potential for self-organization and release into the environment of small particles is seen as a warning signal for an existing risk-zone (Steinfeldt et al., 2004). Thus, the process of mastering and control of forces is an issue.

Moreover, ambivalence arises from the dual-use[6] nature of new technologies. But it is not yet justified to fix up protocells as a potential in a worst-case scenario. There is a range of applications in modern "white"[7] biotechnology, mostly aiming to enhance existing industrial processes or create intermediary products, not carrying any of those problems.

Ambivalence should be regarded as an integrated effect of scientific discovery and technological application. Ambivalence is produced in different areas of society, and not only as a problem of perception and practice of misuse. Thus, it makes sense to talk of ambivalence as a result of inconsistency on several levels, caused by embedding into different action frames and values. Context-laden science and technologies include inhomogeneous relationships between instrumental and noninstrumental values (Feenberg, 2003, p. 45), content values and context values of science (Longino, 1990, p. 5).[8] Ambivalence can also occur between the goal of a technology and its resulting applications, as an effect of interest groups and dependency. A technical object can serve more than one purpose (Joerges, 1996, p. 22). In addition, recent discussions of ambivalence indicate that troublesome social effects arise because applications of new biotechnologies make it much more difficult to draw a clear line between beneficial and harmful applications (Jonas, 1984, p. 51; Dando & Nixdorf, 2004).

It is important to explore social and cultural effects on the development of society and technology. From the social perspective, protocell technology needs to find a way of robust integration into society. Under process aspects, it seems necessary to deal with ambivalences without being blocked in or neglecting risk (Wehling, 2001). Ambivalence, in this sense, has the role of provoking reflections about aims, purposes, and actions, and thus slowing down a dynamic process. I intend to show how technology affects the cultural boundaries we set up, how the demand for ethics plays out in different fields, and how ethics work assesses risks in a legitimate way. Therefore, I select fields of interest on the global, regional, and institutional levels, assuming a relevant influence of shaping integration of applications. Ethical demands and the character of responses are part of these contexts.

Shifting Boundaries: A Link Between New Technologies, Society, and Ethics

Contemporary public debates concerning biotechnology, artificial life, and genomics often articulate fears in terms of risk, violation of nature, maintaining the sanctity of life, or "playing God." Despite variations in terminology and horizons of thought, the basic issues are problems of limitation. More than one voice (Latour, 2001; Viehöfer, Gugutzer, Keller, & Lau, 2004) is interpreting these reactions to the applications of new biosciences, as a result of crossing existing boundaries between nature and society.

Our historical and intercultural knowledge shows that many key boundaries— such as those between body and mind, human and nature, or life and death—are not clearly fixed. New technologies and medical applications challenge in new ways the placement, and even the very existence, of these boundaries. Examples include controversy around technologies at the beginning of life (e.g., insemination and abortion) and the end of life (e.g., organ transplantation); questioning identity and kinship in the light of cloning and reproductive medicine; questioning values of inheritance in a world with genome analysis and alteration; allowing medical therapy for cosmetic enhancement in addition to illness; questioning the value of a single life by taking an instrumental view of man as a resource. These boundary- and value-challenging technologies clash with religious and traditional conceptions of human bodies as sacred and challenge current practices and legal definitions. Controversies also show that different criteria are being simultaneously applied[9] and that decisions about validity cannot be given by science alone (van den Daele, 2001). They also affect our relationship with the world around us as we question involuntary exposure to risk, and humans' relation to the natural environment, for example, climate change and genetically modified food. Furthermore, the potential use of biotechnical and bioinformatical nanomachines in a human body challenges sensitive differences between enhancement, prosthetics, and "natural" processes. These boundaries cannot be drawn easily, and they render decisions about acceptable behavior contentious and subject to continuous revision. This continuity contrasts with the clear distinction between "evil" applications of a technology, such as atomic weapons, and "beneficial" ones, such as nuclear power plants. One more illustration is the discourse over steroid use in sports, and the related difficulty of defining what are natural and unnatural substances (Viehöfer et al., 2004).

Homo faber,[10] with his inventions and constructions, is always seeking to widen the realm of human artifice. Since industrialization, men have channeled these forces into producing a manmade world.[11] These controversies underline that the technological impact of protocells on the foundation of modern biotechnology for the human community will be to allow man to have finer "channeling" between nature and society in man and outside of man.

The blurring of boundaries between human and nature indicated by discourses, artifacts, and technologies is accompanied by the opening of societies to the world, new political cooperation, and shifts in the geopolitical order. Intensifying international exchange of people and goods indicates increased mobility, a foundation on which new transnational organizations and communities develop. However, along with this increased mobility come blurred distinctions (between, e.g., war and peace, national and international, private and public) and the removal of traditional societies from their native lands. Sociologists claim that a general change modern societies undergo is the loss of ties to their cultural past. This cultural state is characterized as *reflexive modernity*, to emphasize that the basic principles belonging to a given self-description of society are no longer given (Beck, Bonss, & Lau. 2004), but instead a reflection of their references.[12] The cultural effects of reflexive modernity can be identified by dissolutions in a cultural web of meaning. This is another ambiguous aspect of protocell research, which makes it difficult to look for an "all-clear" signal, since the culture shock caused by living technology merges with the experience of the loss of cultural and social certainties.

The comparison of the transition today with the rise of the modern age in the seventeenth century seems obvious. A shift in horizon and perspective[13] was caused by advances and changes in science and technology (Ahrendt, 1958, p. 258ff). Along with this were the beginnings of the global discovery of the human dwelling place and the social process of expropriation and industrial production. Pestre (2005) draws a parallel to the social effects caused by practices of privatization of intellectual property in science. Although it is too early to make a proper decision, a certain kind of parallel would indicate a disturbing dynamics for a world on its way to a global society.

Transgression in the social world is one form of the phenomenon of mobility. Aspects of this phenomenon are the increasing migration and mobility of people in the last few decades and the migration of "cultural objects" (Sorokin, 1959) and artifacts, including the transference of meaningful objects from one location to another without cultural context. Mobility in science refers to inter- and transdisciplinary thinking. We all witness shifts in geopolitical centers and values, and experience the shrinkage of distances through technological communication, media, and world travel. The main psychosocial dimension of mobility is the experience of "becoming unfamiliar" (i.e., estrangement from things, selves, or realities). This process of becoming unfamiliar can be painful. So the challenge of how to preserve a sense of control over one's life in a bewildering world of scientific and technological complexity in an ongoing process of globalization (Nowotny, 2005a, p. 23) can be seen as an indicator of public disturbance.

Since traditional ethics is bound to consideration of the near future, Jonas (1984) started from a position of an experienced lack of differentiation. Space and time still seem to be crucial points of ethical consideration. The old Platonic questions "What kind of life do we want?" and "What do we want to pay?" seem to require significant consideration to be sufficiently explained. Who are the "we" being addressed, since globalization is not identical with the development of a "world system" (Giddens, 1996, p. 119), nor does a vision of a world community bring forth its existence? It looks like the need for ethics and regulations arises out of this vacant position. Further observations must include an integration in a new framework of global governance (Willke, 2006).

New political systems constructed during globalization have ethical and moral pathways that are visibly embedded in different traditions. Therefore, cultural scientists argue that "culture" today cannot be regarded as a homogeneous framework with collective norms and values, but rather as a limited part of a larger transcultural framework (Heller, 2004).

These differences exist within a single culture, too. This interpretation is reflected in the characterization of today's ethics as *conflict ethics* (Dabrock, 2000). It is difficult to find an instance of universally acknowledged justice that is common to diverse societies. Jonas's question of whether ethics has the tools to meet the problems of space and time remains relevant.

Kant's notion of the categorical imperative emphasized the two sides of ethics: an interpersonal practice from the perspectives of a general law and the conduct of life. The challenge for ethics in a world characterized by innumerable differences is to find criteria that can be accepted beyond specific traditions. The concept of the dignity of man brings to our attention a kind of prelegal adjustment of social regulating behavior. This concept can be considered as a reminder to maintain a common ground of mutual understanding and respect when forming rules of interaction with different traditions and sensibilities in a tightly connected world.

Roles of Ethics in Different Fields

Although the demand for ethics is already publicly visible, neither the general role of ethics nor the activities demanded of it are easy to define. The transnational production chains created by transnational companies produce a kind of global integration. But an integration of common interests and public goods is weakly represented in the institutional world of a global society. It seems that ethical responsibilities formulated by UNESCO and other international organizations help strengthen awareness of common public regulations.

The creation of protocells aims to reveal mysteries of life. This draws a line of continuity to losses in a religiously grounded belief system. It may be that the hidden culture shock of losing one's conceptual "home," familiar beliefs of a world of mystery and religious truth, promotes both ethics as a "mediator" for experienced value conflicts and the need for social rules. It may be that the renaissance for ethics is nourished by the desire to avoid a world-dominating conflict of fundamentalism.

Regarding ethics as a "text," its intention is to advise for a good practice of daily life. It articulates the hope[14] for good character traits—consistency, moderateness, friendship—during difficult experiences, which are necessary to undergo for learning.

Thus, a situation of open boundaries and lost orientation might offer a chance to become aware of our responsibility of preserving certain things and of our potential for not only building but destroying. Regarding ethics as a "text" might reveal another aspect of meaning. An equivalent of the wish to be part of a cosmos greater than man's world, articulated through religious belief, can be articulated through ethics as Wittgenstein (1930) demonstrated: Being amazed by nature is a possible attitude toward nature that, as a category of feelings, expresses a sense of "being touched" in a passive way. In this context, the insistence of a difference between the manmade world and nature may no longer be merely the expression of anti-modern traditions.

Shifts in Institutional Relations and Effects on Ethical Demands

Gernot Böhme (1997, pp. 9f) distinguishes three realms in which ethical issues are found. *Science-formed philosophy* is practiced mainly at universities,[15] *ethics as participation in general regulations of society* refers to the general public,[16] and *ethics as a moral involvement in matters of daily life* refers to the conduct of personal life.[17] The changing relationships among science, markets, and politics (Dabrock, 2000; Nowotny, Scott, & Gibbons, 2001; Weber, 2003) are the subject of vehement discussions in European societies. Shifts in the characters of these institutions are described as a heterogeneous set of changes toward hierarchization, instrumentalism, and organization of dependency between previously more separated institutional action frames. These changes are settled within a wider framework of change, such as the state's policy providing public services with responsibilities of its own distinctive domains. These changes between institutions challenge ethical sensibilities on issues like autonomy, freedom, and security.

The term *techno-science* (Dabrock, 2000; Weber, 2003) marks today's closer connection among science, markets, and politics. This connection was accompanied in Germany and in other European countries by new financial and legal adjustments (e.g., research embedded in industrial application; universities as economic actors specializing in patent production) and such far-reaching changes in the values of the institutions themselves as science produced mainly as intellectual property rather than purely a "public good" (Buss & Wittke, 2001). This has consequences for institutional organizations and reward mechanisms in academic institutions, including the changing role of scientists toward merchandisers of knowledge.

The present ethical discussion refers to conflicting value systems brought about by this change in institutional context. It is doubtful that truth and objectivity can persist in the face of the interests of the worldwide bioindustry. Increasing commercial autonomy of the producers of science is accompanied by changes in the organization and funding of scientific research, which seem likely to force a shift in research focus from truth to economic efficiency (List, 2004a, p. 31). Investment policies push players to reduce the time allotted to research, as they force them to produce visible, working products (Dabrock, 2000). The amount of money invested creates reasons for new cycles of technological applications, with the tendency to overestimate the real potential of special technologies or, even worse, to neglect risks. In biomedical research, this point becomes more relevant as the relationship between promise and realization comes into focus. More than one author in recent publications talk of the "genetic illusion," which is merely a fantasy of an incredible, questionable market (Sicard, 2003).

One ethical line of conflict deriving from contradictory values arising out of technoscience can be seen in the extremely different time frames for the production of applications versus the assessment of risks. This is a point of dissent in the field of green biotechnology. Other lines of conflict are the movement of controversial research like cloning from one society where they are ethically questionable to another (Wade & Sang-Hun, 2006), and the advantage taken by firms that push novel treatments into markets where the legal procedures regarding them are less strict.[18]

Another dimension of tensions accompanying the closer connection among science, application, and market is articulated in intellectual fields, as tendencies toward *intellectual reduction*, which undermines scientific reasoning. Many discussions are centered on the role of genetic determinism, on genetic research (Propping, 2001; Keller, 2003) and the economy, on the difference between the manipulation and understanding of nature, and on the growing differences between public and scientific understandings of genetics (Keller, 2003). The information available to the public suggests a future of easy and privately consumable advanced technology, and seems to be compatible with weakening of existing good practices in dealing with risks.

Examples are found in green biotechnology as well as in genetic medicine. An incidence of the latter became visible in the reaction to the Gelsinger case by the U.S. National Institute of Health (NIH; Paul, 2003, p. 34).[19] In an international competitive power system, you find a proliferation of promises that are doomed to success. Cases for research are made in terms of the quest to "possess the book of life" or to buy a full analysis of one's personal genetic makeup, and then purchase designer therapy[20]—a kind of "boutique medicine." Here the use of language, or the *Sprachspiel* (Wittgenstein), indicates that scientific knowledge, barriers, and technological competence are mismatched to the elegant promises made. Only specialists can judge and fully reveal the limits of such research. Thus, the fear that public service ethics is being replaced by business interests searching for an adequate business ethic as its correlate (Nowotny et al., 2001, p. 24) cannot be rejected. Ethics itself is thus in danger of being used by particular interests, or becoming part of a new type of symbolic politics, and may become the "opiate of the masses" (Viehöfer, 2006).

Because ethical debates deal with genomics and biotechnology in health care, implications of a shift to molecular medicine are central. No longer is the phenotype of illness the starting point, but rather the genotype and its connection to resultant dispositions. This reversal of an existing framework of meaning—some talk of it as a change of paradigm (Peltonen, 2003)—will become the basis of diagnosis and definition of illness (Honnefelder et al., 2003), with many implications in technological (Lengauer, 2001), organizational, legal, and social fields. Although there is progress in a different view of genes and the genetic code (Honnefelder, 2001), ethical discourses in biotechnology and biomedicine refer to the dignity of man (and living creatures in general) as an appeal to respect others as living beings, to grant the individual right of decision making. The concepts of *person* and *dignity* in the bioethical debate are centered on self-determination and freedom (Haker, 2003), which indicates new conflicts between treatment requirements by an expert culture and personal expectations of medical treatment. This conflict might become even more precarious if medical treatment is done by self-organizing, individualized living artificial cells.

Conflicts and Solutions in Fields of Biotechnological Applications

Artificial cell technology is open to a variety of potential applications (environmental, health care, industrial technologies, and products). Controversy and its resolution help the public to trust and accept as well as indicate awareness of technological assessment of artificial cells. In the following section, I look at conflicts between promises, legitimization, and outcomes in existing biotechnological applications.

Besides the scientific goal of discovery, the potential to decode and change (human) genomes was socially legitimized by the promise to enhance or improve health (Böhme, 1997, p. 207), and minimize preexisting social difference and poverty.[21] Thus, besides the scientists' motivations to discover new things and create and participate in technological innovation, the technological artifacts constructed should serve peoples' needs. Experience with promises and resultant applications show a failure to make good on promises, for green as well as for red biotechnology.

Applications and practices of green and red biotechnology elucidate that technological development promoted and legitimized by the goal of enhancing human life or making it more comfortable turn out not to automatically improve the life of human beings in general. The global distribution of poverty and wealth between countries indicates stagnation or an even deeper gap (Hauchler, Messner, & Nuscheler, 2002, pp. 74ff). Another indicator is the Human Development Index,[22] representing health differences, increased for many developing countries by hygienic protection and better nutrition of mothers and their children, as a result of practiced care.

The influence of new technologies is not easily accounted for in this process. It seems that advanced technologies like biotechnology mainly affect the countries in the global north (Castells, 2004, pp. 132ff). But decreasing differences in effect can be expected in the near future, since emerging economies in Asia and South America are bridging gaps of education, laboratory facilities, know-how, and infrastructure.

Since the beginning of green biotechnology regulation, various conflicts have signified the violation of people's rights. The main issues have been conflicts between public and private property, many caused by the hijacking of intellectual property through patenting naturally grown fauna and flora in nonindustrialized countries. Property conflicts and violations of rights also arise within red biotechnology in proposals for the genetic screening of whole populations. New conflicts continue to arise between mismatched potentials for diagnosis versus treatment. Here, negotiation is still ongoing.

Green biotechnology has proceeded to international conventions and treaties for protection of natural resources, public goods, and diversity of species, whereas the regulation and control of red biotechnology[23] is still in its infancy. As to the use of biotechnology in armed conflicts, international regulations such as the Bioweapon Convention (BWC) have existed since 1975, and nations have implemented laws concerning biosecurity and processes of adapting to new problem zones.[24]

New conventions[25] and legal instructions include protection of the "dignity of human beings" and "human rights." This is regarded by some promoters of regulations as a minimal agreement of mutual acceptance (Honnefelder et al., 1999).

In spite of existing regulations, tensions over the distribution of harms and benefits are already visible in the health sector. Here, engaged people are working to change the model of informed consent into one of "informed contract" (Sass, 2001; Schröder, 2004). The patient's position in a new commercial system of biotechnical medicine[26] has to be strengthened, because applications of genetic medicine have the potential to violate rights and harm the patient (Paul, 2003).

At the moment, industrial applications of nanotechnology,[27] another field intersecting with artificial cell technology, seem to be in accordance with their promise of white biotechnology: Components and new materials, as initial and intermediary industrial products, play an important role in applications and are becoming important factors in the leading markets (Luther & Malanowski, 2003, pp. 155ff). Conflicts involving the dignity of man in these important fields do not play a role at the moment, though there is potential for conflicts in future applications affecting (not only) informational self-determinism of the individual (e.g., "intelligent" clothes including chips, nanotechnological diagnosis, and recognition methods).

Ethical Issues from the Perspective of Risk Assessment

Protocells are embedded in broad discussions about hybrid and artificial beings. Although the creation of cyborgs (Haraway, 1995) resulting from human-machine integration is a fantasy, the risk of such pursuits dominates the discourse. Despite this focus, these discussions do not reject science and technology in general, as Schulze-Fielitz's (2005) evaluation of data from several European surveys shows.

Looking at the state of the art in the development of protocells, one might argue that since their realization is in the relatively the distant future, one should let sleeping dogs lie. But timelines of development are incredibly fast, and visions of protocells are in fact likely to be realized far more quickly than one might imagine. From an ethical perspective, we should consider science fiction approaches as a way to imagine the prospects of risky outcomes in the sense of Jonas's obligation to an ethic for the future (Jonas, 1984, pp. 61ff). Science fiction can articulate our fears of being threatened or overwhelmed by science, and as such constitutes a warning signal for possible areas of risk that need to be substantiated systematically through a more rational and objective approach to risk assessment. Approaches to discovery arise in three main areas: genomics, synthetic biology, and biomimetic design.

The approaches differ in the way that researchers bridge the gap between living and nonliving matter. Construction of living technology in genomics is practiced by inserting minimal yet functional genetic material from one natural cell into another one with its genetic material removed. The minimal cell approach practiced by Craig Venter is an attempt to build minimal self-replicating entities. Synthetic artificial cell creation—embedded in a broader approach of synthetic biology—is working with vesicles that use natural material such as enzymes. Biomimetic design[28] includes techniques such as using vesicles in a microfluidic enzyme-free environment to explore cell functions, under the assumption that gaining scientific knowledge and understanding of powerful natural cycles through imitation for a special purpose will yield useful insights and applications in the areas of ecology, power production, health care, and material or process technology.

Our ambivalence toward such living technology comes out of the very factors on which the advantages of such technology rely. The use of quasi-natural functions, the tiny scale on which these organisms exist, and the ability to self-organize and flexibly adapt to their environment are all factors that will make them useful for tasks in dangerous or hard-to-reach places. A broad consensus seems to exist among experts in the field that the risk presented by such technology depends on its eventual applications. Within an application, risk depends on our mastery of the technology, which includes our ability to minimize or neutralize risk.

The quality of being like natural forms in material and process creates the greatest risks. At this point, problems of social seductiveness start with all the fears of manipulating, dual–use, and changing human identity.

Protocells' capacity to reproduce increases risk, as it raises the question of our ability to influence and control our creations, especially in an open system. For genomics, it is also the differences created in an organism, not just the organism itself, that cause risk. As with green biotechnology, the changes made to one organism could escape into others, and the genomes that we intended to be confined could escape our control and be released into the greater population.

Establishing Risk Assessment as a Process

Considering the different approaches to protocell development reported here, one might argue that pursuing protocells by a route that avoids crossing biological boundaries would be less contentious in a field filled with other potential risks. A decision to take this route may also be ethically sound for science, as it would constitute a refusal to work with existing cells and (human) genomes in our present stage of understanding. Thus, bottom-up research will give us a way to analyze potential risks of protocell research by gaining a better insight, while avoiding the risks involved with developing protocells that are close to nature.

The underlying argument (Jonas, 1984, p. 28) for mastery of process, tool, or technology is that mastering a process and a tool within a process requires knowledge much wider and deeper than constructing that tool and process.

Pursuing artificial cells through biomimetic design might be a relatively lower-risk strategy than other options. But this does not guarantee that there are no risks.

The social aspect of knowledge production in society is that scientists have to deal with a regime of heterogeneously distributed knowledge production (Rammert, 2003), including controversies about political, environmental, technological, and ethical aspects of science. All of these discourses underline the fact that rapid knowledge growth is accompanied by an increasingly sensible perception of not-knowing, not only in a temporary sense, which can be overcome by new research efforts, but also in a more essential sense (Wehling, 2004).

Implementing process observation and learning loops might thus be part of the ethically responsible pursuit of new technology, as the knowledge that we gain through mastery of a technology could help us avoid the risks of implementing it. Experiences with science assessment show that there is no "best moment" (Boeschen, Lau, Obermeier, & Wehling, 2004, p. 143) for influencing innovation and technology development. This is another argument for a process-dependent conception of risk assessment.

Our ambivalence toward technology implies that one must consider technological development through three lenses: as (a) the advancement of science and discovery; (b) change in society, institutions, and social embeddings; and (c) a process of personal/individual decision making. This complex setting, in the first instance, shows that scientific researchers, values of scientific organizations, and principles of action are only part of a net of decision making that depends on institutions and negotiations. Although resolving our ambivalence seems to be a Sisyphean task, two fundamental questions arise. First, what are scientists able to do to influence such a complex field? Second, which kinds of cooperation between actors and actions are effective in a specific situation?

Research on artificial cells has to cope with the influence of antecedent and related discourses on public opinion, including the means of triggering public opinion. The experiences with ELSI (Ethical, Legal, and Social Issues) programs in the United States show some obstacles, such as active communication outside scientific communities being mostly far beyond individual scientists' capabilities (see Yesley, 2005). The successes of advocates such as Bill Joy (2000) indicate that strategies appealing to dramatic overstatement are more effective than rational arguments in inflating warnings about technological developments in a contemporary media landscape.

Anchoring process as a key feature of this field implies not only looking at the risks of a product itself but also evaluating the whole process of production, deployment, and value creation. To undertake this task, we need risk assessment tools that address sustainability and balancing of processional and special risk and benefit according to different technologically appropriate criteria (e.g., ecological balance, health balance, energy balance, information balance, and security balance). We also need standards that observe social criteria and admit dignity as a baseline for interaction (e.g., are products tested appropriately in the light of existing conventions and controls? Is self-determination realized? What are the real and possible distributions of harm and risk?). The development of instance-specific criteria is an innovative aspect of technological trajectories that is not yet worked out (Steinfeldt et al., 2004).

The ambivalence created by the mix of positive and negative possibilities stemming from a technology can be managed. Especially in the area of potential misuse, tools for managing ambivalence include visible standards such as good practice and general scientific rules and practices promoted by national scientific societies and publishers. On international, transnational, and national levels, regulations and conventions control even the largest companies (e.g., Novartis; see Schreiber, 2005).

Preventive control of risk on a personal level is now considered a standard of conduct for scientists in biotechnology. These efforts encourage and standardize honesty, fairness, and trust between scientists and the public in a difficult and competitive field, and they clearly define legal and illegal applications of high-risk technologies. However, it should be added that research on and development of a global society are embedded within a worldwide economy of illegality and corruption.[29] The overlap between criminal and legal activities makes the problem especially difficult and dangerous. These shady businesses become attractive in our institutional world because they are forbidden,[30] and the consequences of misuse are considerable in this case.

How ambivalence regarding technology can be managed without succumbing to destructive tendencies is an open question. Some contemporary developments seem to strengthen destructive forces, such as harmfully wide gaps between values and practices. Exclusion of too many groups, regions, and ways of thinking from the development process and its benefits might undermine our trust in the advantages of technological know-how and expert skills. The problem-solving qualities of technology and engineering can be undermined if short-term, strongly application-oriented economic interests are dominating scientific value orientation.

Let us change perspective now to areas where scientists might be better able to control the process itself: their own environment of technological advancement. Adopting a cautious courage principle (Bedau & Triant, ch. 3, this volume), emphasizing innovation but also calling us to deal responsibly with already known and estimated risks and precarious actions, gives us the possibility to create a new balance between innovation and precaution. Using the cautious courage principle makes us sensitive to the risks and harms of our own activities, especially if we examine or reanalyze our decision making in more than a mere technological dimension. Under the conditions of a technoscientific market, it seems wise to be sensitive to pressures from dependency on others, the institutional context, and scientists' career aspirations. Thus, scientists should be open to questioning: How can we identify factors in a decision field appropriately? Is our sensitivity to ourselves well-trained enough to recognize subtle shifting in our tendencies toward seductive paths during hard research processes? A reexamination of an existing negative example involving the treatment of genetic disease (the Gelsinger case mentioned earlier; see Simon, 2004) shows weak points that helped lead to failure:

- a failure to carefully prove that the risk model was adequate
- ignoring or suppressing contrary positions and evidence
- excessive collection of power in a small group of people

It is interesting to consider these short descriptions of the failure to address risk reported by the Columbia Accident Investigation Board (CAIB), as an example of an accident in another equally high-risk and high-tech area. This very detailed report is written from the perspective that "complex systems almost always fail in complex ways" and "it would be wrong to reduce the complexities and weaknesses associated with these systems in some simple explanation" (CAIB, 2003, p. 6). It is apparent that the main failures resemble those of the Gelsinger case mentioned earlier.[31] The report closes with the statement that "we cannot explore space on a fixed-cost basis" (CAIB, 2003, p. 202), highlighting the importance of long-term analysis of cost, efficiency and safety tradeoffs, and drawing attention to the ambiguous motivation to simultaneously benefit mankind, maximize the return of honor and glory and reduce costs in a tight budget.

Conclusion

This chapter has discussed multiple levels of risk assessment for high-tech projects trying to create living entities *de novo*. It has also described a methodology of continuous consideration of ethical implications during the process of setting these (not yet existent) artificial life forms into operation. It turns out that ethical considerations are vital for protocell research and development because of the many social and economic interactions of this new technology with humankind, and the idea deeply embedded in society that the whole endeavor need not be condemned if ethics are intertwined with the process of scientific research.

Acknowledgments

I greatly appreciate the steady help of Mark Bedau and Emily Parke, especially the patient proofreading of my manuscript. Also I would like to express my thanks to the European Center for Living Technology in Venice and the PACE (Programmable Artificial Cell Evolution) project.

Notes

1. There is more than one ethical discourse in history. Remarkable discussions from the seventeenth century on the difference between humans and machines were started by Descartes,

Pascal, and others. More recently, in the nineteenth century, ethical discussion was centered on the discovery of evolution. Social Darwinists such as Herbert Spencer used the implications of survival of the fittest as an ethics of evolution. In 1890, Thomas Huxley wrote that there is an unfortunate ambiguity between the meanings of "fittest" and "best" (Huxley, 1893; republished by Ridley, 2004, p. 418).

2. Especially the construction and detonation of nuclear bombs in World War II and, later, quasi-nuclear weapons in the Iraq war.

3. For instance, the explosion of the nuclear reactor at Chernobyl.

4. For example, *Faust* (1808) by J. W. Goethe and *Frankenstein* (1818) by Mary Shelley.

5. Ahrendt refers to nuclear energy as an example here.

6. Dual-use technologies have applications in both military and civil spheres; see Liebert, Rilling, and Scheffran (1994) for further discussion.

7. In Germany new technologies are named after the field of application: green = agricultural science, red = pharmaceutical science, gray = environmental remediation, and white = nanotechnology and chemistry (white also connotes "clean").

8. Longino distinguishes between science as "content" governed by constitutive values and science as "context" governed by values of persons belonging to a particular social and cultural environment.

9. That is, from different ethical perspectives: If you value people's right to have their lives saved, you could argue that organs for life-saving transplant should be taken from donors at the point of brain death. On the other hand, if you value purely the right to life and to die with dignity, the process of death is the central focus, which could be defined at a point beyond brain death, i.e., when breath and circulation cease (see Sellmayer, 2004, pp. 150f).

10. Ahrendt distinguishes between *homo faber* and *animal laborans*. Homo faber is the prototype of a tool maker. In contrast to *animal laborans*, who only wants to ease the labors of its own life process, *homo faber* wants to build a world according to his ideas (continuing Plato's *eidos*, or shape and form), using, destroying, interrupting, and killing an existing life process. His image is that of a creator-god (Ahrendt, 1958, pp. 119ff).

11. Instead of keeping them outside or using them as a protection of the human artifice against the forces of nature.

12. The term *reflexivity* has different meanings: One meaning (e.g., Giddens, 1990, p. 39) refers to knowledge and knowledge applications. According to Beck, it relates to the fact that the "second modernity" has to constantly react to problems caused by modernity, and the problems and solutions caused by the problem solving of society, rather than naturally given problems (Beck & Holzer, 2004).

13. Beginning with Galileo and his invention of the telescope, Ahrendt described the "discovery of the Archimedian Point," i.e., the growing consciousness of being bound to Earth with the metabolism of the body and the possibility of leaving to cosmic horizons by mind.

14. Aristotle's *Nicomachean Ethic* represents its strictly secular character as an advice and training of interpersonal relations practices (among "free men," the citizens); sanctity as a religious sensation was an aspect of nature at that time, whereas the "gods" were family and

humanlike. All religions (Judaism, Christianity, Islam) that originated in the Middle East were centered on ethics as rules for social behavior in the community, which was thought of as a transcendent and triadic relation between men and God.

15. With a special character of analytical and rational procedures and problem solving, mostly without connection to personal engagement and involvement.

16. Concerning problems of public interest, i.e., moral decisions about public problems. This feature is well characterized by Kant as *Weltweisheit* (Böhme, 1997, p. 10).

17. Protestant ethics has been well-known as a *Lebensform* since the writings of Max Weber.

18. A recent sad example is the application of TGN 1412 in London by TEGENERO, which led to severe reactions in the test subjects.

19. The NIH insisted, according to existing regulations, on an immediate report of unwanted side effects and interferences from scientific and clinical working groups using genetic treatment. It followed that 691 serious problematic side effects were reported, of which 652 were not previously known to the NIH. Only 6 percent of the accidents were reported correctly.

20. Example by Craig Venter (2001), cited in Brand, Danbrock, Paul, and Schröder (2004, p. 11).

21. Leading words in genetic technology are the same for plants, animals, and humans.

22. For the latter, an increased survival rate for children is an important cause.

23. Controversial actual and potential practices under discussion include changing the human genome with a strong impact on selection, matters of pregnancy and childbirth, buying and selling raw material (e.g., a "designed child") for organ substitution of a sibling, or producing a special disease-free child.

24. The BWC was set up in 1972 and enacted in 1975; 154 states signed. New regulations have emerged since then, including: national legal actions including laws punishing criminal actions and national biosecurity laws in 2002; improved capabilities in 2004 to respond to causes of suspicion, strengthen the efforts to control, and detect, diagnose, and fight infectious diseases; and, in 2005, a code of conduct for scientists. Information from Kathryn Nixdorf, TU Darmstadt, Germany.

25. For example, the UNESCO Declaration (1997) for Protection of Human Genome and Human Rights; European Guidelines for Biotechnological Patentation; Convention for the Protection of Human Rights and Dignity of Human Beings with Regard to the Application of Biology and Medicine; or the Prohibition of Cloning Human Beings (1998; Prohibition of Genome Intervention) (see Honnefelder et al., 1999, p. 11).

26. See, e.g., the case of Gelsinger. The Principal Investigator, founder, and main shareholder of Genova (a biotechnology company), in his function as a doctor, advised his patients through parent counseling; together with another clinical adviser, he experimented on patients; as a scientist he got public funding for developing viral vectors and clinical studies; as an expert he advised the government on which kind of genetic projects are worth funding and which studies are approved; as an expert and editor of a journal, he determined what received publication and created and promoted the topic of genetic medicine; and his University was the main shareholder of his company.

27. Nanotechnology is difficult to define clearly, because it exists on the borderline between disciplines, but one key component is that behavior, surface, and molecular recognition are important in describing special effects at nanoscale (Luther & Malanowski, 2003, pp. 18f).

28. This approach is being developed by the European Union FP6 project PACE (Programmable Artificial Cell Evolution).

29. The UN Economic and Social Council, in 1994, estimated the global turnover by organized transnational crime to be about $750 billion; $500 billion resulted from drug economy alone. In 1999, the IWF roughly estimated global money laundering at between $500 billion and $1 trillion a year; others estimate it at about 5 percent of the global GIP (Castells, 2003, p. 178). It is indeed very difficult to reach valid estimations, because of method dependencies. A recent IMF report on shadow economy (Schneider & Enste, 2000, 2002) gives an overview of different methods. The shadow economy has its legal aspects as well as criminal activities. The authors' empirical findings underline a worldwide increase of these activities during the final decade of twentieth century. Involvement in shadow economy differs from 5 percent to nearly 80 percent of the gross domestic product of a national economy. Data show 40 to 70 percent in Africa/South America/Asia; about 20 to 60 percent in transition countries; and about 5 to 20 percent in OECD (Organization for Economic Cooperation and Development) countries.

30. Drugs are the main area of illicit trade; also important are weapons, nuclear material, humans, human organs, prostitution, illegal workers, gambling, secret information, and technology (Castells, 2003, p. 176).

31. Unresolved conflicts, including those of power transparency, had a dynamic impact on withdrawal of safety issues; Effective communication of technical problems was blocked by hierarchies, signals were overlooked, people were silenced, and useful information and dissenting views on technical issues did not surface at high levels (CAIB, 2003, p. 201). Responsibilities were transferred to contractors, which increased dependence on the private sector for safety functions and risk assessment (CAIB, 2003, p. 202).

References

Ahrendt, H. (1958). *The human condition*. Chicago: University Press.

Aristotle (reprinted 1983). *Nikomachische Ethik*. Stuttgart: Reclam.

Beck, U., Bonss, W., & Lau, C. (2004). Entgrenzung erzwingt Entscheidung. Was ist neu an der Theorie reflexiver Modernisierung? In U. Beck & C. Lau (Eds.), *Entgrenzung und Entscheidung* (pp. 13–64). Frankfurt: Suhrkamp.

Beck, U., & Holzer B. (2004). Reflexivität und Reflexion. In U. Beck & C. Lau (Eds.), *Entgrenzung und Entscheidung* (pp. 165–192). Frankfurt: Suhrkamp.

Beck, U., & Lau, C. (2004). *Entgrenzung und Entscheidung*. Frankfurt: Suhrkamp.

Bedau, M. (2003). *Broad perspectives on artificial cells*. (Working paper). Dortmund: University ECAL workshop, Bridging Living and Non-Living Matter.

Bedau, M. (2005). *Social responsibility and protocell technology*. Presentation at The Social and Ethical Implications of Protocell Technology, PACE Workshop, European Center for Living Technology, Venice, Italy.

Böhme, G. (1997). *Ethik im Kontext.* Frankfurt: Suhrkamp.

Böschen, S., Lau, C., Obermeier, A., & Wehling, P. (2004). Die Erwartung des Unerwarteten. Risk Assessment und der Wandel der Risikoerkenntnis. In U. Beck & C. Lau (Eds.), *Entgrenzung und Entscheidung* (pp. 123–148). Frankfurt: Suhrkamp.

Brand, A., Dabrock, P., Paul, N., & Schröder, P. (2004). *Bio- und Gentechnologie* (report). Berlin: Friedrich Ebert Foundation.

Buss, K.-P., & Wittke, V. (2001). Wissen als Ware. In Bender, G. (Ed.), *Neue Formen der Wissenserzeugung* (pp. 123–146). Frankfurt; New York: Campus.

Castells, M. (2003). *Jahrtausendwende: Das Informationszeitalter III.* Opladen: Leske & Budrich.

Castells, M. (2004). *Der Aufstieg der Netzwerkgesellschaft: Das Informationszeitalter II.* Opladen: Leske & Budrich.

Columbia Accident Investigation Board (CAIB) (2003). Report, Volume I. Limited first printing, Washington DC. Available online at: http://caib.nasa.gov/news/report/volume1/default.html (accessed July 2007).

Crichton, M. (2002). *Prey.* New York: HarperCollins.

Dabrock, P. (2000). *Wer büßt für die 'Sünden' der modernen Wissenschaft und Technik? Nachdenkliches zur Entwicklung der neuesten Entwicklung der Bio- und Reproduktionstechnologien.* Working paper. Meschede: Fachhochschule.

Dando, M. R., & Nixdorf, K. (2004). *Science and technology considerations at the 2011 seventh review conference of the BTWC: Will the convention have been bypassed?* Bradford: University (UK); Darmstadt: Technical University (GER).

Feenberg, A. (2003). Heidegger und Marcuse: Zerfall und Rettung der Aufklärung. In G. Böhme & A. Manzei (Eds.), *Kritische Theorie der Technik und der Natur* (pp. 39–54). Muniche: Wilhelm Fink.

Giddens, A. (1996). *Jenseits von Rechts und Links.* Frankfurt: Suhrkamp.

Grimm, D. (2005). Preface. In H. Nowotny, D. Pestre, E. Schmidt-Assmann, H. Schultze Fielitz, & H. H. Trute (Eds.), *The public nature of science under assault: Politics, markets, science and law* (p. v). Berlin; New York: Springer.

Haker, H. (2003). Feministische Bioethik. In M. Düwell & K. Steigleder (Eds.), *Bioethik: Eine Einführung* (pp. 168–183). Frankfurt: Suhrkamp.

Haraway, D. (1995). *Die Neuerfindung der Natur: Primaten, Cyborgs und andere Frauen.* Frankfurt: Campus.

Hauchler, I., Messner, D., & Nuscheler, F. (2002). *Global trends 2002.* Frankfurt: Fischer.

Heller, A. (2004). Cultural Studies im Wandel: Zur Modellfunktion der American Studies. In E. List & E. Fialla (Eds.), *Grundlagen der Kulturwissenschaften. Interdisziplinäre Kulturstudien* (pp. 39–53). Tuebingen: Francke.

Honnefelder, L. (2001). Was wissen wir, wenn wir das menschliche Genom kennen? In L. Honnefelder & P. Propping (Eds.), *Was wissen wir, wenn wir das menschliche Genom kennen?* (pp. 9–28). Cologne: DuMont.

Honnefelder, L., Mieth, D., Propping, P., Siep, L., & Wiesemann, C. (2003). *Das genetische Wissen und die Zukunft des Menschen.* Berlin; New York: Walter de Gruyter.

Honnefelder, L., Taupitz, J., & Winter, S. (1999). *Das Übereinkommen über Menschenrechte und Biomedizin des Europarates: Argumente für einen Beitritt.* (Report 171/1999). St. Augustin: Konrad Adenauer Foundation.

Huxley, T. H. (1893, reprinted 2004). Evolution and ethics. In M. Ridley (Ed.), *Evolution* (pp. 418–420). Oxford: Oxford University Press.

Joerges, B. (1996). *Technik, Koerper der Gesellschaft.* Frankfurt: Suhrkamp.

Jonas, H. (1984). *Das Prinzip Verantwortung: Versuch einer Ethik für die technischeZivilisation.* Frankfurt: Suhrkamp.

Joy, B. (2000). Why the future doesn't need us. *Wired,* April 2000.

Keller, E. F. (2003). Genetischer Determinismus und das Jahrhundert des Gens. In L. Honnefelder, D. Mieth, P. Propping, L. Siep, & C. Wiesemann (Eds.), *Das genetische Wissen und die Zukunft des Menschen.* (pp. 15–26). Berlin; New York: Walter de Gruyter.

Latour, B. (2001). *Das Parlament der Dinge: Für eine politische Oekologie.* Frankfurt: Suhrkamp.

Lengauer, T. (2001). Bioinformatik an der Schwelle der postgenomischen Aera. In L. Honnefelder & P. Propping (Eds.), *Was wissen wir, wenn wir das menschliche Genom kennen?* (pp. 56–61). Cologne: DuMont.

Liebert, W., Rilling, R., & Scheffran, J. (1994). Die Ambivalenz von Forschung und Technik und Dual-Use Konzeptionen in der Bundesrepublik Deutschland. In W. Liebert, R. Rilling, & J. Scheffran (Eds.), *Die Janusköpfigkeit von Forschung und Technik: Zum Problem der zivil-militärischen Ambivalenz* (pp. 12–30). Marburg: BWI.

List, E. (2004a). Institutionen des Wissens: Zur Frage nach dem Ort der Kulturwissenschaften im Wissenshaushalt der Moderne. In E. List & E. Fialla (Eds.), *Grundlagen der Kulturwissenschaften* (pp. 13–38). Tuebingen: Francke.

List, E. (2004b). Interdisziplinäre Kulturforschung auf der Suche nach theoretischer Orientierung. In E. List & E. Fialla (Eds.), *Grundlagen der Kulturwissenschaften* (pp. 3–21). Tuebingen: Francke.

Longino, H. (1990). *Science as social knowledge: Values and objectivity in scientific inquiry.* Princeton: Princeton University Press.

Luther, W., & Malanowski, N. (2003). *Innovations und Technikanalyse: Nanotechnologie als wirtschaftlicher Wachstumsmotor* (report on future technologies, 53). Düsseldorf: VDI (Verband Deutscher Ingenieure).

Mooney, C. (2002). Nothing wrong with a little Frankenstein. Commentary in *The Washington Post,* December 1, 2002.

Nowotny, H. (2005a). The changing nature of public science. In H. Nowotny, D. Pestre, E. Schmidt-Assmann, H. Schultze Fielitz, & H. H. Trute (Eds.), *The public nature of science under assault: Politics, markets, science and law* (pp. 1–29). Berlin; New York: Springer.

Nowotny, H. (2005b). *Unersättliche Neugier: Innovation in einer fragilen Zukunft.* Berlin: Kadmos.

Nowotny, H., Scott, P., & Gibbons, M. (2001). *Re-thinking science, knowledge and the public in the age of uncertainty.* Cambridge: Cambridge University Press.

Paul, N. W. (2003). *Auswirkungen der Molekularen Medizin auf Gesundheit und Gesellschaft*. Bonn: Friedrich Ebert Foundation.

Peltonen, L. (2003). Zukunft und Perspektiven der Humangenomforschung. In L. Honnefelder, D. Mieth, P. Propping, L. Siep, & C. Wiesemann (Eds.), *Das genetische Wissen und die Zukunft des Menschen* (pp. 27–35). Berlin; New York: Walter de Gruyter.

Pestre, D. (2005). The technosciences between market, social worries and the political: How to imagine a better future. In H. Nowotny, D. Pestre, E. Schmidt-Assmann, H. Schultze Fielitz, & H. H. Trute (Eds.), *The public nature of science under assault: Politics, markets, science and law* (pp. 29–52). Berlin; New York: Springer.

Propping, P. (2001). Vom Genotyp zum Phänotyp: Zur Frage nach dem genetischen Determinismus. In P. Propping & L. Honnefelder (Eds.), *Was wissen wir, wenn wir das menschliche Genom kennen?* (pp. 90–102). Cologne: DuMont.

Propping P., & Honnefelder, L. (2001). *Was wissen wir, wenn wir das menschliche Genom kennen?* Cologne: DuMont.

Rammert, W. (2003). Zwei Paradoxien einer innovationsorientierten Wissenspolitik: Die Verknüpfung heterogenen und die Verwertung impliziten Wissens. *Soziale Welt, 54*, 483–508.

Sass, H. M. (2001). A "contract model" for genetic research and health care for individuals and families. *Eubios Journal of Asian and International Bioethics, 11*, 130–132.

Schätzing, F. (2004). *Der Schwarm*. Cologne: Kiepenheuer & Witsch.

Schneider, F., & Enste, D. (2000). *Shadow economies around the world: Size, causes, and consequences*. International Monetary Fund Report no. IMF WP 00/26.

Schneider, F., & Enste, D. (2002). Hiding in the shadows: The growth of the underground economy. *IMF Economic Issues, 30,* available online at: http://www.imf.org/external/pubs/ft/issues/issues30/index.htm (accessed August 2008).

Schreiber, H. P. (2005). Transcultural bioethics: Corporate ethics as a vehicle for global bioethics. From the Expert Conference of the Federal Ministry of Education and Research (Berlin). Bioethics in the context of law, morals and culture. September 14–15, 2005.

Schröder, P. (2004). Das Ende des "informed consent." *Das Gesundheitswesen, 66*, 8–9.

Schulze Fielitz, H. (2005). Responses of the legal order to the loss of trust in science. In H. Nowotny, D. Pestre, E. Schmidt-Assmann, H. Schultze Fielitz, & H. H. Trute (Eds.), *The public nature of science under assault: Politics, markets, science and law* (pp. 63–86). Berlin; New York: Springer.

Sellmayer, S. (2004). Entscheidungskonflikte in der reflexiven Moderne: Uneindeutigkeit und Ahnungslosigkeit. In U. Beck & C. Lau (Eds.), *Entgrenzung und Entscheidung* (pp. 149–164). Frankfurt: Suhrkamp.

Sicard, D. (2003). Illusionen und Hoffnungen der Genetik. In L. Honnefelder, D. Mieth, P. Propping, L. Siep, & C. Wiesemann (Eds.), *Das genetische Wissen und die Zukunft des Menschen* (pp. 47–56). Berlin; New York: Walter de Gruyter.

Simon, P. (2004). Entwicklung, Risiken und therapeutischer Nutzen der Gentherapie (report 2, Analysen zur Zukunft der Biotechnologie). Berlin: Friedrich Ebert Foundation.

Sorokin, P. (1927, reprinted 1959). *Social and cultural mobility*. New York; Ontario: Free Press.

Stehr, N. (2004). *Biotechnology: Between commerce and civil society*. New Brunswick, NJ: Transaction Publishers.

Stehr, N., & von Storch, H. (1999). *Klima, Wetter, Mensch*. Munich: Beck.

Steinfeldt, M., von Gleich, A., Petschow, U., Haum, R., Chudoba, T., & Haubold, S. (2004). *Nachhaltigkeitseffekte durch Herstellung und Anwendung nanotechnologischer Produkte* (report 177/04). Berlin: IÖW.

van den Daele, W. (2001). *Dealing with the risk of genetic engineering as an example of "reflexive modernization"?* Unpublished paper. Berlin: WZB.

Viehöfer, W. (1997). *"Ozone thieves" and "hot house paradise:" Epistemic communities as cultural entrepreneurs and the reenchantment of the sublunar space*. Unpublished doctoral dissertation, University of Florence, Italy.

Viehöfer, W. (2006). *Totem ohne Tabu: Das Ungeborene in modernen Mythen über den Beginn des menschlichen Lebens* (working paper 2/2004). Munich: University of Munich.

Viehöfer, W., Gugutzer, R., Keller, R., & Lau, C. (2004). Vergesellschaftung der Natur— Naturalisierung der Gesellschaft. In U. Beck & C. Lau (Eds.), *Entgrenzung und Entscheidung* (pp. 65–94). Frankfurt: Suhrkamp.

Wade, N., & Sang-Hun, C. (2006). Researcher faked evidence of human cloning, Koreans report. *The New York Times*, January 10, 2006.

Weber, J. (2003). Vom Nutzen und Nachteil posthumanistischer Naturkonzepte. In G. Böhme & A. Manzei (Eds.), *Kritische Theorie der Technik und der Natur* (pp. 221–246). Munich: Wilhelm Fink.

Wehling, P. (2001). Jenseits des Wissens? Wissenschaftliches Nichtwissen aus der soziologischen Perspektive. *Zeitschrift für Soziologie, 30*, 465–484.

Wehling, P. (2004). Reflexive Wissenspolitik: Öffnung und Erweiterung eines Politikfeldes. *Technikfolgenabschätzung—Theorie und Praxis, 13* (3), 63–71.

Willke, H. (2006). *Global Governance*. Bielefeld; manuscript in preparation.

Wittgenstein, L. (1930, reprinted 1989). Vortrag über Ethik. In J. Schulte (Ed.), *Vortrag über Ethik und andere kleine Schriften* (pp. 9–19). Frankfurt: Suhrkamp. First published in 1965 as Wittgenstein's lecture on ethics, *Philosophical Review, 74*, 3–12.

Yesley, M. (2005). Development of bioethics and ELSI-grants in the USA. Berlin: Expert Conference of the Federal Ministry of Education and Research: Bioethics in the context of law, morals and culture. September 14–15, 2005.

III

Ethics in a Future with Protocells

13

Open Evolution and Human Agency: The Pragmatics of Upstream Ethics in the Design of Artificial Life

George Khushf

Recently there has been a qualitative change in human ability to understand and directly manipulate living systems at the molecular level. In diverse fields like bionanotechnology, synthetic biology, and artificial life, this new knowledge is being used to radically reengineer existing systems for a broad range of purposes, including the synthesis of new materials and diverse medical applications. In each of these fields, a new kind of science is taking form, one that is itself a hybrid of a more conventional science and engineering. Older distinctions between pure and applied science are no longer helpful in categorizing this research. Although fundamental questions are still addressed, the mindset and methods are novel, often involving simulations and technological products as answers, rather than more traditional modes of theory formation and hypothesis testing. When engineering or medical goals are advanced, they rarely involve a straightforward application of some basic science.

In this chapter, I consider how this new science deeply challenges our approach to ethical issues. I first consider in some detail how a traditional ethic involves notions of practical rationality that are closely aligned with more traditional understandings of science. It is assumed that a contemplated activity can be discretely regarded, that ends can be clearly distinguished from the means advancing them, and that we can assess the impact of a proposed activity by modeling it and projecting how it will interact with other parts of its environment. Such a traditional ethic has generally been *post hoc*, following after the development of some science or technology. Responsible agency is then associated with a process of normalization, in which a presupposed activity is harmonized with other activities integral to a flourishing community.

I next consider a representative example of some research in artificial life associated with the design of a self-assembling membrane, and show how this new, hybrid science undermines the conditions for a traditional post hoc, downstream mode of ethical reflection. Such research is entangled with other scientific and technological projects in multiple, complex ways, and it is impossible to isolate and discretely address the relevant issues in the manner required by a traditional ethic. Instead, ethical reflection must be incorporated up front, as a vital part of the research itself. It will involve a series of continual, iterative adjustments in the research process, and can be adequately addressed only if the research team is attentive to the ethical challenges posed by their own work.

Finally, I close by considering a radical implication of this proposal. If ethics and science are deeply intertwined, then ethical considerations could inform concepts integral to the science. By reflecting on the iterative adjustments integral to an upstream ethic, I consider how top-down and bottom-up approaches to creating protocells might be coupled and explore how such a coupling might inform the way some problems in open-ended evolution might be addressed.

Norms Integral to the Responsible Conduct of Scientific Research

Ethics, Responsible Agency, and Human Flourishing

Two prominent themes are integral to ethical reflection.[1] The first concerns free agency. Unlike principles of inertia or gravity, ethical norms depend on an agent who can choose to be governed by them. We thus experience the ought of ethical norms as intimately related to freedom (classic arguments are in Kant, 1956, 1985; see also Mele, 1995; Allison, 1997; Dennett, 2003). Contemplation of action involves possible worlds, and norms provide guidance for actualizing these possibilities. Ethics thus arises within a nexus of choice. Without such choice, we might have some governing biological or social principle, and this principle might explain a large complex of behaviors and social patterns, but we would not have ethics.[2]

The second prominent theme is that of a flourishing or good life (classic arguments are in Aristotle; see also Paul, Miller, & Paul, 1999; Nussbaum, 2000; Foot, 2001). Ethical norms represent types of practices and social structures in which humans flourish. Some of these norms might be relatively transient, applying in only a small niche; for example, a small, short-lived clique might have idiosyncratic norms about greetings. Other norms, like prohibitions against murder or adultery, are more enduring, reflecting general patterns of kinship and even necessary features of a well-functioning community. (All flourishing communities end up having some variant of these more enduring norms; see Wilson, 1993.) When closely wed to the conditions of human flourishing, norms have their own truth conditions: If they enhance life, they are valid; if they diminish it, they are pathological.[3]

In ethical theory, one of these themes is often emphasized at the expense of the other. Thus, for example, a Kantian ethic gives primary weight to the questions of agency, free choice, and practical rationality, whereas a utilitarian or perfectionist ethic gives greater weight to the good and the conditions for realizing it. Attempts to integrate these strands have been notoriously tricky, although most (especially the more nuanced thinkers) appreciate the need to do so. Freedom without flourishing becomes a kind of brute, arbitrary spontaneity, or it degenerates into a mere formalism (as in some interpretations of Kant's ethic). On the other hand, flourishing without freedom degenerates into a pseudoscientific, social engineering project. The agents involved—those who do the social engineering—provide grand schemes for assessing cost and benefit, or for maximizing some good, but they do not carefully reflect on their own context, on the internal dynamics of their deliberations, or on the openness of the future they seek to determine.

In notions of responsibility, the questions of free agency and human flourishing are linked (Jonas, 1984; Taylor, 1985, 1989; Schoeman, 1987; Sen, 1999). We can clarify this by considering what distinguishes responsible agency from irresponsible agency. Consider some contemplated action, for example, driving a car. I am aware of a capacity or power to drive. I know how to locate a key, start up a car, and then manipulate the gas pedal and brake to make it move. I thus have a sense of agency, that is, an ability to do something within the world. This sense of agency is linked to specific material and social conditions, and it involves know-how, which uses a technology or technique to act within the world.[4] At the same time, I am aware that what I do has an impact on myself and on my environment. I could be an agent in the sense of initiating a causal stream within the world, but I could fail to accomplish any of my goals and could also cause a broad array of unintended harms. Agency can thus be spontaneous, brute, and irrational. But I can also take the vague sense of being able to drive—the brute ability to spontaneously jump in a car, start it, and crash aimlessly ahead—and refine it. An episode will consist of some discrete use of this ability, with a clear starting and stopping time. During this episode, I might drive in a way that either accomplishes or does not accomplish my goals, and I might do this in a way that has a positive, negative, or neutral impact on my environment. This refinement assumes I have goals, and that driving will serve them. If I have learned the appropriate rules for driving and developed the required skills, and then drive in an alert manner in conformity to the rules and exhibiting these skills, we say I am doing this responsibly. By contrast, if I drive before I am ready or when I am in some way impaired (as when drunk), then we say I am irresponsible. Responsible driving thus involves a broader set of skills and abilities than just driving. This broader set enables me to use the car in such a way that I do not hinder or interfere in inappropriate ways with others in my environment. Responsible driving thus involves a kind of partition for the driving episode. By means of the appropriate exercise of knowledge and skills, I can get on the road, go where I want, and return without adversely impacting the ability of others to do the same thing. The relevant norms that make driving responsible therefore concern the modularizing of the driving episode, the assurance of an appropriate means-end linkage, and the harmonization of any specific instance with other instances and activities. Responsible agency is a kind of agency that orders and harmonizes otherwise disconnected modules or episodes so they cohere and are integrated in a complementary, coordinated manner.[5]

From this example, we can see that responsible agency presupposes a framework for using abilities to appropriately realize broader ends, simultaneously ensuring there are no adverse consequences for others. The background norms make driving "responsible" (i.e., normalize it) by situating it in a broader context of individual and social activities. In this way, the basic activity is channeled so it does not disrupt human flourishing.[6] By contrast, irresponsible agency involves a careless, unreflective action, which might have unintended harmful effects on others. The activity is an arbitrary, brute eruption of a causal stream within the world. It remains disconnected from the broader web of other activities, and thereby puts at risk the dynamic of a flourishing community.

I can now frame this notion of responsibility in a normative way by stating that a person should consider the broader context and consequences of any action, not just on herself, but also on others and on the environment. The ethical analysis presupposes that you have a desirable activity that arises in some other, preethical context.[7] As with the brute ability of driving, we have some sketch of a possible project or action, and we seek to optimize it. Within that action, there is some implicit or explicit end. The ethical analysis takes place at a second level. It models the contemplated action, situating it within the context of broader notions of human flourishing. As with the norms that govern driving, the second level of analysis fills in the content of the proposed activity or project, normalizing it so the result fits with other social activities and projects.

The analysis of ethical norms becomes more complex when we recognize that norms are themselves open. The agent does not just choose whether or not to follow a pregiven norm, but also partly chooses what the norms are. The conditions of flourishing are thus themselves open. Humans create new norms that can either enhance or diminish life. The development of norms thus introduces a highly complex individual and social context. People must assess whether and how norms might advance a flourishing life. This assessment is partly a question of fact, concerned with the actual conditions of a flourishing life, and partly a question of choice, concerned with the kind of life that will flourish. In some ways, the assessment of the conditions of flourishing precedes the choice about the kind: One selects that pattern of life most conducive to flourishing. But in other ways, the choice precedes the assessment: Different kinds of life will flourish in different contexts and involve different norms. In practice, we often have an iterative adjustment between these two.

An appreciation of these complexities allows for considerable ethical pluralism, as well as recognition that norms will be historically, culturally, and socially conditioned. However, it is also clear that some norms are more appropriate than others. As with physiological norms, considerable variations exist, but so do pathologies.

When material or social conditions are altered, previously appropriate norms might become maladaptive. With such shifts, older norms might take on new meanings. Thus, for example, a greeting originally designed to immobilize weapons— shaking with the sword-wielding hand—might become a purely symbolic gesture. However, other norms might, in fact, put the community at greater risk of harm or make it unable to function effectively in the altered context. Consider, for example, older prohibitions against usury (Jonsen & Toulmin, 1990). In largely agricultural or feudal societies, such a norm protects those who are more vulnerable against exploitation. In a market economy, however, this prohibition hinders the effective use of capital and thereby undermines the conditions that allow for accrual of wealth and advancement in social welfare. If a culture lacks the resources for the appropriate development of norms, a change in context could be deeply destabilizing, leading to a radical disruption of communal equilibrium. We can thus distinguish between first-level norms, which are subject to development, and a second level of normative reflection. At this second level, more general metanorms govern the scope, character, and development of first-level norms. First-level norms, together with their associated practices, are then analogs of the preethical given; they reflect a relatively stable, discrete domain. The function of second-level normative reflection is to normalize those pregiven norms (and their associated practices) by bringing them into harmony with other domains to sustain flourishing. Such second-level normalization is a continuous, ongoing process.

Earlier we considered how norms might regulate the activity of driving. A brute ability to drive constituted the first, preethical given, and norms then regulated this activity so that one person's driving does not harm others, and so the goals of driving might be realized. But these norms of driving are not static, and we cannot view the systems of traffic as fully separable from other social systems, such as those of commerce or defense. At the second level, we have social and political processes for critically reflecting on and revising the norms that govern driving. These include mechanisms of representative government, deliberative civil discourse, and related systems for revising our laws, as well as the judicial mechanisms for contesting norms that are deemed to be unjust, such as norms governing search and seizure, taxation for supporting roads and infrastructure, and so on. Since second-level norms govern the process of first-level norm formation and revision, they could be viewed as process norms. (An important subset of these second-level norms is embodied in canons of procedural justice.) When such process norms are formulated at a highly abstract level, they reflect ideals of practical rationality (e.g., Kant's categorical imperative; see also Milgram, 2005).

Ethics and Technology

New technologies can make new forms of human life possible. In doing this, they alter the conditions for human flourishing. Technological development thus has an impact on the truth conditions of norms, and it can precipitate an ethical crisis (Ellul, 1964; Jonas, 1984; Postman, 1993). However, technology does not drive social change in a purely deterministic manner. Technologies are themselves regulated. Norms have emerged that mitigate the disruptive and life-diminishing consequences of technological change, channeling development so it enhances rather than diminishes life. Such norms normalize technological development, and they involve regulatory structures that function at multiple levels—economic, social, political, and cultural. Often these second-level norms are embodied in explicit policies and protocols that govern technological development; for example, in environmental regulations related to the testing and use of a new pesticide or FDA regulations governing introduction of new pharmaceuticals. Changing material and social conditions are thus themselves regulated processes, and ethical norms govern these processes for changing norms. At the most general level, there are norms for critical reflection on norms. These general norms embody canons of responsible agency, and they take both social form, as in a jurisprudence, and individual form, as in the logics of practical rationality.

The second-level, process norms are usually more stable and enduring than first-level norms. When an emergent technology is disruptive, it generally problematizes the first-level norms. Many of our regulatory processes are constructed so that new norms can coemerge with the technologies they regulate. The development of these norms depends on the higher-level practical rationalities. However, at times a kind of technology emerges that more radically challenges all levels of ethical reflection. A prototypical technology of this kind is writing. When this developed, the deepest patterns of thought and social organization were undermined, and completely new modes of reflection and action became possible (Havelock, 1963, 1986; Eisenstein, 1980; Ong, 1982). Primarily oral cultures were unable to resist the literary cultures that internalized the possibilities created by the technologies of writing. I refer to this more fundamental crisis posed by some new technologies (often surrounding a key development, like that of writing, the printing press, or more recently, the computer) as a *radical ethical crisis*, and to the technology that precipitates it as a *radically disruptive technology*.

Norms governing processes of development and embodied in canons of practical rationality have the same truth conditions as the first-level, material norms. They are valid if they advance patterns of human flourishing, and maladaptive otherwise. However, in them we see more clearly a certain paradoxical character integral to all ethical norms (Canguilhem, 1991). Such norms presuppose change, but they also are formulated to prevent certain kinds of change; namely, that which undermines human flourishing. Second-level, process norms govern change, but they must also presuppose something invariant: an enduring pattern of life that is supposed to be nourished and preserved in the midst of change. When faced with a radically disruptive technology, we thus face a fundamental ambiguity: The transformations brought about by the technology can be viewed as altering the truth conditions of the antecedent norms, thus calling for revision. Or they can be viewed as exactly the kind of disruptive development that the norms were formulated to prevent.[8] Some deep transformations thus pose a question at the heart of ethics: Does this emergent technology reflect changing processes of development, and thus call for the development of alternative forms of practical rationality and human agency? Or does it reflect a pathological development that undermines the conditions of responsible human agency?

In periods of radical ethical crisis, the truth conditions of process norms have a strange status. If a potential pathway of technological development is prohibited from developing in one context, as, for example, genetically modified organisms (GMOs) were in Europe, and allowed to develop in another, as they were in the United States, then conditions emerge for determining whether more permissive or prohibitive norms were more appropriate. In these cases, one does not just assess the material norms governing the emergent technology (e.g., GMOs). One also assesses process norms, such as those of more progressive versus precautionary stances with respect to the emergent technology. In cases where a technology is completely cut off or universally embraced, however, one loses the conditions for assessing the alternative. The enduring, preserved features of human life become a reflection of the adjustments associated with previous, contingent choices.[9]

One way to summarize the paradoxical character of process norms is to say that the truth conditions of such norms cannot be fully disentangled from the mechanisms for assuring conformity to those norms and the choices made about what must be preserved. Truth conditions of norms thus depend on two things:

1. The choices made by individuals and communities. When we speak of flourishing, we must speak of the flourishing of this or that thing. The definition of "self" is, in an important sense, prior to an analysis of the conditions under which the self flourishes. Humans are somewhat plastic; they determine who they are and will be, and this determination is reflected in their practices, social dynamics, and

interactions with others and the ecological landscape. The choices made regarding human nature, then, determine what conditions foster the flourishing of that life.
2. The actual material and social conditions of human flourishing. Once a specific form of life is chosen, there are constraints on how it might be advanced. What fosters life? What diminishes it? These conditions are not simply made up. Science discovers them. We can get them right (or wrong), and we are constrained by ecological and biological realities.

We thus have two factors: a choice and, for any specific form of life, a descriptive, scientific account of the conditions that foster or undermine flourishing. Usually we hold one of these factors fixed and discretely consider the other. Thus, we might consider the implications of different possibilities, and then decide among them. Such deliberation and choice come under the domain of ethical deliberation. Alternatively, we might ask how to accomplish a certain end. Once the end is specified, we draw on the tools of applied science to realize it.

At times, however, we are unable to neatly distinguish the ethical and scientific, and are forced to engage in a complex, spontaneous kind of adjustment between means and ends, where the logics of choice and scientific technique are fundamentally entangled (Schon, 1983). A radically disruptive technology brings one to such an entangled place, precipitating a radical ethical crisis. In such contexts, we lose footing for assessing whether the emergent technology is adaptive or maladaptive, because the answer (at least partly) depends on a choice regarding the nature of the life in question.[10] The intermingling of science and ethics itself poses a deep challenge to both fields. Our conventional notions of practical and scientific rationality assume a sharp distinction between fact and value, science and choice. We have different institutions and cultures for these different worlds. If they mingle too deeply, both science and ethics are undone.

Science and Ethics

This chapter is concerned with the ethical assessment of a scientific activity; for example, with the ethics of artificial life research. "Science" is thus the protoethical activity, analogous to the brute ability to drive. The task of ethical analysis is to place this activity within a broader social context. To do this, the scientific activity must be modeled, its consequences anticipated, and the activity refined. Such a modeling activity is itself scientific: a second-level science of the science, which might be called *philosophy of science* or *science studies*. In this way, we move from what might be called *brute science* to *responsible science*. I will call this second-level modeling activity the *science integral to ethics*. Beyond this, I can also reflect in a more general way on the whole process; for example, I can ask whether the second-level modeling activity is of the same kind as the first (Latour & Woolgar, 1986, pp. 252f). Do both follow a method? Do they reflect a common scientific ideal? When I reflect on the second-level reflection, as I do in this chapter, then I can speak of a third level. Here we have the *science of ethics*. When, for example, I seek to provide an account of the development of norms that links up with evolutionary theory, then I am conducting my investigation at this third level of analysis.

Later, I wish to consider what happens to these three levels when a radically disruptive technology precipitates an ethical crisis. However, before considering this, we need to consider how these three levels might be conceptualized for cases of normal science.[11] As an example of normal science, consider the research and testing associated with the development of a new pharmaceutical (Brody, 1995).

Level 1:   The protoethical science. Here we might include things like initial genetics work to identify a mutation with interesting properties, subsequent work to identify a protein for which this gene codes, and then attempts to artificially synthesize some chemical that mimics the properties of the protein. Various computer simulations might be used to explore a range of possible chemical combinations, and promising candidates can then be synthesized and tested. Finally, appropriate animal models can be used to test the candidate drug.

Level 2:   The science integral to ethics. Ethical reflection places the protoethical science in a broader context of social oversight and use. By means of such oversight, we have developed a network of protocols and policies that make clear what responsible development and use of a new drug must entail. This includes initial testing for toxicity, followed by a sequence of tests for efficacy. Detailed standards have been worked out for how tests are conducted on human populations, how research participants' consent must be obtained, how participants in a drug trial must be selected, and so on. We also have standards for when drugs can be prescribed, who can purchase them, and how markets for them may and may not function.

Level 3:   The science of ethics. After a system of norms emerges around some practice, I can reflect on the history of its development and consider how well the norms function in regulating the practice. Thus, with respect to the norms governing pharmaceuticals, I can study FDA policy, look at the functioning of institutional review boards in regulating human subjects research, examine how commercial interests might influence assessment of efficacy or prescription patterns, and so on.

Several observations can be made with respect to these three levels and how they function in relation to one another.

1. *The distinction between levels is pragmatically determined.* Ethical analysis of science and technology usually considers some domain that has its own integrity. The protoethical domain (level 1) is taken as given. That means it must be relatively stable and already have coherence that allows it to be critically analyzed. In reality,

all such activities are already highly norm-governed, and these norms reflect a long history of previous ethical reflection and adjustment. If we decompose the protoethical activity, we could identify subdomains, take these as level 1, and ethically analyze them. In this case, features that were intrinsic to the activity on our first account become extrinsic. From this we see that the lines between levels 1, 2, and 3 depend on pragmatic considerations. A historical account of the development of the activity would document a sequence of steps in which a more primitive practice is increasingly augmented by norms that regulate the practice and integrate it into a broader web of social activities.

2. *The distinction between levels depends on the norms that can be taken for granted in the protoethical activity.* Here we can distinguish between intrinsic norms and ends, which are integral to the level-1 activity, and extrinsic norms and ends, which will be integral to the level-2 analysis, but may or may not be explicitly affirmed by those involved in the level-1 activity. In our previous example, we assume that those involved in the research and development of the pharmaceutical are already responsive to standards of appropriate scientific conduct, that they appreciate the need to develop new drugs that can mitigate the suffering associated with disease, and so on. At the second level, additional norms are introduced such as those of informed consent (a research subject should never be regarded as a mere means), appropriate balance of risks in trial design (beneficence), and fair selection of research subjects (justice).[12] Regulation of prescriptions assumes that clinical oversight is needed to monitor appropriate drug use. In these norms, we see standards that might be insufficiently regarded by those involved in the development and promotion of a drug.

3. *Higher-level analysis presupposes that the lower levels can be isolated from their own environments (i.e., taken as modules) and decomposed into parts, which function in relation to one another in a clearly specified manner.* When a complex practice like scientific research is taken as the object of ethical analysis, we must first frame the activity as discrete. Then we can distinguish between internal norms of the practice and the external reflection that nests the practice within a broader social context. As in any system hierarchy, we can isolate components at a given level and then consider how these are composed of lower-level components.[13] What we identify as sub and supra depends on the level at which our primary analysis takes place. In ethical analysis certain kinds of concepts are used to decompose the activity in question. These include the distinction between means and ends, and also involve concepts for categorizing the ends. In the case of scientific research, a fundamental distinction is often made between practices that lead to technologies or involve an intent to do something within the world (applied science) versus those that lead to theoretical products such as factual descriptions or theories (fundamental science). For drug development, we thus can distinguish between more fundamental scientific research, which often takes place at universities, and the applied research that leads to a new pharmaceutical. Usually, fundamental science is held accountable only to

its internal norms; in fact, we express great concern when interests and ends external to science are allowed to influence how it is conducted. However, technologies are different. Since they directly influence the lives of all, a much broader set of considerations is regarded as relevant in their regulation.

4. *Ethical analysis arises downstream. Downstream ethics moves up among hierarchically nested levels.* Our cultures of scientific research and technological development assume we will have a relative independence of science from regulatory oversight, and that more stringent oversight only takes place when a technological product is created. During the initial phases of research and development, science is governed by its internal norms; for example, those associated with open and honest reporting of data, fair attribution of credit, peer review, and so on. If the research leads to new kinds of technology or involves a broader liaison with industry, then ethical issues are seen to arise at the stage when a technology moves out of the lab and into the market (study of this is often called *translational research*). In our example, ethical analysis often black-boxes pharmaceutical research and development, assuming that the primary norms of interest during that stage are the internal norms of science, together with standard norms of industry concerning things like worker health and safety. When some technology pops out, then, the second level of ethical analysis comes into play, and it largely concerns the assessment and use of this technological product. *Each higher level presupposes and supervenes on the lower levels. This post hoc character often takes the form of a temporal sequence: First the scientific research is conducted, then the ethical reflection and oversight takes place.*

5. *The concepts integral to science in no way depend on ethical analysis.* These scientific concepts are governed by the demands of knowing truly and intervening effectively. Ethical analysis will be informed by scientific concepts and analysis, using these to assess the impact of the science and technology. We thus see an asymmetry between science and ethics, an asymmetry that is integral to the post hoc character of ethical analysis. When a new drug is developed, appropriate facts and scientific background must be included in the protocol for the human subjects research. This science informs assessments of cost/benefit ratios.

6. *Irresponsible science arises when any of the preceding conditions are violated.* If, for example, the concepts of a science are informed by the interests and ends of a particular group, it is distorted. Similarly, if a technology is promoted for uses that were not properly assessed, or if commercial interests influence and distort the regulatory process, then we have irresponsible science. Notions of responsible agency thus depend on our ability to isolate and assess the impact of a given activity.

From these six observations, we see that traditional ethical analysis is closely intertwined with a specific view of science and technology. Science comes in discrete episodes, and it involves specific kinds of products. If the product is a factual description or a component of a theory, then we have fundamental science. This is largely governed by internal norms. If the product is a technique or technology, then we can evaluate its impact. As science is discrete, so too is ethics. Ethics places the activity of science and its technological products within a broader context, models its impact, and then advances norms that mitigate harms and optimize benefits. Viewed in this way, ethics is itself a kind of social technology (Ellul, 1964). It provides the techniques for normalizing a protoethical activity. Norms normalize by integrating an activity into a broader social web (Canguilhem, 1991). The kind of modularity and discrete character we have identified is thus an outcome of the ethical activity, as well as its presupposition.[14] We could view the episodic character of the normal science as a kind of constructed modularity. Regulatory oversight ensures that the modularity is sustained.

Critical Reflection on the Traditional Ethic

In the preceding analysis, I sought to make explicit broadly held assumptions regarding what the responsible conduct of science entails. I am hoping that these standards are sufficiently obvious that they do not need to be justified here. They are, for example, reflected in current standards of clinical research (National Commission for the Protection of Human Subjects, 1978). They are also implicit in various criticisms of the pharmaceutical industry and in accounts of evidence-based medicine (see, e.g., Angell, 1997, 2005; Salek & Edgar, 2002). Godin (2006) has documented the history and continued influence of the associated linear model of research and development, with its distinction between pure and applied science.

All this does not mean that a more fine-grained philosophical, historical, or social scientific analysis would sustain the assumptions. In any given instance of drug development, for example, we might find a complex, nonlinear web of interactions between all levels of R&D and the regulatory institutions and policies that govern testing, approval, and use of the drug. In fact, careful analyses of specific instances of pharmaceutical development and regulation document such a complex web, and thus the inadequacy of the linear model (Brody, 1995; Rabinow & Dan-Cohen, 2005). This results in a disconnect between assumptions integral to standards of ethical conduct and the actual processes of research, development, and ethical oversight.

In considering the dissonance between standards and practice, we need to make some careful distinctions. Ideals of practice might be effective and appropriate even when they can never be fully implemented (Khushf, 2004). Assumptions integral to the ideals might serve as useful approximations to contexts of practice, and inform a regulative ideal that serves the intended normative function in aligning the covered activity with other social practices and ends. In fact, for many new pharmaceuticals, medical devices, pesticides, and so on, the standards have effectively performed such a function. We could thus speak of them as "valid" in these domains, even if a careful historical analysis would reveal a far more complex picture of the actual course of research and development in any specific instance.

We can now more carefully specify what occurs when a radically disruptive technology precipitates an ethical crisis. Although the linear model and its associated pure-applied distinction may be problematic in all cases, it nevertheless had a heuristic function and appropriately informed standards of ethical analysis in cases of normal science. In an ethical crisis, this heuristic function breaks down, and the conventional division of labor between scientific and ethical analysis can no longer be sustained. In such contexts, the background tensions between the account of norms, on one hand, and the historical, philosophical, and social analyses, on the other, come forward as problems that must be explicitly addressed. This calls for a fundamental revision of the standards. However, such revision is itself implicated in the crisis. The task of revision occurs at a nexus of description and choice that, by its nature, violates the conventional canons of both scientific and ethical discourse.

At this stage, I will consider only two aspects of this crisis. First, let us assume that in practice the pure-applied distinction is breaking down, and that the activity of science is infused with interests and values. What do we make of this? Does this signify something bad, a distortion that we need to counter? Are we witnessing a kind of decay? Or are we moving to a more honest kind of science, one that owns up to an ideological dimension that was always there? Should we advance greater social oversight, or is there something about the independence of the research activity that we want to protect? If we still affirm in some form the older ideals of science and ethics, then we may wish to reconstruct these ideals in a more appropriate form, one that better maps to the realities of practice. But we will not want to give them up. Alternatively, if we think the older ideal was itself inappropriate, then we might be willing to radically alter the norms, completely rethinking our notions of science and ethics. The approach we take is partly a function of choice, but it also depends on the role we give science in society and what we see as the constraints on human flourishing. An appreciation of the dissonance between our ideals and current realities provokes a broad range of questions and challenges, but leaves us floundering for an answer.

Second, if I start "from above," at level 3, and attempt to rethink something like the post hoc character of ethical reflection, then I seem to pull the rug out from under myself. Rethinking ethics at level 3 presupposes that I have some relatively stable account of science and ethics at levels 1 and 2. In this case, levels 1 and 2 become my level 1, and level 3 becomes level 2. My attempt to rethink these thus presupposes exactly the kind of asymmetry and post hoc character that I now claim to question.

Our notions of ethics and responsible agency are so closely wed to the accounts of science and practical rationality I sketched earlier that, when we critically reflect on them, we do not view this as part of an ethical activity. For this reason, the science of ethics is not usually regarded as part of a normative ethics. We might, for example, try to situate the development of normative reflection in the context of evolutionary theory, or we might find a rich sociological account of some specific technological development. But in these accounts, we lose touch with responsible agency. Instead, the force of the account seems to be a debunking of notions of agency. The subject as agent is lost. And whenever we try to recover a normative ethic, we find that the three levels settle out again, and we implicitly draw on exactly those notions that are problematized in our evolutionary or sociological accounts.

On the Cusp of Volitional Evolution: The Challenge of Radically Disruptive Technologies

Precautionary Framing of Upstream Ethics

A host of developments associated with nanotechnology, biomedicine, information technology, and cognitive science deeply challenges our ideals of science and ethics, and with them, our notions of responsible human agency. The challenge is seen in the very form of ethical reflection that is required. Instead of our traditional down-stream ethics, we see a consensus emerging on the need for so-called upstream ethics; namely, on the need for ethical reflection that takes place at the beginning of the R&D process, and that is closely intertwined with the developing science.[15] Such a consensus is seen in international calls to address ethical issues as a vital part of major programs in areas like genomics, proteomics, metabolomics, or bionanotechnology.

Although it is commonly recognized that ethical reflection is needed upfront, the character of such an ethic is far from clear. In fact, we find an ambivalence of the kind I associate with a radically disruptive technology. For many, upstream ethics entails an attempt to normalize technological development, and thus mitigate its potentially disruptive character. Such an approach is especially apparent in more precautionary approaches to ethics (Jonas, 1984; Fukuyama, 2002; McKibben, 2003). Advocates of such an approach attempt to inhibit technological developments until epistemic and social conditions can be established that approximate what is necessary for conventional science and ethics.

The precautionary character of upstream ethical reflection is nicely shown in E. O. Wilson's writings (1998), and a brief review of his argument helps clarify the kind of challenge posed by recent technological developments.[16] Wilson notes how scientific knowledge has greatly expanded our ability to sustain our lives and alter our environment. However, the more we alter the environment, the more we depend on new knowledge and technology to mitigate the effects of our alterations. This problem is especially apparent in our growing dependence on medicine. By enabling those with disease and disability to survive and reproduce, medicine mutes the stabilizing role of natural selection. Wilson thinks this leads to a short-lived second stage in evolution, one which will rapidly be displaced by a third, more radical stage, when we can directly reengineer the human genome so any defects are permanently repaired. Thus, on the near-term horizon, "the rules under which evolution can occur are about to change dramatically and fundamentally" (p. 299). Wilson refers to this third stage as *volitional evolution*:

Homo sapiens, the first truly free species, is about to decommission natural selection, the force that made us. There is no genetic destiny outside our free will, no lodestar provided by which we can set course. Evolution, including genetic progress in human nature and human capacity, will be from now on increasingly the domain of science and technology tempered by ethics and political choice. . . . Soon we must look deep within ourselves and decide what we wish to become. Our childhood having ended, we will hear the true voice of Mephistopheles. (p. 303)

Wilson counsels a conservative approach to such hereditary change, and he justifies such an approach by appealing to evolution. He argues that a biologically informed, naturalistic account of *Homo sapiens* recognizes human fragility and the delicate character of any ecological balance. Like all animals, humans gravitate to those environments in which our own epigenetic rules emerged. Those who wish to alter these too radically, who see humans as *Homo proteus*, might in some sense make themselves better, but they would no longer be human. In this radical step, they would defy the knowledge that arises from an appropriate biological regard for our own existence.

Wilson's account is especially instructive because he does not generally espouse a traditional ethic. When he shifts from scientific to normative reflection, however, we see traditional notions of responsible agency. The fundamental question of free agency is understood in terms of the possibility of a brute, arbitrary choice: a fundamental remaking of ourselves. To do this responsibly, we would need to understand how human life self-assembled. Wilson thinks we should pursue such knowledge, and he expresses confidence that we will eventually attain a reductionist account. Until we do so, however, any deep technological alteration of human life will be irresponsible, and we should advance an "existential conservativism."

To the extent that we depend on prosthetic devices to keep ourselves and the biosphere alive, we will render everything fragile. To the extent that we banish the rest of life, we will impoverish our own species for all time. And if we should surrender our genetic nature to machine-aided ratiocination, and our ethics and art and our very meaning to a habit of careless discursion in the name of progress, imagining ourselves godlike and absolved from our ancient heritage, we will become nothing. (p. 326)

With these arguments, Wilson reaffirms the traditional distinction between pure and applied science. In the technological alteration of human nature and our environment, we find artificial processes that are of a different kind than normal evolutionary processes. Volitional evolution involves, for him, a change in the rules—a new, radically different phase of evolution. When Wilson presents the challenge in this way, and when he advances his precautionary, conservative ethic, he reaffirms traditional accounts of responsible human agency. Ironically, to do this, he needs to deny the possibility of a genuinely free choice when it arises, seeing this as a negation of ethics and of science.

Into the Abyss: The Depth of the Challenge Associated with a Progressive Stance
I think it is possible to frame an alternative to Wilson's precautionary approach. A genuine alternative would need to provide guidance for the radical technological developments Wilson contemplates. It would need to account for a technology that in some sense runs ahead of the science, and address the entanglement of science and ethics that arises.

Wilson's account provides a valuable clue for how the alternative might proceed. For him, volitional evolution involved a radical break with the evolutionary processes that brought us to this stage. The next stage was to be a function of "science and technology tempered by ethics and political choice." This was viewed as discontinuous with natural selection, which was now "decommissioned." Also, for Wilson, free choice was discontinuous with science and evolutionary process. Thus, if we wish to advance an alternative, we need to place technological change and free agency in the context of evolution, rather than view these as radically discontinuous. To do this, we need a richer account of evolution than Wilson provides, one that can integrate natural selection, self-assembly, and developmental processes, and one that is broad enough to encompass biological and cultural/technological change (Oyama, 2000; Gould, 2002; Harms, 2004). Since we do not have a reductionist account of the self-assembly responsible for our own emergence, the needed account must also provide some alternative to a reductionist science, one that can do justice to higher-level dynamics without needing to decompose them into their lowest-level components and local interactions.

Beyond this richer account of evolution and development, the needed alternative must also be in some sense continuous with antecedent notions of ethics, practical rationality, and responsible human agency. Although these need to be fundamentally rethought, we cannot jettison them and begin *de novo*. We thus need a science of ethics that involves a critical but constructive development of the kind of ethic we discussed in the first part of this chapter.

Finally, our needed alternative cannot just be worked out "from above." It cannot take the form of a Hegelian science of science. Instead, we need to consider examples of the emergent science, and bring reflection on this concretely developing science into relation with more general and abstract reflection. This will involve integration of the self-assembling science "from below" and the critical reflection on science and ethics that arises "from above." This would bring together top-down and bottom-up processes of normalization.

To sketch this alternative, I now turn to an example in artificial life research. Many other examples would equally well illustrate my argument.

The Design of Artificial Life

Artificial life is a vibrant, new field that uses a synthetic method to investigate living phenomena. In a prominent overview, the central goal is said to involve development of "a coherent *theory* of life in all its manifestations" (Bedau et al., 2000, p. 363; my emphasis). This makes artificial life a kind of theoretical biology, "foremost a scientific rather than an engineering endeavor" (p. 364). Though this definition captures something important, I think it is also deficient. Rodney Brooks (2001, p. 409) is closer to the mark when he highlights the unconventional, hybrid character of this science, defining it as a "mixture of science and engineering." In fact, we might take it as a quintessential example of technoscience. It is concerned with fundamental questions. But in place of more conventional tools of hypothesis formation and testing, it uses a synthetic method to evoke living forms in various material substrata. Models are simultaneously technological products. Knowing and making are thus deeply intertwined, and with this entanglement the pure versus applied (or science versus engineering) distinction breaks down.

Mark Bedau (2003) provides a helpful taxonomy of artificial life research. He identifies three basic kinds, depending on the material substrata researchers use. (1) "Soft" A-life involves computer simulations; living patterns are evoked in the space of computer memory. (2) "Hard" A-life seeks to implement living patterns in various kinds of hardware. It includes things like reconfigurable hardware, robotics, and various swarm-sensing systems. (3) "Wet" A-life seeks to synthesize life from inanimate, biochemical components. This includes top-down strategies that seek to reengineer current living systems or parse DNA to the minimum necessary to sustain its life, as well as bottom-up protocell research. Bedau sees these three domains as relatively independent, each with its own criteria and research agendas, but he also recognizes that a fully satisfactory account of life might require some kind of deep integration of all these types. In "bridging the gap between non-living and living matter . . . we can expect 'wet' artificial life to become inseparably intertwined with 'soft' and 'hard' artificial life . . ." (Bedau, 2003, p. 511).

When we combine the technoscientific character of artificial life with this entanglement of problem sets and research domains (and the associated inability to modularize specific activities and discretely reflect on them), then we get exactly the kind of context that undermines conventional modes of post hoc ethical reflection. However, it is tricky to properly exhibit the entanglement in any case description: Attempts at representing the phenomenon also involve parsing and subsequent expression in piecemeal, serial form, and these conditions are undermined by the phenomenon being represented. We are thus faced with the challenge of representing the conditions under which representation itself fails.

To exhibit the interconnection of artificial life domains, I start with an influential assumption in early, soft A-life, namely, that life is form, involving patterns that can be abstracted from any given material substratum. Research associated with this assumption provoked criticisms based on a perceived "reality gap" between computer simulations and the kind of "wet" Earth systems traditionally studied by biology. One can then view wet A-life as an answer to these criticisms: By evoking emergent patterns in a biochemical medium, the simulation becomes a demonstration of the conditions necessary for wet systems typical of actual life on Earth to emerge from nonliving components. After considering a specific case of wet A-life research, I then return to the ethical issues and show why conventional modes of ethical reflection are no longer possible.

"Life as Form" and the "Reality Gap" in Early Artificial Life Research

Claus Emmeche (1994, p. 18) nicely states a prominent assumption in much A-life research:

All life is form. Neither actual nor possible life is determined by the matter of which it is constructed. Life is process, and it is the form of this process, not the matter, that is the essence of life. One can therefore ignore the material and instead abstract from it the logic that governs the process, taking it out of the concrete material form of the life we know. Hence, one can thus achieve the same logic in another material "clothing" or substratum. Life is fundamentally independent of the medium.

On the basis of this conviction, researchers can study life by means of a synthetic method that evokes living forms in the space of computer memory. Computer simulations are, according to this view, not just inauthentic, partial representations. The complex patterns associated with, for example, cellular automata are genuine manifestations of life. These simulations show how complex, lifelike patterns such as bird flocking or fish schooling behavior can emerge from simple, lower-level rules. Such a bottom-up construction of complex phenomena is believed to be the key to the emergence and development of all living forms (Camazine, Deneubourg, Franks, & Sneyd, 2001).

Some of the novel features and problems of artificial life can be seen in attempts to engineer complex patterns by using bottom-up self-assembly. In such research, one rarely understands exactly how the higher-level phenomenon arises from simple, lower-level states and rules. Instead, one experiments with different media and rules, and then observes certain patterns that spontaneously emerge. "Emergence" refers to higher-level dynamics that arise from simpler, lower-level elements and local rules of interaction. For many artificial life researchers, there is a sense that many of the most important, higher-level patterns will never be reduced. This does not necessarily mean that they do not arise from lower levels. (Such "strong emergence" is generally rejected, since it conflicts with assumptions about bottom-up self-assembly.) Rather, in many cases, the complexity is so great that we could never model the emergent pattern; the simulation would require a complexity comparable to that of the actual system in which the pattern emerged (Bennett, 1986; Chaitin, 2005). This recognition poses a fundamental challenge to traditional reductionistic science: If you cannot reduce the higher-level dynamic, how can you be sure that any purported explanation accounts for the full range of factors integral to the phenomenon? Artificial life simulations can be taken as one way of answering this (for an extensive discussion and alternative answers, see Camazine et al., 2001). By replicating lifelike phenomena in alternative media, researchers demonstrate that the observed complexity can be accounted for in terms of specific kinds of elements, even if they can never completely work out the exact pathways.

Many artificial life researchers believe that the "autonomy" and "spontaneity" associated with wet, biological Earth systems (and the wonder evoked in the face of this spontaneity) is exactly the same in kind as the spontaneity of emergent patterns found in soft artificial life simulations (Emmeche, 1994, ch. 2). Thus, by means of these simulations, we gain an understanding and control over living systems that we previously did not have (Bedau, 2003). But this is a new kind of understanding and control, one that does not involve the complete transparency and mastery we associate with traditional reductionistic explanations. One "controls" by providing rules for an artificial universe. (As Gould, 2002, p. 925, notes, "the results depend crucially on the human mental protoplasm that sets the particular rules and idiosyncrasies of the virtual system.") One then iteratively refines and tunes initial conditions and rules. But one does not directly construct the emergent patterns. These are novel, arise indirectly, and could not be anticipated in advance. This notion of control is more akin to the management of an estate or an ecosystem than the traditional top-down engineering design. It is in this emergent notion of control that we can find the kernel of an alternative notion of responsible agency.

Some obvious objections immediately arise in response to these claims about the equivalence of soft and wet A-life (Emmeche, 1994, ch. 6; Gould, 2002, pp. 922–928). There seems to be a double "reality gap" between simulations of evolution and self-assembly, on one hand, and actual biological processes, on the other. First, even the simplest natural systems cannot be simulated by *ab initio* methods (Laughlin, 2005). If it is to be useful, any simulation of a natural system requires all sorts of ad hoc adjustments. This divergence seems to call into question the degree to which artificial simulations appreciate the principles integral to natural system dynamics. Second, and perhaps most problematic, there seems to be something different about the messy, actual systems with their complex combinations of physics, chemistry, and so on, on one hand, and the rarefied computational world integral to the computer simulations, on the other. Something about the coupling of the analog and digital in actual systems seems lost in the artificial life simulations.

Design of a Minimal, Self-Replicating System

One can view some recent developments in wet artificial life research as a way of addressing this reality gap. If one could use techniques similar to those found in soft artificial life research to evoke living forms in biochemical media, artificial life would be brought much closer to actual living systems (for a review of these projects, see Emmeche, 1994; Szostak, Bartel, & Luisi, 2001; Pohorille & Deamer, 2002; Noireaux & Libchaber, 2004). Such simulations would be creations, not inauthentic imitations. The products of these simulations could then be used to address fundamental theoretical questions about the origin and development of life.

Not all wet artificial life research has the same potential for addressing this reality gap. There are many reasons for conducting wet A-life research. If one, instead, focuses on engineering or medical goals, different features of wet artificial life research might become important. Consider, for example, the top-down design strategy integral to Craig Venter's work. He starts with a simple bacterium, *Mycoplasma genitalium*, and then identifies the minimal genome necessary for the proper functioning of its living processes. By reverse engineering such a system, he hopes to isolate the essential components and processes of life, and then use these for various technological projects. As Szostak and colleagues note (2001, p. 387), this top-down approach is less helpful for addressing fundamental questions about origins: "Any 'stripping-down' of a present-day bacterium to its minimum essential components still leaves hundreds of genes and thousands of different proteins and other molecules. We must look to simpler systems if we hope either to synthesize a cell de novo or understand the origin of life on Earth."

If we wish to address origin-of-life questions by synthesizing life, our technological product (the protocell) would need to be a special kind of creation: It would need to re-present the spontaneous emergence of life from a nonliving substrate. Here we get a new kind of representation that is commensurate with the new kind of control. Instead of directly representing the phenomenon by means of a conventional model, one reduplicates conditions that cultivate the phenomenon of interest. Representation now has a double aspect, and it is indirect. First, one directly re-presents conditions that mimic those in which life first emerged. This first level (representing the conditions) can occur in different media; for example, computer memory, hardware, or a biochemical medium. For wet artificial life, one selects biochemical media. Second, the patterns of interest arise from these conditions. The result is then an actual technological product: a minimal, self-replicating system that can evolve. But this technological product can be taken as the instrument of a new kind of theoretical biology, providing the needed confirmation of hypotheses regarding the way life on Earth first emerged.

A Genetic Algorithm for Optimizing Self-Assembly of a Protocell Membrane
Even if we narrow our focus to bottom-up wet artificial life research, there are still different design strategies, all entangled with other kinds of research in a host of complex ways. Consider, for example, the Programmable Artificial Cell Evolution (PACE) project, an ambitious U.S. and European Union effort aimed at designing a protocell. The U.S. team, based at Los Alamos, is using a more direct rational design strategy that attempts to encapsulate an artificial analog of DNA (lipophilic peptide nucleic acid, or PNA) within a self-assembling membrane (a micelle) and drive its metabolism by an artificial photosynthetic reaction (Rasmussen et al., 2004). European research includes groups in Germany, Spain, and Italy that are also using more indirect design strategies, including artificial evolution. To illustrate the kind of research being conducted, I will now focus on some work being done by a Venice-based team on the self-assembly of a membrane that could encapsulate the PNA.

Natural membranes can be viewed as a complex kind of soap bubble. They consist of amphiphiles, which are chemical compounds that have both a polar, hydrophilic component and a nonpolar, hydrophobic component. Within an aqueous solution, the amphiphiles organize themselves into a double layer, with the polar end facing outward and the hydrophobic end facing inward. By combining different kinds of amphiphiles in different quantities, their self-assembly can be optimized to yield membranes with desired properties. The self-assembly process associated with such membrane formation is highly complex, depending on multiple chemical and physical properties. As the membrane forms, further development also depends on interaction between the supramolecular structure and its environment. At present, scientists are not able to fully understand and control this self-assembly. Much research proceeds by trial and error.

To optimize the recipe for the formation of such a membrane, the Venice-based team used artificial evolution. In the 1960s, John Holland (1975) developed a design technique referred to as a genetic algorithm (GA). In its original form, the algorithm operated on subroutines. Each subroutine is given a task, such as the solution of a traveling salesman problem. A criterion is put forward to assess the relative fitness of the solutions. The most fit subroutines are then selected and some rule is used to generate offspring from these programs. These now solve the same problem and their results are similarly assessed. By iterating this process, an evolution-like search procedure can be used to discover an approximate optimum to a problem that is too complex to determine the exact solution.

The novelty of the work by Theis and colleagues (2007) involves using a GA to search for an optimal recipe for a self-assembling membrane. In doing this, they overcome the reality gap associated with the artificial simulations. More significantly, they directly couple evolution and self-assembly, two key ingredients responsible for the emergence of natural living forms. The intractability of the problems of self-assembly are overcome by using the process nature used. This mimics evolution; the natural becomes a technique of artifice (I will return to these issues of biomimesis later). To accomplish their task, Theis et al. used a library of sixteen amphiphiles. A recipe consisted of five volume units from this library. The possibility space involved 15,504 possible recipes. Recipes were mixed in a ninety-six-well clear plastic plate, and a spectrophotometer was used to read the turbidity of each sample. Turbidity provided a rough measure of the fitness of the self-assembled structure. Thirty recipes were tested in each generation (with three samples of each recipe). Initially, recipes were randomly determined. The twelve most fit samples were then selected and a given algorithm was used to generate child recipes. The process was then iterated over four generations, and the fitness distribution of the GA-selected samples was compared to that of a random sample. The results clearly documented the effectiveness of the GA in optimizing the recipe.

Preliminary Assessment of the Theis et al. Research

Initially the research by Theis and colleagues seems relatively modest in scope: An established algorithm is used to optimize recipes for a product in supramolecular chemistry. However, further reflection makes clear that their work does not fit easily within traditional schemas. In fact, when this research is placed in its broadest context, its radically disruptive character becomes more and more apparent. Consider the following features of this experiment:

1. *The goals of this research are ambiguous, and they are not fully transparent in the publication.* With the exception of an acknowledgment of a grant at the end of the essay, the link to the larger PACE project is not discussed or referenced within the Theis et al. publication, and relatively little is said about the relevance of self-assembling amphiphilic systems for the formation of artificial or natural membranes. There is also little about how this research fits within the larger artificial life agenda, although reference to the reality gap and a concluding comment about the transition in the study of complex systems to practical methods makes clear that this is a significant contribution to the artificial life debate. Even if we focus on the more proximate goals, it is not clear whether this research is about the self-assembly of amphiphilic systems (an important research area in supramolecular chemistry and bionanotechnology) or about the more practical application of the genetic algorithm (the conclusion seems to indicate that amphiphilic systems were simply used to provide a "proof of principle" regarding the practical utility of a GA for optimizing the properties of a real chemical system). In the protocell project and the larger artificial life agenda, these two components are inextricably linked. In fact, the goal is to understand the deep connection between self-assembly and evolution in the emergence of living forms. However, in the research publication, the GA functions primarily as an optimization technique, one that only accidentally draws on evolutionary principles. When we reflect on this research, as I do here, its breadth and complexity become apparent. Those within the artificial life communities will appreciate the full range of concerns. However, much of this will be opaque to other researchers. An ethicist who comes from the outside and looks at this research publication (or others in the protocell project) would completely miss all of the important issues.[17]

2. *The research cannot be discretely regarded.* For the reasons already indicated, we cannot simply isolate this one research project and situate it within a larger social context, as traditional ethical reflection would require. Some might try to solve this problem by expanding the focus to include the protocell project, instead of just focusing on the self-assembly/GA project. But the protocell project is just as ambiguous and multivalent. In multiple, complex ways, it overlaps and interfaces with projects in supramolecular chemistry, bionanotechnology, embodied cognition, computer science, and other fields. The protocell project attempts to address one of the "grand challenges" posed by the larger artificial life community (i.e., the first one, associated with "constructing a life form in the laboratory from scratch;" Bedau et al., 2000), so one would need to expand the focus still further to include the larger artificial life agenda. But the grand challenge is also oriented toward answering a fundamental question in biology—namely, how life emerged from nonlife—so we could expand the focus to include the biological sciences more

generally. Each expansion of focus loses some features as it gains others. This complex nesting and overlap of activities and goals characterizes all science, but in more traditional science (such as that associated with the development of a new pharmaceutical) we have found ways of modularizing the activities and products so they are amenable to piecemeal ethical and policy analysis. That is what is lacking here, and it motivates the upstream ethical engagement. However, the very features that demand early ethical reflection also seem to undermine the conditions for it.

3. *The research is both fundamental and applied.* It leads to technological products; for example, a new optimization technique or strategies for self-assembly of membranes that might be used for new pharmaceuticals. Theis et al. list fundamental scientific contributions associated with the understanding of cell membranes together with technical applications that include "drug delivery, medical imaging contrast agents, cosmetics, and food processing" (p. 2). The larger protocell project involves a fairly radical technological intervention within the world; namely, the creation of a new, non-DNA- or RNA-based life form.[18] But this activity is clearly oriented toward answering fundamental scientific questions, and these scientific questions primarily motivate many of the researchers. Here, knowing and making are deeply intertwined, and it would be impossible to tease out and separate potential technological products from the theoretical ones. The creation of a minimal, self-replicating system that can evolve is the means by which these researchers seek to answer the fundamental biological questions about the origin and character of life.

4. *This research is part of a highly novel and radical technocratic intervention within the world, but the "intervention" is also a "development" that takes place apart from any specific, intentional research endeavor.* We can reflect on protocell research in two ways. First, we can consider it as a contingent, intentional endeavor. We can then ask about the goals, how these might be realized, and so on. However, we can also attempt to situate this research in a broader social and historical context, and ask whether certain essential features of the project arise naturally from the kinds of historical processes that are operative. From this second perspective, we can recognize a broader process of technological evolution, and observe that artificial life research is similar to other kinds of research in areas like synthetic biology, bionanotechnology, artificial intelligence, and cybernetics.

Something about the current state of technological development evokes these projects, although each takes a slightly different form and raises its own unique issues. This leads us to ask whether and how these new forms of research might reflect broader and deeper processes of social and technological development and evolution. Here, the emergence of artificial life is as much a happening as it is a research project. Within technological evolution, we have a further evolution of evolvability, enabling generation of new kinds of novelty on which selection can operate. What we are calling "artificial" life can then be situated in the context of this "natural" process, and regarded as an example of novelty that now might arise from it. We need to consider how this broader technological development might take more or less appropriate form. To address this ethically, language needs to shift from top-down "control" (which we do not have) to a more nuanced account, one that seeks to appropriately manage an evolutionary process already underway (Norton, 2005). The actors are situated within the process; they do not stand above it, looking down. Researchers are agents involved in intentional action, but they and their endeavors are also products of broader processes. Ethics thus needs to be interstitial, advanced from a nested perspective, at a "middle place" that reflects the true place of human agency.

5. *Any attempt to apply a precautionary principle does not make sense, because we lack the conditions for modularizing the research in question.* Consider the protocell project. This involves exactly the kind of technological intervention that advocates of a precautionary principle are most worried about. All the concerns raised about genetically modified organisms are intensified. We do not know what risks such an artificial organism might pose. Could it propagate in a natural environment? What concerns with containment might arise? How might this knowledge be misused? Though it might be possible to discourage funding for the current PACE research endeavor, that would not in any way eliminate the project. It would only make the endeavor less transparent, and thus less open to reflection and refinement.

The protocell project is not one project, but many, and each of its subcomponents is part of broad, mainstream science. The genetic algorithm for optimizing self-assembly of amphiphilic systems will be funded under programs for nanotechnology or information science, and they will still be developed by artificial life researchers interested in creating minimal self-replicating systems. The artificial photosynthetic process will be developed by people working on alternative fuels. And the more these various streams of research advance, the easier they link up with each other in novel ways. Integrated research projects seem to naturally self-assemble out of the current broader research context. Advocates of a rear-guard precautionary stance might try to pop these bubbles as they arise, but fairly soon there will be so many of them that they would need to halt all science to succeed.

6. *When we attempt to ethically situate the research, we find a strange blurring of scientific and ethical reflection.* It is virtually impossible to distinguish internal norms from external norms that arise by situating this research in a broader social context. In the traditional downstream ethics, internal norms could be taken as relatively stable; this enabled the demarcation of the research endeavor, with the subsequent situation of that within a broader social context.

In contrast, the internal and external norms associated with emerging sciences such as artificial life or bionanotechnology are all continually in flux. To socially situate the research by Theis et al., we need to do what I have been trying to do in this chapter, namely, see how it fits within protocell and artificial life research, understand how this is part of a broader trend in the development of science and technology, consider possible implications for our systems of regulation and oversight, and so on. We cannot even make sense of the ethical issues until we do this. But at what stage do I move from internal to external norms? All attempts are deeply entangled within the webs of the evolving research endeavors. Questions at levels 1, 2, and 3 are intertwined; the levels do not seem to settle out. The "ethics work" might lead to provisional assessments and proposals, but these will then feed back to those in the research community, be refined by them, and perhaps inform some developments, slightly altering the goals or leading to constraints that might then condition the design of some subcomponent of some project. This ethics work will depend on the expertise of the researchers who enable outsiders to understand these developments, and thus must proceed as a collaborative endeavor, rather than as one of external regulatory oversight.

7. *The techniques integral to the research can be generalized as strategies for integrating this research with other components of the protocell project, and for advancing alternative forms of communal interaction.* Traditional ethical reflection is informed by traditional scientific concepts. Reductionist science informed a specific ideal of ethical oversight and control. As the meaning of scientific understanding and control shifts, we see alternative patterns of social understanding and control emerge, eventually leading to the possibility of alternative forms of practical rationality. Thus, for example, the Theis et al. genetic algorithm might be extended as a technique for optimizing the self-assembly of research teams that work on components of artificial life projects.

In fact, some scientists are already advancing such sociopolitical generalizations of their insights. In one especially interesting example, leaders in the artificial life community asked how they might build consensus and clarify conflicts among other members. They observed that "traditional science and engineering approach the problem of predicting, controlling, and optimizing the behavior of a system through a premeditated linear method" that considers "how the system's behavior depends on the interconnected functioning of its components and their connections with the environment" and then uses a "model to figure out how to control those aspects of the system that bring about predefined desired goals" (Rasmussen, Raven, Keating, & Bedau, 2003, p. 209). However, some systems—such as the system of artificial life researchers—are far too complex for this kind of modeling and control. They thus needed another strategy to address the challenges of communal development:

One source of inspiration for solving this problem is to take a page from nature and ask how complex distributed biological systems, such as immune systems or evolving populations, achieve self-regulation and continual adaptation using a completely different method, one that involves no premeditation or central control. This distributed control in such systems combines emergent dynamics with the existence at all times of a broad diversity of potential solutions. The system's overall dynamics yield the selection of candidate solutions. If no adequate solution is found within the current diversity of current candidates, the system fails and "dies." If a solution emerges, then the system dynamics will further optimize it, not by sacrificing the diversity of solutions but by changing the relative dominance of their subspecies. Herein lies the thrust of our approach to social decision-making and control. (Rasmussen et al., 2003, pp. 209–210)

Using this general insight, they then designed a system for using the Internet to clarify the core accomplishments and challenges the artificial life community faces, and also to address ethical and regulatory concerns (such as the formation of a formal professional society, and editorship of a journal). In this example, we see how alternative notions of understanding and control integral to ethical reflection might emerge out of scientific developments.

8. *Ethical reflection would have to arise in the context of the iterative adjustments made by researchers as they continually revise their goals and local design strategies in response to unfolding successes and failures of provisional projects.* Since the meanings of success and failure will depend on both the goals (as conceptualized at any stage) and external feedback arising from necessarily partial attempts at situating this project in a broader social context, we find a highly complex, nonlinear dynamic characterizing the development of the research. The emerging scientific understanding of such complex, nonlinear dynamics enables us to understand the character of ethics in a way that is in closer accord with sociohistorical accounts of how norms emerge and develop. However, unlike those descriptive sociohistorical accounts, the science of complex system dynamics enables us to preserve a sense of agency.

A Radical Implication: Emergence of an Alternative Form of Practical Rationality at the Cusp of an Open Evolutionary Process

Reflection on artificial life research has shown us why a new account of ethics is needed. We have also seen how emerging sciences (e.g., complex systems science) might inform such an ethic. But in these deliberations, we still have the traditional asymmetry between science and ethics. One of the more radical implications of disruptive technological developments is that even this asymmetry is altered. Science and normative reflection are deeply intertwined, meaning that normative reflection can and ought to inform the science, just as the science informs the ethics. In these closing reflections, I provide one rather speculative way in which ethics might do this.

Let us start by imagining that researchers succeed in creating a PNA-based self-replicating system that can evolve, and that they used artificial evolution and self-assembly to indirectly design this life form. Once this happens, we can ask whether and how this life form fits within the larger history of the emergence and evolutionary development of life on Earth. Was this act somehow different, discontinuous with the processes that have led to other kinds of life, as E. O. Wilson suggests? Or is it continuous with the broader development?

We can ask about this continuity of life in both a scientific and ethical way. Scientifically, we can consider whether and how the new life form fits within the broader evolutionary history of life on Earth. The PNA life form would be unique. Unlike other DNA-based life forms, it would be designed by humans. However, because the humans will have used techniques of self-assembly and evolution that directly mimic the broader biological process, these processes will be equivalent in form, although the physical and chemical substrate will be to some extent novel. Further, human technological development also follows evolutionary principles of development, and the emergence of research projects that seek to create life seems to be a natural outgrowth of our current stage of development. We could thus view the new, "artificial" life as an example of the kind of novelty that now arises as a result of an ongoing process of evolving evolvability. The new life form is thus both continuous with current life (as the product of an ongoing evolutionary process on Earth) and discontinuous (as the product of human design). To fully explain the emergence of this novel life form, we need to view the event of protocell emergence as a specific, recent transition in evolutionary development. This explanation complements an alternative, engineering explanation that simply recounts the techniques humans used in designing it.

In the scientific, descriptive account, the question of continuity concerns the fit with *past* evolutionary development. We view the new life as the cusp of an open evolutionary process, and then seek to understand the emergence of this novelty. If we now shift from scientific to ethical analysis, the focus turns to the *future*. Instead of ending with novelty, we begin with it and then ask whether ongoing life processes will continuously unfold or whether they might be undermined or radically disrupted. Can the PNA-based life form replicate outside of the laboratory? Could it put at risk existing life by competing for resources? If not, could it be modified by other people who are less responsible? From an ethical perspective, the question of continuity concerns the trajectory of life. We take the emergence of novelty as the preethical given, and ask how this might be harmonized with all other life.

Now let us shift back again and ask how we should view protocell research prior to the creation of an actual protocell. In this case, we are not describing an existing stage in evolution. Instead, by our efforts we seek to initiate such a stage. In thus reflecting on our own activity, the process of evolution is, as it were, self-aware. Choices about how to advance this project regard the future character of the macroevolutionary process, of which we are a part. This means the process is not fully determined; it is still open, and our current reflection and practices help specify how it will turn out. When we look ahead and contemplate the emergence of new life, the scientific and ethical features that relate this contemplated life to existing life processes do not yet settle out as distinct.

We can now consider how, within our own awareness, this transition we create takes form. And here is the key, speculative insight: If we currently bear witness to a transition that is in some deep sense similar to those transitions that have previously taken place (i.e., equivalent in form), then our reflection on our current project of creation gives a kind of privileged access to the inner workings of this historical process. We thus have clues that might help us answer some of the more pressing scientific challenges. Normally we can only model from the outside. But now questions of human agency (as seen by the actors "from within") provide access to broader evolutionary processes of development (as seen by scientists who study this "from without").

When we regard our current activity in this way, I think a host of interesting questions and observations arises, not just for ethics, but also for a broader science of the self-assembly, development, and evolution of life. Here I am only going to mention two of these promising lines of investigation.

The first concerns the relation between bottom-up and top-down processes of self-assembly. In the Theis et al. paper, bottom-up self-assembly is combined with a kind of artificial evolution, which optimizes the fitness of recipes for membrane formation. The research is framed to imply that the "outcome" is obtained by bottom-up methods. But that conclusion requires that we ignore how the researchers established the conditions under which self-assembly takes place.[19] If we bring into view the researcher's activity, it becomes clear that both top-down and bottom-up processes are involved. Initially, the researchers configure a design process that involves a specific kind of possibility space, and launch a search algorithm that optimizes the fitness of recipes within this space. The initial construction of this design process can be viewed as a kind of developmental differentiation of an initially amorphous process for membrane design.

If we now focus on the full range of protocell research activities, we find that the researchers are involved in a complex, iterative tuning activity, by which broader theoretical understandings of self-assembly and evolution are instantiated within a specific research design. These broader understandings have arisen by extensive study of natural systems, and this knowledge emerged over a long period of time and assimilates interaction with a broad array of systems. We could thus view the humans involved as a kind of resonant system that transduces macro-level processes within a microprocess. This is a kind of differentiation. Successes and failures in the different attempts to replicate natural design, then, lead to iterative adjustments that inform future attempts. Knowledge gained enables better understanding of the macro-level processes, and this leads to new attempts at design. If protocell researchers eventually succeed, when we consider this evolutionary transition, we will need to explicitly consider the complex, iterative resonance between the macro-

and microprocesses, and we will need to discuss how the human agents enabled this replication of the broader natural process within a new chemical substrate. We will need to consider not just specific techniques, such as the refining of the genetic algorithm, but also how these techniques were generalized in communal structures that lead to novel and more effective ways of coordinating and integrating research components. And, just as we moved from microtechniques (such as the modified GA) to new social institutions, so also we moved from novel social formations (as worked out in new Internet-based systems for mediating knowledge) to new micro-level techniques. Once we appreciate how broader techniques of social interaction resonate with techniques of scientific research, we have a matrix for transducing information in both directions.

These observations suggest that weak notions of emergence (associated with bottom-up self-assembly) might be insufficient for both the design of new life and the explanation of how existing life emerged. Strong emergence, involving top-down causal paths and a resonance between macro- and micro-level processes, might be needed to understand the self-assembly integral to the emergence of living forms. This might also be needed to better conceptualize the character of research that seeks to reduplicate the natural processes leading to such form. Guided by such an observation, attention might turn to possible natural systems that could provide such a resonant coupling of higher- and lower-level dynamics.

The second promising line of investigation suggested by our speculation concerns the meaning of responsible human agency. Humans, by their agency, modulate and tune the top-down and bottom-up processes in artificial life research. If they do this just right, something happens. As with earlier artificial life computer simulations, there is clearly a kind of control. But it is a new kind of control, in which conditions are established to foster novel, spontaneously emerging forms that transcend all accounts of the mechanisms involved. To understand this kind of agency, we need to see it as always nested, always interim, in the middle, and on the way.[20]

In place of older notions of control that assumed full transparency, we now must advance a newer account of responsible management. Ethics also involves a tuning, and such activity cannot be neatly distinguished from the agency of the scientists. In both cases, success depends on a subtle, delicate resonance between a broad array of systems at multiple levels. Scientific and ethical reflection will be attentive to a different range of systems and concerns. In the context of the ongoing activities of research, these different dynamics will need to continually come into alignment with one another. If they get out of alignment, the regulatory systems might inappropriately hinder or distort exactly those scientific developments that would have exhibited the greatest promise for appropriately configuring the next steps of technological development. Or, alternatively, the scientific systems might lead to happenings that are discordant with current living systems, and with this, a degeneration in the subtle balance integral to human flourishing. Here ecological and systems models and metaphors seem more appropriate, but we do not get the precautionary stance so often associated with such models, because we never stand outside the developing systems.

With all this, we see that upstream ethics cannot be viewed as a simple adjunct to science. The new science calls for a whole new kind of ethic, one that involves a new form of responsible agency. Somehow this transformation of agency is intertwined with the next transition in an open evolutionary process. To further specify the content of such responsible agency, we must explicitly take up the scientific projects of design, but in a new, responsible way. In doing this, our science and ethics must become deeply intertwined. Only in this way will upstream ethics take an appropriate form.

Acknowledgments

Research for this chapter was supported by the National Science Foundation award EEC-0646332—Complexity, Systems and Control in Nanobiotechnology: Developing a Framework for Understanding and Managing Uncertainty Associated with Radically Disruptive Technology. The views expressed in this chapter are my own, and do not necessarily reflect those of the National Science Foundation.

Notes

1. In these sections I attempt to summarize a large literature on ethical theory, hopefully clarifying the overlap between recent developments in ethical theory and those in the philosophy of science. For most of this, it is outside the scope of this chapter to sufficiently nuance and reference the full range of background literatures. However, some might find helpful a brief sketch of the kind of literatures I have in mind, if only to clarify my overly ambitious aspirations. The two prominent themes I discuss are standard fare in overviews of Anglo-American ethical theory, with the first strand represented by deontological and the second by consequential and perfectionist traditions. (An equally valuable account might further distinguish utilitarian and more Aristotelian strands, with the Aristotelian traditions representing a more holistic, integrative account, e.g., Hauser, 2006; for an influential review of analytical ethics and its open problems, see Darwall, Gibbard, & Railton, 1992).

Following Ross and more recent principlists such as Beauchamp and Childress, we can see each of these traditions as reflecting prominent features of any ethic, although in a one-sided manner; for example, freedom and autonomy are central for deontologists, whereas consequentialists focus on realizing some good. Through my account of responsible agency, I seek to integrate these strands, drawing on recent philosophical and social scientific research on practical rationality. (Representative work in this area includes Harms, MacIntyre, and Nusbaum.) However, beyond this, I seek to do two additional things. First, I want to show how recent developments in science and technology problematize this notion of responsible agency. Thus, while in one sense the account I give of responsible agency is novel, carrying forward current agendas in ethical theory, it must simultaneously be representative of a whole tradition of ethical reflection, and in this sense is not novel at all. My account of responsible agency must bring to language the implicit constellation of assumptions that informs debates over ethical theory. If it does not do this successfully, then the breakdown of this notion could not reflect a deeper breakdown of traditional approaches to ethics.

The second additional concern in my presentation relates to how this breakdown is historically situated. Two vast literatures consider how recent science and technology challenge our conventional accounts of ethics. One attempts to situate ethics in the context of broader theories of biological evolution and development (Oyama, 2000; Harms, 2004). Another more "continental" strand is concerned with how technologies (including techniques of social organization) alter the human condition, including practical rationality and ethics. (This second strand would include so-called science, technology, and society (STS) research, exemplified, for example, in the work of Gibbons et al. (1994) and Zieman, as well as in broader phenomenological, hermeneutical, and neostructuralist writings, which follow Merleau-Ponty, Gadamer, Foucault, and Jonas, to name just a few.) Unfortunately, there is too little communication between these traditions of reflection; they represent Snow's two cultures, and integrating these strands can be viewed as one of our deepest philosophical challenges. As a step in that direction, I tried to frame the current ethical crisis in a way that is informed by, and speaks to, core problems in each of these different traditions of reflection.

2. Here we might distinguish between a descriptive and normative ethic, or between ethics and a science of ethics (or metaethic). In normative ethics, we presuppose an agent that is faced with a choice, and the task is to provide some guidance. This agent might be an individual, but we might also have communal and political agents; thus, for example, a normative ethic might provide guidance for how a policy should be formulated. If, in describing ethics, we end up with a story about how there are no subjects or agents, and how norms emerge in the context of complex sociohistorical processes, then we lose the vital context of ethics. Such accounts might inform an ethic, but by themselves they do not provide an ethic. An especially prominent example of this is found in the work of Rabinow. He has provided a series of insightful accounts of the forms of life that arise in the context of developing biotechnology. These explore the discrepancy "between scientists' self-representation and the representations of scientists by those who study them." For Rabinow, this social scientific work constitutes an ethic: "While this discrepancy is of little consequence for practicing scientists (most will never have heard of its existence), it provides much of the subject matter and the authority of the social studies of science. Actually, the identification of the discrepancy is itself normative, a practice of producing knowledge. The fact that the self-image of the biosciences is not 'true,' i.e., fully adequate to its own norms, is precisely what makes it an appealing subject for anthropological study. I define practices as norms in context and in process. In that light one can identify contemporary inflections in the process of fashioning forms at least partially adequate to scientific norms understood as practices, with all the hesitations, conflicts, and failures attendant to such efforts" (Rabinow, 1996, p. 14).

Although such accounts provide a deep appreciation for complex, emergent dynamics of a scientific practice, they become normative only when they can provide specific guidance for the re-formation and further development of those practices. To do that, a different kind of discourse is needed, one that specifically takes up the question of agency and addresses how an open future might be determined. If there is no open future, then one description simply goes over into another, and we can reflect on these descriptive accounts (as social scientific descriptions of social scientific description) in the same way they reflected on their lower-level science. We get descriptions of descriptions of descriptions all the way out, but with no horizon for orienting the development. If that is the case, then we have the end of ethics, not a specific kind. The challenge is thus to integrate the social scientific framing of the discrepancy (taken as a point of departure) with a future-oriented activity of adjustment, which brings discordant trajectories into alignment. Later in this chapter, I define upstream ethics in terms of such an iterative, tuning activity.

3. My account of the truth conditions of norms is an attempt to integrate two prominent philosophical strands. The first was well framed by Kitcher (1994) in his response to E. O. Wilson's attempt to work out an evolutionary ethic. Kitcher sees an explanation of ethics as a clarification of the truth condition of ethical norms. An insightful recent attempt to provide such an account, based on the evolutionary design process, is given by William Harms (1999, 2004).

The second prominent strand is perhaps best represented by Canguilhem's *The Normal and the Pathological* (1991). Canguilhem shows how physiological norms are only specified together with pathologies, and then extends this to a broader understanding of "technological normalization." Here, normalization is standardization. "[A] decision to normalize assumes the representation of a possible whole of correlative, complementary or compensatory decisions." "[I]n a social organization, the rules for adjusting the parts into a collective which is more or less clear as to its own final purpose—be the parts individuals, groups or enterprises with a limited objective—are external to the adjusted multiple. Rules must be represented, learned, remembered, applied, while in a living organism the rules for adjusting the parts among themselves are immanent, presented without being represented, acting with neither deliberation nor calculation. Here there is no divergence, no distance, no delay between rule and regulation." "In society the solution to each new problem of information and regulation is sought in, if not obtained by, the creation of organisms or institutions 'parallel' to those whose inadequacy, because of sclerosis and routine, shows up at a given moment. Society must always solve a problem without a solution, that of the convergence of parallel solutions. Faced with this, the living organism establishes itself precisely as the simple realization—if not in all simplicity—of such a convergence. As Leroi-Gourhan writes . . . 'all human evolution converges to place outside of man what in the rest of the animal world corresponds to specific adaptations,' which amounts to saying that the externalization of the organs of technology is a uniquely human phenomenon" (Canguilhem, 1991, pp. 257, 250, 254–255).

4. Here, an action is defined in terms of some type. This reduces the action to a technique or technology, e.g., driving, which is then to be harmonized with other activities similarly defined. We thus presuppose an already determinate, pregiven module. Alternatively, we could begin with a complex system of interacting components, and consider normalization as a process of differentiation that leads to modules that settle out as distinct. In that case, the sparsely defined modules (given as a type) are the outcome of the normalization process, rather than the presupposition of it. In a more complete account, we should consider how normalization involves both of these: There is a harmonizing of tentatively specified modules, which arises from the bottom up, and there is also a differentiation, in which somewhat plastic protomodules are further restricted, so they take on very specific functions.

5. The norms integral to good driving guard against failures, but initially the failure is not a failure to drive. Both responsible and irresponsible driving are instances of driving, and

that is accomplished. The failure thus regards a higher-level action, which situates the episode of driving in the broader context of individual and social projects. But even here a refinement could (and should) be introduced. We could view the failure as both (a) a failure to drive and (b) a failure in which driving disrupts flourishing. The key is to link these two. We then get two notions of driving: (1) the brute activity associated with the know-how; and (2) the more refined activity, which situates the know-how in the social contexts and practices where driving is used, and where it serves broader goals.

Note that driving here has the same ambiguity that we find in analyses of biological functions. In the second sense of driving (we can call it "responsible driving"), the brute, modular activity is integrated with other life processes, so that this possible activity is only "called up" when it can function in harmony with other social activities, and thus when it does not pose a threat to individual and social homeostasis/equilibrium. When an activity becomes normalized, as in "normal science," it becomes modular and is harmonized with a broad range of other activities that are all harmoniously related with a broader system.

6. The classic account of channeling in developmental biology is found in Waddington (1957). Pomper (1985, pp. 34ff) shows how these same concepts apply to psychological development. He summarizes: "[F]ated path and stable flow signify the channeling of development in the presence of exogenous factors—factors external to the genetic material. The exogenous factors, which are introduced over time, affect but do not normally prevent the achievement of normal development and a final state of maturity. The epigenetic principle therefore governs a sequence of stages of normal development" (p. 35). This account applies equally to the development of regulatory mechanisms, although some rethinking of the analog of genes is needed.

7. I have been speaking of a desired activity like driving that is initially taken as a protoethical given. But we could also, following Frankfurt (1971), speak of a protoethical, first-level desire. We can then posit a second-level desire that situates this first-level desire in the context of a broader life project. In both cases, practical rationality involves a kind of critical reflection on an initially given content. This, in fact, provides one way of clarifying the core insight in Kant's notion of universalizability. You initially have some given, presented in the form of a maxim of action (arising from the lower will). You then ask whether this maxim could be a universal law, thereby harmonizing it with other maxims that are similarly extended and situated. Such an interpretation moves beyond the purely formal accounts of Kant's categorical imperative (focusing on logical contradictions like those involved in universalizing a lie), thereby clarifying his notion of practical rationality and also accounting for the full range of concerns Kant addresses in his metaphysics of morals.

8. There is a striking parallel between this problematic ambiguity and that of the "immune self," which similarly must distinguish threats and preserve self. For an outstanding discussion of these issues, see Tauber (1996).

9. Such accidental and contingent factors are a part of any lineage, and they inform any natural history. An epistemic uncertainty corresponds to ontic contingency. This is one reason why adaptation is such a confounded concept, and why an irreducible, speculative aspect will always be associated with the reconstruction of any evolutionary lineage.

10. Ellul (1964) provides a profound but one-sided account of how the logics of scientific description and ethical practice can be intertwined. For example, he considers how techniques

of economic description lead to specific kinds of interventions and planning (pp. 158–178), which, in turn, lead to a redefinition of human nature so it conforms to economic planning (pp. 218–227). Here, the process is unidirectional: The "technological intention" (p. 52) leads to a redefinition of human nature, so that it becomes the kind of life that flourishes by means of economic planning. "Economic man" is a result of economic planning, rather than a purely theoretical assumption of economic analysis. Ellul thinks sociological analysis can present only such a deterministic logic, and thus cannot consider the other side of the dialectic, where individual and social choice regarding human nature is framed ethically, leading to a reassessment of economic techniques and the technological intention (pp. xxvii–xxxiii). This may be true of sociology, although that would simply highlight the deficiency of such analysis, pointing to the need for a richer way of framing the fundamental issues, so that the full texture of the crisis and the possibilities for addressing it come into view.

11. Here I follow Kuhn, who views normal science as "research firmly based upon one or more past scientific achievements, achievements that some particular scientific community acknowledges for a time as supplying the foundation for its further practice" (1996, p. 10). Normal science can be taken as a relatively stable, protoethical activity, although it is also already highly normalized, and thus reflects previous activities that harmonize an earlier core with broader theories and social practices.

12. The Belmont Report (National Commission for the Protection of Human Subjects, 1978) frames the ethical issues integral to human subject research in these terms. An influential ethical theory based on these principles can be found in Beauchamp and Childress (2001).

13. Nice reviews of developmental systems theory can be found in Koestler and Smythies (1969), Weiss (1971), and, for a more recent example, Robert (2004). The basic idea is that living systems are hierarchically organized. Study of such systems can focus on any level. Once a level of the hierarchy is chosen, there will be a scale of time and space appropriate for elucidating the central dynamic of the system at that level. At that scale, certain modules (or *holons*) will settle out as relatively stable, and their interaction will constitute the dynamic of interest. There will also be a higher level, which provides the environment for this dynamic of interest, and a lower level, where the modules can themselves be viewed as systems consisting of even lower-level components. In practice, scientific and ethical analysis enters at any level (in the middle), and finds ways to parse the system into the appropriate subcomponents that can be taken as relatively stable modules. To this extent, a radical reduction is not needed and, in fact, never obtained. Analysis of any level usually can go only so far down before a different logic and different modules are utilized. Similarly, synthesis can proceed up only a level or two before one needs to abandon the relevant "atoms" and select new ones more appropriate to the level of analysis.

14. Here I frame the constructed modularity as the outcome of an ethical activity. However, we can also see it as an outcome of scientific activity (Latour & Woolgar, 1986). The question then is how these two forms of construction are related. Much of this chapter involves the attempt to appropriately frame this question and appreciate how its current form deeply challenges our models of both science and ethics.

15. A nice overview of the reasons for upstream ethics can be found in Wilsdon and Willis (2007). See also Royal Society and Royal Academy of Engineering (2004) for an account of upstream ethics in nanotechnology.

16. In an earlier essay, I considered how Wilson's reductionist view of science breaks down when considering bionanotechnology (Khushf, 2004). Wilson's view of science and his conservative ethic are deeply intertwined.

17. For this project I am indebted to the PACE researchers for allowing me to participate in a week-long workshop at Los Alamos, and especially to Mark Bedau for invaluable feedback.

18. For the sake of this discussion, I assume that the new life form would be PNA-based. I can then distinguish it from current life on Earth by saying it is not DNA/RNA-based. However, the Venice-based work on protocell membranes could also be used with DNA- or RNA-based protocell research. There is no necessary link between it and the PNA protocell project at Los Alamos.

19. There is a parallel between our question about whether self-assembly is bottom-up and the question about whether development is determined by genes. Robert (2004) provides a strong criticism of accounts that one-sidedly emphasize the bottom-up paths. For example, studies of genes generally involve heuristics that hold an environment fixed, and then modify genes (e.g., by knock out), concluding that the knocked out gene then determines some trait. However, this is an artifact of the study design. One could also hold genes fixed and modify the environment, concluding that a given environment determines some trait. In the case of the Theis et al. study, one is tempted to ignore the top-down process that establishes the conditions present in each reaction chamber of the experiment. When this becomes invisible, the process is seen as simply bottom-up. Robert (2004, p. 9) rightly notes that this conclusion arises from a systematic bias: "[C]ontext simplification is biased toward lower explanatory levels, so simplifying the environmental context stems from, and leads to, focusing on simple components of a system. Higher-level components of systems, and higher-level systems, are legislated out of epistemological and methodological existence in favor of lower-level systems and their components. . . . we may be prone . . . to draw unjustified causal inferences; it is remarkably easy to fall into the trap of generating causal stories about genes against a constant environmental background (which itself exists only in the laboratory)."

20. Since the time of the Greeks, people have sought an ethical theory that begins from scratch, with simple axioms, as in Euclid's geometry. But such an ethic does not reflect the human condition. In life, we always start in the middle: We find ourselves in a context, with goals and desires, already governed by norms and by a process for reflecting upon and revising them. "Ethics" then concerns the living activity in which we harmonize a broad range of goals, desires, and preexisting norms, addressing tradeoffs, and making adjustments so that our lives may flourish. We can thus view the new, *interim ethic* as one that better reflects the human condition. It is an ethic that brings us closer to the normative processes that already govern our lives.

References

Allison, H. A. (1997). We can act only under the idea of freedom. *Proceedings of the American Philosophical Association*, 71 (2), 39–50.

Angell, M. (1997). *Science on trial: The clash of medical evidence and the law in the breast implant case.* New York: W. W. Norton & Company.

Angell, M. (2005). *The truth about the drug companies: How they deceive us and what to do about it.* New York: Random House Trade Paperbacks.

Aristotle. (2003). *Nicomachean ethics.* J. A. K. Thomson (Trans.). New York: Penguin Classics.

Beauchamp, T., & Childress, J. (2001). *Principles of biomedical ethics,* 5th ed. New York: Oxford University Press.

Bedau, M. (2003). Artificial life: Organization, adaptation and complexity from the bottom-up. *Trends in Cognitive Sciences, 7* (11), 505–512.

Bedau, M., McCaskill, J., Packard, N. H., Rasmussen, S., Adami, C., Green, D. G., et al. (2000). Open problems in artificial life. *Artificial Life, 6,* 363–376.

Bennett, C. H. (1986). On the nature and origin of complexity in discrete, homogeneous, locally interacting systems. *Foundations of Physics, 16,* 585–592.

Brody, B. (1995). *Ethical issues in drug testing, approval, and pricing: The clot-dissolving drugs.* New York: Oxford University Press.

Brooks, R. (2001). The relationship between matter and life. *Nature, 409,* 409–411.

Camazine, S., Deneubourg, J.-L., Franks, N. R., & Sneyd, J. (2001). *Self-organization in biological systems.* Princeton, NJ: Princeton University Press.

Canguilhem, G. (1991). *The normal and the pathological.* New York: Zone Books.

Chaitin, G. (2005). *Meta math: The quest for omega.* New York: Pantheon Books.

Darwall, S., Gibbard, A., & Railton, P. (1992). Toward Fin de Siecle ethics: Some trends. *Philosophical Review, 101* (1), 115–189.

Dennett, D. (2003). *Freedom evolves.* New York: Viking.

Eisenstein, E. (1980). *The printing press as an agent of change.* Cambridge: Cambridge University Press.

Ellul, J. (1964). *The technological society.* New York: Vintage Books.

Emmeche, C. (1994). *The garden in the machine: The emerging science of artificial life.* Princeton, NJ: Princeton University Press.

Foot, P. (2001). *Natural goodness.* New York: Oxford University Press.

Frankfurt, H. (1971). Freedom of the will and the concept of a person. *Journal of Philosophy, 68* (1), 5–20.

Fukuyama, F. (2002). *Our posthuman future: Consequences of the biotechnology revolution.* New York: Farrar, Straus and Giroux.

Gibbons, M., Limoges, C., Nowotny, H., & Schwartzman, S. (1994). *The new production of knowledge: The dynamics of science and research in contemporary societies.* London: Sage Publications.

Godin, B. (2006). The linear model of innovation: The historical construction of an analytic framework. *Science, Technology & Human Values, 31* (6), 639–667.

Gould, S. J. (2002). *The structure of evolutionary theory.* Cambridge, MA: The Belknap Press of Harvard University Press.

Harms, W. (1999). *The evolution of normative systems*. Available online at: http://billharms .home.comcast.net/Evonorms6.pdf (accessed September 2007).

Harms, W. (2004). *Information and meaning in evolutionary process*. Cambridge: Cambridge University Press.

Hauser, M. D. (2006). *Moral minds: How nature designed our universal sense of right and wrong*. New York: HarperCollins.

Havelock, E. (1963). *A preface to Plato*. Cambridge: Cambridge University Press.

Havelock, E. (1986). *The muse learns to write: Reflections on orality and literacy from antiquity to the present*. New Haven, CT: Yale University Press.

Holland, J. H. (1975). *Adaptation in natural and artificial systems*. Ann Arbor: University of Michigan Press. (For further information, see note 91, p. 183 in Emmeche, 1994).

Jonas, H. (1984). *The imperative of responsibility: In search of an ethics for the technological age*. Chicago: The University of Chicago Press.

Jonsen, A., & Toulmin, S. (1990). *The abuse of casuistry: A history of moral reasoning*. Berkeley: University of California Press.

Kant, I. (1956). *Critique of practical reason*. L. W. Beck (Trans.). Indianapolis: The Bobbs-Merrill Co.

Kant, I. (1985). *Foundations of the metaphysics of morals*. L. W. Beck (Trans.). New York: Macmillan.

Khushf, G. (2004). Systems theory and the ethics of human enhancement: A framework for NBIC convergence. *Annals of the New York Academy of Sciences, 1013*, 124–149.

Kitcher, P. (1994). Four ways of "biologicizing" ethics. In E. Sober (Ed.), *Conceptual issues in evolutionary biology* (pp. 439–450). Cambridge, MA: MIT Press.

Koestler, A., & Smythies, J. R. (1969). *Beyond reductionism: New perspectives in the life sciences*. Boston: Beacon Press.

Kuhn, T. (1996). *The structure of scientific revolutions*, 3rd ed. Chicago: University of Chicago Press.

Latour, B., & Woolgar, S. (1986). *Laboratory life: The construction of scientific facts*. Princeton, NJ: Princeton University Press.

Laughlin, R. B. (2005). *A different universe: Reinventing physics from the bottom down*. New York: Basic Books.

McKibben, B. (2003). *Enough: Staying human in an engineered age*. New York: Times Books.

Mele, A. (1995). *Autonomous agents: From self-control to autonomy*. Oxford: Oxford University Press. (Cited in Dennett, 2003, p. 281.)

Milgram, E. (2005). Practical reason and the structure of actions. Stanford Encyclopedia of Philosophy. Available online at: http://plato.stanford.edu/entries/practical-reason-action/ (accessed September 2007).

National Commission for the Protection of Human Subjects (1978). *The Belmont report: Ethical guidelines for the protection of human subjects in research*. Washington, DC: DHEW Publication.

Noireaux, V., & Libchaber, A. (2004). A vesicle bioreactor as a step toward an artificial cell assembly. *Proceedings of the National Academy of Sciences of the United States of America, 101* (51), 17669–17674.

Norton, B. G. (2005). *Sustainability: A philosophy of adaptive ecosystem management.* Chicago: University of Chicago Press.

Nussbaum, M. (2000). *Women and human development: The capabilities approach.* New York: Cambridge University Press.

Ong, W. (1982). *Orality and literacy.* London: Methuen.

Oyama, S. (2000). *The ontogeny of information: Developmental systems and evolution,* 2nd ed. Durham, NC: Duke University Press.

Paul, E. F., Miller, F. D., & Paul, J. (1999). *Human flourishing.* New York: Cambridge University Press.

Pohorille, A., & Deamer, D. (2002). Artificial cells: Prospects for biotechnology. *Trends in Biotechnology, 20* (3), 123–128.

Pomper, P. (1985). *The structure of mind in history: Five major figures in psychohistory.* New York: Columbia University Press.

Postman, N. (1993). *Technopoly: The surrender of culture to technology.* New York: Vintage Books.

Rabinow, P. (1996). *Making PCR: A story of biotechnology.* Chicago: University of Chicago Press.

Rabinow, P., & Dan-Cohen, T. (2005). *A machine to make a future: Biotech chronicles.* Princeton. NJ: Princeton University Press.

Rasmussen, S., Chen, L., Deamer, D., Krakauer, D. C., Packard, N. H., Stadler, P. F., & Bedau, M. A. (2004). Transitions from nonliving to living matter. *Science, 303,* 963–965.

Rasmussen, S., Raven, M. J., Keating, G. N., & Bedau, M. A. (2003). Collective intelligence of the artificial life community on its own successes, failures, and future. *Artificial Life, 9,* 207–235.

Robert, J. S. (2004). *Embryology, epigenesist, and evolution: Taking development seriously.* Cambridge: Cambridge University Press.

Royal Society and Royal Academy of Engineering (2004). *Nanoscience and nanotechnologies: Opportunities and uncertainties.* London: The Royal Society.

Salek, S., & Edgar, A. (Eds.) (2002). *Pharmaceutical ethics.* New York: John Wiley & Sons.

Schoeman, F. (Ed.) (1987). *Responsibility, character, and the emotions.* New York: Cambridge University Press.

Schon, D. A. (1983). *The reflective practitioner: How professionals think in action.* New York: Basic Books.

Sen, A. (1999). *Development as freedom.* New York: Knopf.

Szostak, J. W., Bartel, D. P., & Luisi, P. L. (2001). Synthesizing life. *Nature, 409,* 387–390.

Tauber, A. I. (1996). *The immune self.* New York: Cambridge University Press.

Taylor, C. (1985). *Human agency and language: Philosophical papers.* Cambridge: Cambridge University Press.

Taylor, C. (1989). *Sources of the self: The making of modern identity.* Cambridge, MA: Harvard University Press.

Theis, M., Gazzola, G., Forlin, M., Poli, I., Hanczyc, M. M., & Bedau, M. A. (2007). Optimal formulation of complex chemical systems with a genetic algorithm. *Proceedings of the European Conference on Complex Systems.* Forthcoming.

Waddington, C. T. (1957). *The strategy of the genes: A discussion of some aspects of theoretical biology.* New York: Macmillan.

Weiss, P. (Ed.) (1971). *Hierarchically organized systems in theory and practice.* New York: Hafner Publishing Company.

Wilsdon, J., & Willis, R. (2007). *See-through science: Why public engagement needs to move upstream.* Available online at: http://www.demos.co.uk (accessed July 2007).

Wilson, E. O. (1998). *Consilience: The unity of knowledge.* New York: Vintage Books.

Wilson, J. Q. (1993). *The moral sense.* New York: Free Press.

14

Human Practices: Interfacing Three Modes of Collaboration

Paul Rabinow and Gaymon Bennett

A congeries of "postgenomic" projects have defined their challenge as taking up the functional redesign of biological systems. One strategy devised to meet this goal is actually a heterogeneous collection of enterprises deploying diverse tactics loosely grouped under the compelling label of *synthetic biology*. Synthetic biology began as a visionary if minimally defined project whose goals were nothing if not audacious. Following in the rhetorical footsteps of the manifesto-like proclamations of the preceding two decades in molecular biology, one version of the program reads as follows:

Synthetic Biology is focused on the intentional design of artificial biological systems, rather than on the understanding of natural biology. It builds on our current understanding while simplifying some of the complex interactions characteristic of natural biology. . . . Those working to (i) design and build biological parts, devices and integrated biological systems, (ii) develop technologies that enable such work, and (iii) place the scientific and engineering research within its current and future social context. (Synthetic Biology Project Web site, 2004)

At the outset, the name was a basically a placeholder, or as some of its critics contend, a brand. Regardless, whereas its chief proponents understand synthetic biology as a process of modularization and standardization, it appears to us to be developing in and renovating a tradition nicely labeled the "Engineering ideal in American culture" (Pauly, 1987). Synthetic biology aims at nothing less than the (eventual) regulation of living organisms in a precise and standardized fashion according to instrumental norms. Unlike the visionaries of the previous decades' genome sequencing projects, and their prophecies of the molecular as the "code of codes," synthetic biologists clearly have a feeling for the organism, albeit the organisms with which its practitioners intend to populate the near future (Fox-Keller, 1984; Kevles & Hood, 1993). Synthetic biology's pioneers work hard at conveying a feeling of palpable excitement that biological engineering will invent and implement technologies that will make better living things. However, exactly what that would mean beyond efficiency and instrumental capacity-building is largely unexamined in any rigorous fashion, and in fact opens up a series of distinctive topics calling for inquiry and deliberation.

Synthetic biology arose once genome mapping became standard, new abilities to synthesize DNA expanded, and it became plausible to direct the functioning of cells. Its initial projects addressed a key aspect of the global crisis in public health—malaria. At the same time, an initial ethical and professional concern that it has had to deal with arises from the risk of bioterrorism. Among its current challenges is a cluster that concerns the production of biofuels. In sum, synthetic biology can be understood as arising from, and responding to, specific challenges and demands. Not all of the problems facing this field are radically new, and not all of the solutions will be either. What they do call for is resourceful solutions and inventive methods of thinking, experimentation, and organization.

As of 2007, at least four strategic tendencies self-identify as synthetic biology. There are two whose goal is to engineer whole cells. One seeks to do this from the "top down" by simplifying existing organisms and then engineering whole genomes or chromosomes, inserting them into the existing cellular machinery so as to orient them to function in a specified manner (see Lartigue et al., 2007). Another "bottom-up" approach, following in the line of earlier efforts at creating synthetic life forms, attempts to build protocells starting with basic nonliving materials (Rasmussen et al., 2008).

The two other variants are the ones we are working most closely with. The distinction between these two variants is an analytic one that we draw from our observations; it is not stated or otherwise formalized. The first variant has been developed primarily by a group of researchers at MIT. It consists of the attempt to engineer, modularize, and standardize working parts on the analogy of industry and prior developments in engineering. The goal of the MIT researchers is to make synthetic biology an engineering discipline in the formal sense (for more information see the BioBricks Foundation Web site). The final strategic tendency of synthetic biology is a variant of this latter approach, one that works in conjunction with the MIT model. It is characteristic of researchers at Berkeley, where there is a stated openness to the goal of standardizing synthetic biology as an engineering discipline, but actual work focuses more on specific functional problems. This approach seeks to develop and use synthetically engineered parts, though not as a goal in itself, or as the demonstration of the power of the subdiscipline. Rather, techniques and work of standardization are taken up insofar as they can eventually contribute to work on specified bioengineering projects. Such projects are not, strictly speaking, limited by the label of synthetic biology.

SynBERC

In 2006 a group of researchers and engineers from an array of scientific disciplines proposed a five-year project to render synthetic biology a full-fledged engineering discipline. Representing major research universities—University of California (UC) Berkeley, MIT, Harvard, UC San Francisco, and Prairie View A&M in Texas—the participants proposed to coordinate their research efforts through the development of a collaborative research center: the Synthetic Biology Engineering Research Center, or SynBERC (www.synberc.org).

SynBERC was designed, proposed, and funded as an effort to invent new venues and research strategies capable of producing resourceful solutions to real-world problems where existing venues and strategies appear to be insufficient. As the Web site puts it, with the typical bravado of an early-stage undertaking: "The richness and versatility of biological systems make them ideally suited to solve some of the world's most significant challenges, such as converting cheap, renewable resources into energy-rich molecules; producing high-quality, inexpensive drugs to fight disease; detecting and destroying chemical or biological agents; and remediating polluted sites."

In addition to its far-reaching research and technology objectives, SynBERC also represents an innovative assemblage of multiple scientific subdisciplines, diverse forms of funding, complex institutional collaborations, a near-future orientation, intensive work with governmental and nongovernmental agencies, focused legal innovation, and imaginative use of media. More unusual still, from the start, SynBERC has built in human practices as an integral and coequal, if distinctive, component.

The SynBERC initiative is designed around four core "thrusts": Parts, Devices, Chassis, and Human Practices. These thrusts, in turn, are designed to meet specified goals. The goal of Thrusts 1 through 3 is to link evolved systems and designed systems, with emphasis on organizing and refining elements of biology through design rules. Thrust 4 examines synthetic biology within a framework of human practices. The term *human practices* was coined to differentiate the goals and strategies of this component of SynBERC from previous attempts to bring "science and society" together into one framework so as to anticipate and ameliorate science's "social consequences." The task of the human practices thrust is to pose and repose the question of how synthetic biology is contributing or failing to contribute to the promised near future through its eventual input into medicine, security, energy, and the environment.[1] The purpose of such a task is to assess this role of synthetic biology through critical examination. The question of how synthetic biology will affect these domains as it develops (not only after it achieves something) constitutes a central, if not unique, concern of Thrust 4.

The SynBERC principal investigators (PIs) have claimed in their grant proposals and made structurally explicit in the initial organization of the center, that the far-reaching promises of synthetic biology cannot be realized under existing conditions and organization of scientific research. If the PIs are correct in their assessment, and if in basic ways the promise of synthetic biology is dependent on new forms of collaboration, then the success of SynBERC will depend as much on changes in work habits and at the institutional level as it will on technical virtuosity.

Given the power differentials between the scientists from a variety of backgrounds, and the existing disciplinary structures of reward that shape and reinforce current practices, there is no guarantee that collaboration will be forthcoming. Indeed, experience suggests that the habits and dispositions of elite scientists, as well as the organization of their laboratories and research objects, will resist change, consciously and tacitly. Certainly many of these scientists are willing to accommodate the Ethical, Legal, and Social Issues (ELSI) mode. They are open to filling out safety forms, to ethical discussions as long as these are periodic and nonintrusive, and to regulation as long as this is downstream of their research. Some are even open to hypothetical discussions about well-meaning social concerns and consequences. In short, some are willing to *cooperate*.

The question remains open, however, whether elite scientists with all the demands on their time are ready to submit themselves to transformative changes in their habits and procedures. Are they willing to contribute to developing *collaboration*? This is a genuinely open question for us, and constitutes a key starting point of inquiry that we undertake in an experimental mode. By experimental mode, we mean that we will monitor the progress (or lack of progress) of this design initiative, and analyze the results.

The Work of Equipment

The goal of our work has been to invent, experiment with, and, if successful, formalize a distinctive form of collaboration between synthetic biology, anthropology, and ethics. The first design parameter takes into account the predominance of cooperation in existing modes of work. As a mode of work, *cooperation* should be distinguished from *collaboration*. A collaborative mode proceeds from an interdependent division of labor on shared problems. A cooperative mode consists in demarcated work with regular exchange; cooperation does not entail common definition of problems or shared techniques of remediation. The first practical challenge, therefore, is to identify a venue appropriate to and capable of such experimentation.

In this chapter, we identify one such venue in which we have begun to work—specifically, SynBERC, mentioned earlier. Our task is to analyze existing modes of interaction and engagement between and among the human sciences, the biosciences, ethics, and organizational forms.

We argue that standard *cooperative* models of science and society, such as those developed under the Human Genome Initiative (HGI) ELSI program, need adjustment and remediation. By adjustment we mean recalibrating the core components of the ELSI program, developed to couple with the early stages of the genome sequencing projects, given the significant changes that have taken place in the biosciences during the last decade. By remediation we mean redesigning formerly cooperative practices to create interfaces among synthetic biologists, anthropologists, ethicists, and others such that they can undertake mutual work on commonly defined problems. The success of such work depends on a number of factors, not least of which is the challenge of introducing new habits and forms of organization into the existing structures and practices of elite science.

As an initial step toward achieving this goal of a distinctive mode of practice, the Berkeley team proceeded with an informed awareness that, to use our technical language, there exists a rather inchoate, if insistent, demand for new *equipment* to reconfigure and reconstruct the relations between and among the life sciences, the human sciences, and diverse citizenries, both national and global. This insight resonates with a year's intensive exploration with members of the Anthropology of the Contemporary Research Collaboratory (ARC; www.anthropos-lab.net), indicating that the demand for conceptually comparative inquiry exists in other emergent domains such as biosecurity and biocomplexity.

Further, this conviction that there is a need to invent new practices and imagine new relationships is buttressed by the demands of the pragmatic situation: The National Science Foundation funds our work. As we shall see, the demands coming from that quarter at times constitute a double-bind: an acceptance of the need to do something different and a pressure to produce immediate deliverables, in an older form, meeting older criteria of relevance. It is worth noting that large-scale programs are underway in Europe demonstrating the possibility and legitimacy of distinctive, post-ELSI approaches. The demands of the security environment in the United States, however, have overshadowed other imperatives, and these new directions are, as yet, largely unknown within the U.S. funding structure.[2]

Our own research and reflection, reading of the relevant literature in science studies, and insightful inquiry from intellectual historians have convinced us that there is—and has been—a level of pragmatic concern and development that lies between technology and method. Settled technologies honed to maximize means-ends relationships abound in our industrial civilization; the social and biological sciences have produced vast reservoirs of methodological reflection to justify and advance their work. Inquiries into past and contemporary situations of change make clear that neither technology per se nor grander methodological elaborations quite cover the terrain of how diverse domains are brought together into a common assemblage. Nor do they sufficiently explain how ethical considerations and demands have been brought into a working relationship with the quest for truth and made to function pragmatically (Shapin, 1995; Barry, 2001).

What is equipment? Equipment is a term (i.e., a word + concept + referent) that, by definition, does not retain a constant meaning. Such variation is a source of its richness and flexibility.[3] Equipment, though conceptual in design and formulation, is pragmatic in use. Defined abstractly, equipment is a set of *truth claims, affects,* and *ethical orientations* designed and combined into practices useful for work on specified problems and objects (see MacIntyre, 1984). In this way, equipment has historically taken different forms, enabling practical responses to changing conditions brought about by specific problems, events, and general reconfigurations (Rabinow & Bennett, 2007b).

Turning equipment into *equipmental platforms* has been a central part of our work. Equipmental platforms are designed to be of general use in a broad problem area (Rabinow & Bennett, 2007a). Whereas equipment identifies basic components given specific problems, a platform distinguishes practices appropriate to pragmatic work on those problems, as well as serving as the basis for the organization of such pragmatic work. The kinds of practices that equipmental platforms distinguish and organize are those relevant to the objects, metrics, and purposes specified by equipment. In order to put equipmental platforms into use, to move from the general to the particular, we must customize them for particular cases. In this chapter, we are concerned with analyzing the modes of practice according to which issues and challenges in synthetic biology have been taken up. These modes can be thought of as cohering with the specifications of distinct equipmental platforms.

For example, through the 1960s concerns arose regarding the capacity of the developing medical and biological sciences to provide adequate means of analysis for understanding and coping with the ethical and ontological consequences of their own advances. A small number of leading scientists took the initiative to invite philosophers and theologians to think about ways in which research might be moving in the direction of transforming or even destroying human life.[4] Out of these and other political encounters, by the middle of the 1970s a new kind of specialist, the "bioethicist," had appeared alongside the life scientist as someone authorized to offer serious truth claims about the relation of science and society. Bioethicists were assigned the task of elaborating principles according to which "good" science could be discerned from "bad" science. Such discernment was intended to provide an ordering and regulating function, ensuring that science would contribute to a healthy society and guard against pathological practices. That is to say, bioethicists were assigned the task of producing equipment. The first step in meeting this demand consisted in articulating principles required for composing appropriate equipmental platforms. In our terms, the actual challenge was to take philosophical principles and render them as design parameters for equipment.

An important example of the early development of such equipment is the work of the National Commission for the Protection of Human Subjects of Biomedical and Behavioral Research. The National Commission was tasked with developing practices appropriate to the protection of human subjects of research. It needed to respond to public outrage over the Tuskegee and Willowbrook experiments (see Jonsen, 2000, pp. 108–109). And it needed to be adequate to the task of preventing the abuse of research subjects in the future. In sum, the National Commission was faced with the task of developing equipment appropriate to specific types of problems under particular historical circumstances and addressing those problems in a distinctive and reconstructive manner.

In like fashion, the challenges surrounding synthetic biology call again for assessing the extent to which existing equipment is adequate for emerging problems, and, where found to be insufficient, for developing new equipment platforms.

Diagnostic

In what follows, we provide a diagnostic of three current modes of equipment. A *diagnostic* has two functions. The first is analytic—to lay out distinctions. A diagnostic serves a critical function; it facilitates the work of decomposing complex wholes in order to test the logic on which composition has been based. In diagnostics, however, the work of decomposition cannot be an end in itself. Rather, analysis must be followed by recomposition. This synthetic work is the second function of a diagnostic. Thus, a diagnostic operates to distinguish and designate, as well as characterize and fashion, categories and elements so as to give them an appropriate form.

We are developing a diagnostic that is designed to be directly helpful for our work in SynBERC, but is also intended to be applicable (with appropriate adjustments) to a range of analogous problem spaces. This approach has helped us analyze more clearly the challenges of how to proceed in organizing and putting into motion this multidisciplinary endeavor. The diagnostic offered in this chapter discriminates the ways in which various *modes of engagement* are designed to manage and respond to qualitatively different kinds of problems. In this way, distinctive modes of engagement can be interfaced and adjusted to each other such that the resulting assemblage is adequate to the kinds of problems SynBERC, and other similar contemporary enterprises, are designed to address.

The challenge, as we see it, is to characterize existing and emergent modes in such a way that they can be constructed as complementary parts of a broader collaborative human practices approach. This chapter analyzes two predominant modes of engagement, the representation of technical experts and the facilitation of "science and society," as well as a third mode, emerging today: inquiry and equipment. A goal of this analysis is to explore the conditions under which existing expertise and "boundary organizations" can be appropriately adjusted and interfaced with synthetic biology, with each other, and with the third mode, which is emergent and in the process of design and experimentation.

To aid the design and construction of such interfaces and the overall project of remediation, we begin with an ideal typical and schematic presentation of these modes so as to determine practices that are helpful and unhelpful, to determine, in turn, more clearly existing limitations and challenges. Our approach is in the line of the construction of "ideal types" proposed by Max Weber a century ago. We build three distinct forms that are constructed to be analytically distinct one from each other. We are fully aware that in the "real world" these divisions are not so neat and compartmentalized. The function of the ideal type, after all, is to highlight distinctions to enable inquiry into the specifics of existing cases. At the same time, of course, these ideal types have been constructed from materials drawn from pre-existing efforts and examples. Hence, there can appear to be slippage between the ideal typical function of producing an analysis and a description of existing configurations. Further, in the case that interests us the most, our own work on synthetic biology, we are engaged in both a projective thought-experiment—a *Gedankenbild*, to use another of Weber's pertinent expressions—and the initial attempts to make this construction operative. Our task is thus both analytic and observational, as well as synthetic and participatory. We hope to keep these points clear in our presentation while realizing that empirical reality is never so stable, clear, or neat.

Mode One: Representing Modern Experts

Mode One consists in inventorying, consulting, and cooperating with experts. The core assumption—often taken for granted and not subject to scrutiny—is that the expertise of existing specialists in one domain is adequate without major adjustment to emerging problems. Of course, in many instances, problem domain and expertise are adequately paired. The vast number of technical specialists trained and supported by the state bureaucracies, corporations, international agencies, and nongovernmental nonprofits of the industrialized world certainly are competent to address many current challenges. It is worth remembering that many of these challenges have been formulated, worked over, and compartmentalized by these experts and the organizational form, practices, and limits within which they operate.

Expert knowledge functions as a means-ends maximizer. Even when such expert knowledge is operative, it gains its very strength precisely from its capacity to bracket purposes or goals. Expert knowledge is structured and functional only when that which counts as a problem is given in advance, stabilized, and not subject to further questioning. In emergent situations, however, neither goals nor problems are settled, and so technical expertise cannot be effectively marshaled without some adjustment. In many instances, obviously, when goals and problems become settled, technical expertise must be given a useful place within an assemblage. Said another way, routinization is normal but qualitatively different from states of emergence or innovation.

Having access to technical competence and successfully deploying it in delimited situations (which need to be identified and stabilized themselves) to effectively address problems are not the same thing. Hence, in addition to technicians, stable organizations need managers or technocrats whose task is to oversee and coordinate specialists and technicians. Such coordination facilitates a *cooperative* mode of engagement by subdividing specializations and assigning tasks. As with technical expertise, it is frequently supposed that the competencies of technocrats are transferable from stable to emergent situations. Thus, in the United States, technocrats and technicians often rotate out of public, governmental, or corporate service into positions as consultants or lobbyists claiming transferable competence.

In Mode One, the role of the social scientific practitioner (MOP) is to identify and coordinate legitimated specialists and technocrats. The MOP is expected to maintain broad overview knowledge of a number of subdisciplines, at least to the extent that he or she can legitimately claim to present a range of candidates as authoritative and available. Candidates are presented and ranked along scales (both formal and informal) of authority, availability, connections, and character. The MOP's authority is based on this work of inventory and ranking. The types of equipmental platform according to which MOPs calibrate their work include those that distinguish kinds of authorized experts and draw these experts into a cooperative frame.

The MOP frequently does not provide a critical analysis of the status of expertise per se, or of existing expertise and its specific functions, but rather understands his or her work as providing an evaluative assessment of specific first-order practitioners, in a first-order mode. The metric of this inventory-making is not a second-order one. We are taking the distinction between first- and second-order observer from the German sociologist Niklas Luhmann. A second-order observer is someone who observes observers observing. This sounds opaque but is actually quite straight-forward. First-order observers take their world as it comes to them (often in a highly mediated form). They then do their work. This intervention in the world is what Luhmann refers to as "observing." Hence, the term is more than perceptual; it is an action, frequently a sophisticated one. A second-order observer observes actors acting. Such a second-order action is neither removed from the world nor given any special privilege. Furthermore, as Luhmann writes, "[a] second-order observer is always also a first-order observer inasmuch as he has to pick out another observer as his object in order to see through him (however critically) the world" (1997, p. 117). We take up this distinction in a nonjudgmental and simple manner: It helps to distinguish different positions and different modes of doing one's work.

The MOP *represents* existing *expertise*; this representation takes a twin form. The MOP literally re-presents existing expertise in a readily comprehensible form (often PowerPoint). The MOP is a representative for the legitimacy of existing expertise. The MOP does not put forth claims of validity concerning substantive issues dealt with by the chosen experts. It follows that under specific circumstances, in funda-mentally stabilized situations, institutions, and problems, Mode One work can provide benefits by identifying, bringing together, and representing existing expertise.

From the outset of SynBERC, it was clear that even in the domains where the existing core of specialist expertise might well be more directly pertinent (e.g., intel-lectual property) than in some others (e.g., ethics), it was certain that start-up companies with whom scientists in SynBERC had direct association as founders or board members (e.g., Codon Devices, Amyris) would have ready access to such experts (e.g., would have already taken great care to address intellectual property [IP] issues). This supposition has been amply supported by evidence. In a word, the small start-up companies associated with SynBERC and other parallel organization have already hired patent lawyers and given priority to related financial matters (or in the case of established organizations such as BP, have whole departments long in place). These counselors are privy not just to the generalities of synthetic biology as an emergent field but to the specifics of the scientific and technological inventions at issue. Further, venture capitalists who have invested in these start-ups provide the contacts necessary for maximizing protection, and insist on enforcing those protections. Finally, SynBERC was conceived within a certain ethos of maximizing the "commons," and from the inception was associated in a working relationship with groups such as Creative Commons with long experience in innovative patent and organizational design.

It is commonly recognized that questions concerning industrial strategies and IP are of fundamental importance to synthetic biology. Work to date has focused on how synthetic biology will have to adapt its open-source goals to existing models of industrial strategies and IP. Our approach is to inquire into what distinctive forms of industrial partnerships and IP can be invented given the objectives of *specific synthetic biology projects*.

Externalities

For each of the modes, after the ideal-typical figure, we present a list of "externalities" and "critical" limitations as a series of talking points. A substantial scholarly and professional literature is available on many of these issues. Rather than giving the impression that we are comprehensively presenting each of these questions, we prefer a schematic form as a means of indicating that we are attempting to think about, explore, and draw lessons from these topics at this initial stage of both our inquiry and the development of SynBERC. At the end of the chapter, we raise a series of challenges that those attempting to work collaboratively must face.

Some immediately identifiable externalities bear on Mode One. The term externality, as we are using it, is taken from neoclassical economics. It refers to factors that "result from the way something is produced but is not taken into account in establishing the market prices."[5] The identification of such limits allows one to pose the question: When and where is it an effective use of limited resources to undertake MOP strategies?

1. In emergent problem spaces appropriate experts do not necessarily exist. This fact falls outside of Mode One operational capacities. Such a deficit, however, does not imply that there is no possible way to adjust and integrate existing expertise. Rather, it simply calls for second-order reflection on this state of affairs.

2. Even when appropriate expert knowledge does exist, its very strength—technical criticism as means-ends maximization—gains its legitimacy precisely from its capacity to bracket purposes or goals. In an emergent situation, such bracketing must itself be subject to scrutiny.

3. In either case, a different skill set is required to move into the contested networks and pathways of what is taken to be the impact, consequences, opinions, of "society" or "the public." The response of MOPs to this challenge is to look for other specialists in surveying opinion, assessing consequences, and preparing for the impact. The reservations of (1) and (2) thus apply here as well.

4. Mode One is based on the modernist assumption that there is a society, that it has been divided into value spheres, that there is a problem of legitimization, and that the challenge is to invent a form of governance in which these issues can be adjudicated through procedure and specialization. These assumptions have been debated and challenged for over a century (Weber, 1941; Boltanski & Thevenot, 1991; Beck, Giddens, & Lash, 1994). And within a new globalized, accelerative,

security-conscious context, it is not obvious (far from it) that MOP presuppositions are defensible.

Critical Limitations

Given these externalities, the question still needs to be addressed: Where expertise is engaged, what are its critical limitations? Answering this question will enable us to pose the question of where and when Mode One experts are useful in an assemblage such as SynBERC. We have identified several critical limitations:

1. In Mode One, the future appears as a set of possibilities about which decisions are demanded (Koselleck, 1979). The range of these decisions is delimited by a zone of uncertainty. The genesis and rationality of such a zone is that Mode One experts operate with a metric of certainty. The ever-receding zone of uncertainty, however, is not fundamentally unknowable, only uncertain. But, precisely because it forms a horizon depending on current decisions, this zone of uncertainty cannot be specified in advance. Uncertainty, however, does not undermine the decision-making imperative of experts. Rather, it compels incessant decisions and affirms that an appropriate form of verifiable certainty (probability series, risk analysis, technical measurements, etc.) can be attained. The authority of experts is not undermined by the oft-demonstrated inability either to forecast the future or to make it happen as envisioned. Rather, this dynamic provides the motor of their legitimization. In sum, a zone of uncertainty is an intrinsic part of this equipmental mode (Pence, 1998).

2. In Mode One, uncertainty is taken up as a boundary condition. It allows Mode One practitioners to move from the generation of verified claims and their delimitation to the coordination of discussion and communication. Rather than deflating the authority of experts or making obvious the need for other modes of inquiry, this move to discussion and communication allows for the rehearsal of the past triumphs of expertise, and renders such past verificational successes as points of reference to orient debate about the present and near future (Deleuze & Guattari, 1991).[6]

3. Uncertainty entails an ever-receding horizon. As such, rather than functioning as a fundamental limitation, uncertainty provides a refinement and corrective such that MOPs can (ostensibly) operate more realistically, and therefore, more effectively. MOPs attempt to factor in "uncertainty" as a parameter in identifying and ranking expertise. What they fail to factor in is the structural insufficiencies of existing expertise, both external and internal.

Mode Two: Facilitating Relations Between Science and Society

We take the distinction between Mode One and Mode Two from the work of Helga Nowotny and coauthors. Nowotny et al. have been part of an active debate and an articulated conceptualization of the strengths and limitations of Mode One (Gibbons et al., 1994; Nowotny, Scott, & Gibbons, 2001). Their book is an elaboration of a report commissioned by the European Commission. In fact, Mode Two has become the norm for official policy in Europe regarding "science and society" (Nowotny et al., 2001). Although there are examples of this mode in the United States (see following section), such instances are dispersed and are not currently normative in an official policy sense. Mode Two arose as a reaction to the perceived arrogance of scientists and technocrats and their lack of professional competence to deal with concerns beyond their direct disciplinary or subdisciplinary questions. Further, as policy makers and civil society activists have discovered and documented, (1) Mode One is unable to include a range of existing social values in planning, (2) neither the purely scientific nor purely technological per se were competent to evaluate consequences and impacts, and (3) by including opinion, through polling and surveying techniques, Mode Two projects meet less resistance and appear more representative.

Mode Two social science practitioners are *facilitators*. Their role *qua* facilitator is to bring various actors (scientists, technical experts, policy makers, law makers, civil society actors, political activists, industry representatives, government and private funders, etc.) together in a common venue. That venue, often created for a particular crisis or event but eventually standardized, is fundamentally a space for *representation and expression*. Stakeholders are encouraged to express themselves, to advocate, to denounce, to articulate, to clarify, and eventually, it is hoped, to form a consensus. Such a consensus is taken to be normative and made to function equipmentally in the organization of research and development programs.

Mode Two is calibrated according to an innovative equipmental platform. This platform takes "social values" as norms for discriminating which activities are appropriate, and elaborates these values to serve as the basis for the organization of such activities. It follows that the challenge for Mode Two practitioners is to develop procedures for identifying significant social stakeholders, discern their opinions and values, and design mechanisms through which such opinions and values become normative for research and development in the sciences. In order to meet this challenge, venues that facilitate boundary organization and their modes of governance must be invented and institutionalized. In both Europe and the United States, the venue for this work has predominantly been the "Center."[7]

Cutting Edge Example: Nanotechnology and Society

A leading example of Mode Two social science is the Center for Nanotechnology and Society at Arizona State University (CNS-ASU). CNS-ASU has been designed to adopt and adjust its organizational practices to the limitations of Mode One by focusing on "emerging problems" and "anticipatory governance:" "Designed as a boundary organization at the interface of science and society, CNS-ASU provides an operational model for a new way to organize research through improved reflexiveness and social learning which can signal emerging problems, enable anticipatory governance, and, through improved contextual awareness, guide trajectories of NSE [nanotechnology science and engineering] knowledge and innovation toward socially desirable outcomes, and away from undesirable ones" (CNS-ASU homepage).

The proposed means of moving toward such socially desirable outcomes is to "catalyze interactions" among a representative variety of publics. The metric of these interactions is not to produce technical expertise per se, but to raise the consciousness and responsive capacities of high-level policy makers, scientists, and "consumers" (Fisher, 2007). Interactions and awareness are facilitated by designed and monitored dialogues on the goals and implications of nanotechnology. This engagement will facilitate the construction of a communications network positioned upstream rather than downstream of the research and development process. Upstream positioning is designed to anticipate and evaluate the impact of nanotechnology on society before "rather than after [its] products enter society and the marketplace" (CNS-ASU homepage).

Equipmental Platform: RTTA and Reflexive Governance

Two sets of strategies are being designed and developed to meet CNS-ASU's goals. The first is a program of "research and engagement" called real-time technology assessment (RTTA). RTTA consists of four components:

"mapping the research dynamics of the NSE enterprise and its anticipated societal outcomes";

"monitoring the changing values of the public and of researchers regarding NSE";

"engaging researchers and various publics in deliberative and participatory forums"; and

"reflexively assessing the impact of the information and experiences generated by our activities on the values held and choices made by the NSE researchers in our network" (CNS-ASU homepage).

The second procedure is a program for "anticipatory governance." Anticipatory governance can be distinguished from "mere governance," defined as "the kind that is always found running behind knowledge-based innovations." Rather, through the facilitation of interfaces between societal stake holders and researchers, CNS-ASU is attempting to develop practices of governance with the capacity to "understand beforehand the political and operational strengths and weaknesses of such tools," and "imagine socio-technical futures that might inspire their use" (CNS-ASU homepage).

Externalities

Some immediately identifiable externalities bear on Mode Two. Identifying such externalities allows one to answer the question: Where and when should Mode Two strategies be undertaken in synthetic biology?

1. Mode Two attempts to factor in and move beyond the limitations of Mode One's focus on existing expertise. However, given built-in funding and legitimacy demands, such a move is frequently hindered. For example, further experts are required to identify and manage the polling of such diversity; yet additional specialists are required to manage these burgeoning classifications, groups, and subgroups. Audit culture expands to meet its own criteria of inclusiveness, accountability, and responsibility (bureaucratic demands of accountability), namely, that the technologies of polling and opinion collection be developed and managed by experts (in polling, in the presentation of results, in public relations, etc.; see Stathern, 2000). In sum, the challenge of moving beyond expertise includes the requirement for new experts.

2. Mode Two supposes that ethical science is science that benefits society, which is made up of stakeholders whose values must be given a venue for expression. This supposition generates two problematic limitations. The first is that various stakeholders are vulnerable to the charge of being ignorant or incompetent: Scientists often believe that lay people are incapable of understanding the details of their work in its own terms (often correct), and hence are not capable of producing legitimate evaluations (often contestable). Policy makers, social activists, and social scientists often believe that the results presented in scientific or technology journals do not correspond to the complexity of social reality. Journalists' attempts to explain science to society are thought to simplify both poles. It follows that charges and countercharges of hype joust with charges of ignorance.

3. The second limitation is that it has become clear in Europe that techniques of producing society's representatives were required, as well as techniques of legitimating these representatives. The legitimating process is frequently challenged by those who consider themselves to be excluded.

Critical Limitations

Given these externalities, the following question needs to be addressed: Once the appropriate venues for Mode Two have been established, what are its critical limitations? Answering this question will enable us to pose the question of where and when Mode Two practitioners are useful in a collaborative assemblage such as SynBERC.

Mode Two is characterized by at least three identifiable critical limitations:

1. Mode Two takes seriously the challenge to respond to society and the public in order to orient research responsibly. However, experience has shown that specifying who exactly one is talking about when referencing society or the public frequently

turns out to be an elusive task (see Jasanoff, 2005). These broad rubrics cover highly diverse actors.

2. Furthermore, two decades of work in science and technology studies and related fields have put into question the very existence of referents to homogenizing terms like *science, society,* and *public.* Sciences are plural when they retain any distinctiveness at all. Society has been increasingly replaced by community and the individual in its neoliberal frame.

3. In Europe, given the bureaucratic framework of the European Union, not surprisingly, the way in which the first critical limitation has been dealt with was through the channels of representation and formal procedures. Proceduralist approaches, however, rarely resolve value disputes, although they may provide means of adjudication in specific instances. Likewise, proceduralist approaches rarely resolve scientific differences. Finally, proceduralist approaches tend to mask power differentials.

4. Regardless of how successfully bureaucratic procedures are designed and implemented, problems remain. As many critics have pointed out, such as the President's Council on Bioethics in the United States and *ATTAC* in France, opinion polling, formal proceduralism, consensus building, and the multiplication of representatives' expression cannot answer the ethical and political question of whether or not a given course of action is good or bad, right or wrong, just or unjust. In fact, proceduralist exercises have no way of posing this question. It follows that representation and expression as modes of organizing scientific and political practice, much like technical expertise, while coherent and valuable within a democratic framework, nonetheless because of their inherent limitations pose serious dangers that must be taken into account.

Mode Three: Inquiry and Equipment

It is not the "actual" interconnections of "things" but the *conceptual* interconnections of *problems* which define the scope of the various sciences. A new "science" emerges where new problems are pursued by new methods and truths are thereby discovered which open up significant new points of view.
—Max Weber (1949, p. 68)

A defining goal and enduring dimension of our Mode Three work is to design practices that bring the biosciences and the human sciences into a mutually collaborative and enriching relationship, a relationship designed to facilitate a remediation of the currently existing relations between knowledge and care in terms of mutual flourishing. The means to inquire and explore to what extent these new relationships will be fruitful consist in the invention, design, and practice of what we have referred to as *equipment*. Equipment, recall, is a technical term referring to a practice situated between the traditional terms of method and technology. The mode of engagement we are developing aids in achieving our goal of designing and synthesizing equipment. If successful, such equipment should facilitate our current work in synthetic biology (understood as a human practices undertaking) through improved pedagogy, the vigilant assessment of events, and focused work on shared problem spaces. An ongoing task is to provide conceptual analysis of these three elements, so as to reflect on their ethical significance and ontological status, as well as to provide equipment that contributes to more responsive and responsible solutions:

1. *Pedagogy.* Pedagogy involves reflective processes by which one becomes capable of flourishing. Pedagogy is not equivalent to training, which involves reproduction of expert knowledge, but rather involves the development of a disposition to learn how one's practices and experiences form or deform one's existence and how the sciences, understood in the broadest terms, enrich or impoverish those dispositions.

Our inquiry, engaged with the practices and experiences of the synthetic biology community, addresses the question: How is it that one does or does not flourish as a researcher, as a citizen, and as a human being? Flourishing, here, involves more than success in achieving projects; it extends to the kind of human being one is, personally, vocationally, and communally. Adequate pedagogy of a bioscientist in the twenty-first century entails active engagement with those adjacent to biological work: ethicists, anthropologists, political scientists, administrators, foundation and government funders, students, and so on. Contemporary scientists, whether their initial dispositions incline them in this direction or not, actually have no other option but to be engaged with multiple other practitioners. The only question is how best to engage, not whether one will engage. Pedagogy teaches that flourishing is a life-long collaborative formative process, and involves making space for the active contribution of all participants.

2. *Events.* A second set of concerns involves events that produce significant change in objects, relations, purposes, and modes of evaluation and action. By definition, these events cannot be adequately characterized until they happen. Past events that have catalyzed new relationships between science and ethics include scandals in experimentation with human subjects and the invention of equipment to limit them, the promise of recombinant DNA and its regulation, crises around global epidemics and significant biotechnological interventions, the Human Genome Initiative and the growth of bioethics as a profession, and 9/11 and the rise of a security state within whose strictures science must now function. Just as scientists are trained to be alert to what is significant in scientific results, we also work to develop techniques of discernment, analysis, and synthesis that alert the community to emergent problems and opportunities as they take shape.

Research in human practices is underdetermined. Past bioethical practices often operated as though the most significant challenges and problems could be known in advance of the scientific work with which these challenges and problems were to be associated. Our hypothesis is that such practices are not sufficient for characterizing the contemporary assemblage within which synthetic biology is embedded. This assemblage is composed of both old and new elements and their interactions.[8] Though some of these elements are familiar, the specific form of the assemblage itself, and the effects of this form, can be known only as they emerge.

We understand *emergence* to refer to a state in which multiple elements combine to produce an assemblage, whose significance cannot be reduced to prior elements and relations. Emergence should be distinguished from uncertainty. Uncertainty operates under a mode of verification; it takes for granted that the future, though uncertain at present, can be anticipated as knowable, following regular patterns. Emergence, by contrast, calls for equipment capable of operating in a mode of *remediation*. It takes for granted that the future, although unknown at present, will have distinctive features that do not depend on the regularities of current configurations.

3. *Problem space*. Events proper to research, as well as adjacent events, combine to produce significant changes in the parameters of scientific work. These combinations of heterogeneous elements are historically specific and contingent. At the same time, they produce genuine and often pressing demands that must be dealt with, including ethical and anthropological demands. In sum, our understanding of the contemporary challenge is to meet what Max Weber calls "the demands of the day," through the design and development of equipment. Such equipment must be adequate to remediating these heterogeneous combinations, the problems raised, and a near future in which it would be possible to flourish.

The challenge of human practices equipment is how best to design and implement interfaces among and between the three modes. This is a daunting challenge, since older patterns of power inequalities and their associated dispositions continue to remain in place. It is even more daunting if human practices attempts to operate (to use two technical terms) according to the metric of flourishing for the purpose of remediation. The term *remediation* has two relevant facets. First, it means to remedy, to make something better. Second, remediation entails a change of medium. Together, these two facets provide the specification of a particular mode of equipment. When synthetic biology is confronted with difficulties (conceptual breakdowns, unfamiliarity, technical blockages, and the like), ethical practice must be able to render these difficulties in the form of coherent problems that can be reflected on and attended to. That is to say, ethical practice remediates difficulties to make a range of possible solutions available.

In the 1970s, bioethical equipment was designed to protect human subjects of research, understood as autonomous persons. Hence, its protocols and principles were limited to establishing and enforcing moral bright lines indicating which areas of scientific research were forbidden. A different orientation, one that follows within a long tradition but seeks to transform it in view of reconstructive and emergent situations, takes ethics to be principally concerned with the care of others, the world, things, and ourselves. Such care is pursued through practices, relationships, and experiences that contribute to and constitute a *flourishing* existence. We note here that flourishing is a translation of a classical term (*eudaemonia*), and thus a range of other possible words could be used: thriving, the good life, happiness, fulfillment, felicity, abundance, and the like.[9] Above all, *eudaemonia* should not be confused with technical optimization, as we hold that our capacities are not already known and we do not understand flourishing to be uncontrolled growth or the undirected maximization of existing capacities.

Understood most broadly, flourishing ranges over physical and spiritual well-being, courage, dignity, friendship, and justice, although the meaning of each of these terms must be reworked and rethought according to contemporary conditions. Such conditions are not constituted by fixed or pregiven forms. If so, they would neither be appropriate to the emergent, nor could they be useful in the work of remediation. Rather, these conditions must be taken up as dynamic and adjustable, calibrated to actual conditions under which the terms of flourishing can be concretely specified. The conditions of flourishing must be specified in a form that is amenable to intervention and amelioration under concrete arrangements.

The question of what constitutes a flourishing existence, and the place of science in that form of life (i.e., how it contributes to or disrupts it), must be constantly posed and re-posed. In sum, human practices equipment is designed to cultivate forms of care of others, the world, things, and ourselves in such a way that flourishing becomes the mode and the purpose of bioscientific, ethical, and anthropological inquiry and practice.

Externalities
A number of immediately identifiable externalities bear on Mode Three. The establishment of clarity about external and internal limits distinguishes warrantable scientific advance from opinion and hype. Identifying such limits allows one to pose the question: When and where is undertaking Mode Three strategies an effective use of limited resources?

1. Mode Three is allied with, but should be carefully distinguished from, the Foucaultian analytic practice of the "history of the present" (Foucault, 1995). When

analysis is undertaken with that goal, its task is to show the lines flowing back from the present into previous assemblages (and elements and lines that preceded those assemblages). Such work clarifies the contingency of current expert knowledge, its objects, standards, institutions, and purposes. The goal is not primarily to debunk or delegitimize such expertise, although a dominant mode of academic criticism habitually does takes the form of denunciation (see Boltanski, Thevenot, & Porter, 2006). Rather, the goal is to make clear how such expertise came about, what problem space it arose within, what type of questions it was designed to answer, how and where it had been successfully deployed, and what blind spots were produced by its very successes. The purpose of analytic work in the history of the present is not necessarily to replace the specialists and managers that already exist. Above all, it aims to open up current practices to critical scrutiny.

2. The habits of elite scientists as well as the institutions and ethos of bioethics orient expectations toward a mode of cooperation, not collaboration.

Critical Limitations

Given these externalities the question needs to be posed: What are the critical limitations of Mode Three? The range of critical limitations of human practices is not yet known. However, two limitations can be identified at the outset:

1. Modes One and Two are designed to work within and be facilitated by governmental, academic, and other stabilized venues. These are legitimate venues when the equipmental demands consist in the regulation or regularization of a problem space. Well-characterized equipment exists for operating in such nonemergent spaces. Adaptations to emergent fields such as synthetic biology and nanotechnology are underway.

2. There will be a repeated and insistent demand for Mode Three practitioners to provide expert opinion, propose first-order solutions, represent opinions, invent and implement a venue for expression, and facilitate consensus. Mode Three practitioners acknowledge the validity of such demands for certain problem spaces, certain actors, and certain venues. Mode Three, however, is designed such that fundamentally it cannot—and should not—honor such requests insofar as it operates on and in emergent problem spaces. There clearly is a price to be paid for respecting this externality. It is the price to be paid for being patient, consistent, and clear-sighted. This consistency may well eventually add value to Mode Three. Its immediate worth, however, is found in its bringing attention to the need for inquiry.

Conclusion: Interfaces

What if, as seems likely given the premises of the strategy of designing and constructing appropriate form, there actually were no experts in emergent domains and problem spaces? At the time of SynBERC's founding, for example, everyone would have agreed readily that there were no specialists in the first three thrusts—Parts or Devices or Chassis—although scientists with diverse skills held potential for such innovation and coordination. Developing venues, modes of practice, technical and other equipment, modes of collaboration, and so on, is, after all, a central goal of the center. The founding strategy was to identify a challenge, make its significance comprehensible, and pursue strategies for addressing it. There was an excited confidence that, with success, many others would follow. A new mode of practice would be launched.

Human Practices and Mode One

Logically, it follows that, as with Thrusts One through Three, so too with Thrust Four. Simply cooperating with technical experts and keeping a watchful eye on the scientists seemed and seems to be an insufficient, even implausible, way to proceed. Indeed, such an approach will likely provide the false assurance of short-term deliverables and the potential for long-term strategic misdirection. Consequently, an obvious initial challenge has been to invent venues within which academic experts-at-a-distance, who might otherwise only share a cooperative relation to emergent hybrid assemblages such as SynBERC, are situated in such a way that their existing expertise can be remediated and redeployed in view of new problems. The claim is not that existing experts have nothing to offer. The question to be explored is: What can human practices provide that existing experts cannot?

One of the distinctive organizational characteristics of SynBERC was its division into thrusts; another was its strategy to include "test-beds" from the start. A test-bed is a concrete research project designed to function as a proof-of-concept for work in the thrusts. Originally there were two of these—bacterial foundries and tumor-seeking bacteria—and then a third, biofuels. The Berkeley and MIT Thrust Four leaders agreed that, informally, the MIT group (and its Mode One collaborative approach) would serve, in addition to its other contributions, as a test-bed for the Berkeley group's experiment in inventing a new type of equipment. With this division of labor, it was hoped, a collaborative approach could be developed. The advantage of this strategy was that the Mode One team would produce immediately recognizable deliverables: workshops, conferences, specific recommendations, organizational advice, network connections in East Coast power centers, and so on.

It was clear that, initially, Mode Three would have no such list of familiar deliverables or modes of delivering them. What Berkeley did have, however, was a keen sense (based on years of anthropological research in the world of biotech and genomics, contemporary reflections on that world, and deep experience in ethical work in the broader political and industrial context) that current modes of practice had built-in structural limits, and, because of the very way they had emerged and been institutionalized, were unlikely to be flexible and creative enough to collaborate effectively within an organization such as SynBERC. We took as an initial task a rigorous diagnosis of what such change might look like, and the initial steps toward actualizing such change. Of course, no one knew in advance if the scientific test-bed form would produce successful collaboration with the separate thrusts. And after one year, the proverbial verdict is still out.

Human Practices and Mode Two

If a primary task of Mode Two is to facilitate representation and expression of stakeholders, this work is likely to be relevant at a subsequent stage. Since fields such as synthetic biology have barely begun to take shape, gain funding and attain a visibility arising from their accomplishments as opposed to the positive or negative hype that surrounds such enterprises, the "public" or "society" may well have no opinion whatsoever, and certainly no detailed opinion or well-informed representatives (none exist) at the early stages of emergent disciplines and assemblages.

There are now professionals at organizing public opinion and alerting stakeholders in other assemblages of possible concerns about developments in related fields. These analogy-professionals' claims to represent broad numbers of people and civil society interests should be examined with care. That being said, these Mode Two professionals have already established funding mechanisms, relations with journalists, functioning Web sites, and networks with heterogeneous civil society groups. It would seem to be a pressing and legitimate function of Mode Two practitioners to assess, sort, adjudicate, and moderate emerging common places and rhetorical thematics.

If preemptive analogizing is both rampant and relatively superficial, futurology is not the answer to emergent things either. There are many versions of predicting or narrating the future, including forecasting. Forecasting refers to the use of quantitative analysis to identify the future trajectories of current trends. The goal of such forecasting is to anticipate small variations from these trends (e.g., variations in oil prices). Forecasting has two built-in limitations. First, it bases its conclusions on the logical outcomes of only one possible future. Second, this one possible future is thought to be a direct and predictable unfolding of current states; as such, it assumes a much greater similarity between the present and the future than usually proves to be the case. Forecasting as a way of dealing with the future requires assembling technical experts that can quantitatively elaborate extensions of current trends. If the future is contingent and emergent, as in fields such as synthetic biology, however, such forecasting has limited value.

Human practices takes up the question of the near future and its bearing on current practices in a different way: scenario thinking. Scenario thinking identifies a range of logically distinct futures. All of these futures are feasible, and yet each one entails dramatically different implications for current and near-future practices and institutional organization. Techniques of scenario thinking help to create a matrix of much more complicated future possibilities than forecasting does. They help tease out and pull apart assumptions about the relation between the present and the near future. This underscores that what is needed is not better predictions about the unfolding of current trends, but the development of capacities for imagining different futures and exercising real-time changes in practice and organization. This work highlights the ways in which current practices and organizations may or may not adequately prepare us to respond effectively to such different futures.[10]

Scenario thinking involves the identification of critical contingencies about the future that might play a formative role in shaping synthetic biology. This approach underscores that the stakes of scientific development cannot be sufficiently known in advance, and that forecasting and prediction by experts is likely to provide false assurance. Critical contingencies can be fleshed out and articulated as variations within specific scenarios. In turn, these alternatives establish a common framework for articulating and working on shared problems.

Human Practices and Mode Three

We do not think that what is distinctive and intriguing about developments in synthetic biology is that they are revolutionary or even cutting-edge. These are modernist terms from a prior historical configuration that draws attention to what is "new" and "radically transformative" as the locus of significance. Our interest and attention is drawn to the combination and recombination of elements, old and new, into a stylized form whose defining diacritic is not its newness per se. Rather, in what has been described elsewhere as "the contemporary" as opposed to the modern, what counts as significant are the forms and possibilities that open up once the quest for the new is moderated and backgrounded (although not ignored). Hence, the basic rules of what counts as good science and engineering in synthetic biology are the traditional or standard ones. What objects are taken up and how they are combined and recombined are themselves part of a larger *Gedankenbild* that is part organizational, part conceptual, part technical—and part equipmental. How such an assemblage might be put together, made to function effectively, cope with breakdown and unexpected occurrences, and discern and address emergent problems both intrigues us and concerns us.

Additionally, well-established modes of engagement are structured by specific metrics. Prominent metrics have included normalization and the protection of dignity. Normalization allows for the regulation and modulation of fields of statistical regularities, such as industrial safety. The metric of dignity facilitates emergency intervention into situations of rights abuse. Though recognizing the worth and utility of these metrics, human practices is designed to discover if it can function according to a different metric—flourishing.

We were oriented toward a reconstructive effort because various research teams at the ARC had been engaged in intensive inquiry on emergent topic areas such as biosecurity and biocomplexity for the preceding two years. For example, we observed in the latter how a rethinking of issues had contributed to a shift from biodiversity as a central approach to a range of environmental concerns, to the emergent field of biocomplexity. The former approach was based on understanding and preserving species as an inherent good, but the latter concentrated more on the types of milieu that would sustain biological complexity to flourish. Hence, a certain range of prior expertise, and prior disciplinary suppositions and ethical commitments, taken as settled and desirable, could well slow or even block the understanding and collection of the data that will be required for the conception of sustainability at work in biocomplexity.

A similar example can be given with biosecurity. It has become clear through our research that recombinations and reconfigurations of existing expertise are required for a biodefense system adequate to emergent problems to be constructed. Although previous Cold War experience can constitute a baseline for thinking about biosecurity today, we find ourselves in a radically different security situation. It follows that a vastly different array of bioscientific understandings and technologies, and new dispositions among security experts (and, for that matter, potential aggressors), are just as vital as new dispositions and approaches among bioscientists.

As an integral component of the overall enterprise, human practices is positioned to take up problems in a way that experts-at-a-distance cannot. For example, problems in industry relations and intellectual property are certainly crucial to how synthetic biology will develop. However, human practices does not need to ask the question of what IP platforms exist and how they can be applied. Rather, human practices is in a position to pose the question of what kinds of objectives are really at stake in specific projects, and how those stakes require rethinking about the interfaces among university labs, government funding, biotech interests, and the like. In this way, the problem of how to leverage existing resources, talents, and technologies to advance the aims of synthetic biology can appropriately be posed. Once posed, these problems can be collaboratively worked on. Such collaboration will require existing experts, to be sure. However, the expertise will need to be interfaced with emergent problems in such a way that experts will be required to think forward rather than reproduce existing insights. In sum, our work is oriented toward understanding how potentially viable design strategies emerge, how these strategies might inform synthetic biology, and what efforts are undertaken to integrate them into a comprehensive approach to the near future.

In the early stages of human practices development, Mode Three has been faced with three primary challenges, one critical and two productive. The first challenge is to accept its positionality as adjacent and second order. Given the positionality of Modes One and Two as consultative and first order, it is not surprising that even sympathetic observers and participants would put forth the demand for first-order and advisory deliverables. Consequently, a primary challenge for Mode Three is to develop a toolkit of responses and practices that temper and reformulate such demands.

A second challenge concerns the form of *collaboration*. Given the emergent character of innovations and practices in synthetic biology, the precise forms of collaboration have not and cannot be settled in advance. Rather, such collaborations will require intensive and ongoing reflection with SynBERC PIs on emergent ethical, ontological, and governance problem spaces within which our work is situated and develops. We have been experimenting with both directed group meetings and having undergraduate and graduate students directly engaged within SynBERC labs as their work unfolds.

A third challenge concerns *reconstruction*. We are giving reconstruction in human practices a specific technical meaning, similar to that put forward by John Dewey: "Reconstruction can be nothing less than the work of developing, of forming, of producing (in the literal sense of that word) the intellectual instrumentalities which will progressively direct inquiry into the deeply and inclusively human—that is to say moral—facts of the present scene and situation" (Dewey, 2004, p. 25).

What is pertinent in Dewey's formulation is that science and ethics are interfaced and assembled in accordance with the demands of "progressively directed inquiry." Such inquiry is not primarily directed at real or imagined consequences or first-order deliverables, although the work of Modes One and Two on these topics is relevant in and of itself as well as being primary data for reflection. Rather, inquiry is directed at the possibility of the invention and implementation of equipment that facilitates forms of work and life. Whether such facilitation will occur, and whether it is efficacious or beneficial, remains to be seen.

Notes

1. For more details see the SynBERC Web site. This chapter only treats the efforts of the fundamental modules of Thrust 4.

2. See the Science and Technology Foresight program of the European Union, http://cordis .europa.eu/foresight/home.html.

3. Mapping and analyzing the distributions of this term would be the work of a much more extended genealogy. How to undertake such an enterprise within the anthropology of the contemporary as opposed to the history of the present is, currently, largely unexplored, and lacking the requisite navigational concepts and methods. "Equipment" takes different forms in the contemporary. This variability stems from several facts. First, the contemporary is neither a unified epoch nor a culture, and consequently there is no reason to expect there would be a single form within it. Further, scholarly works in present history have shown multiple facets to even a settled problematization and thus, it would be logical to assume that multiple solutions follow, requiring diverse equipment.

4. It should not be overlooked that, with the Belmont Report, ethicists were, for the first time, made part of the U.S. government, despite the increasing turn to moral discourse as the site of truth distinctions since 1950.

5. Microsoft Word dictionary definition.

6. Badiou (1995) writes that one calls "opinions les représentations sans vérité, les débris anarchiques du savoir circulant. Ou les opinions sont le ciment de la socialité. . . L'opinion est la matière première de toute communication" (p. 46).

7. See, e.g., the BIOS Center, London School of Economics, www.lse.ac.uk/bios, and the Center for Bioethics and Medical Humanities, University of South Carolina, http://www.ipspr.sc.edu/cbmh/default.asp.

8. For more on this technical use of the term *contemporary*, see ARC (www.anthropos-lab.net).

9. We will address these issues at more length in another article.

10. See Global Business Networks at www.gbn.com.

References

Badiou, A. (1995). *L'éthique: Essai sur la conscience du mal.* Paris: Hatier.

Barry, A. (2001). *Political machines: Governing a technological society.* London: The Athlone Press.

Beck, U., Giddens, A., & Lash, S. (1994). *Reflexive theory of modernization.* London: Polity Press.

The BioBricks Foundation. Available online at: www.biobricks.org (accessed September 2007).

Boltanski, L., & Thevenot, L. (1991). *On justification: Economies of worth.* (K. Portor, Trans.). Princeton, NJ: Princeton University Press.

Boltanski, L., Thevenot, L., & Porter, C. (2006). *On justification: Economies of worth.* Princeton, NJ: Princeton University Press.

Center for Nanotechnology and Society at Arizona State University (CNS-ASU) Web site. Available online at: http://cns.asu.edu/ (accessed September 2007).

Deleuze, G., & Guattari, F. (1991). *What is philosophy?* (H. Tomlinson & G. Burchell, Trans.). New York: Columbia University Press.

Dewey, J. (2004). *Reconstruction in philosophy.* Mineola, NY: Dover Publications.

Fisher, E. (2007). Ethnographic invention: Probing the capacity of laboratory decisions. *NanoEthics.* Available online at: www.cspo.org/documents/Fisher_ ProbingLabCapacity_Nanoethics-07.pdf (accessed September 2007).

Foucault, M. (1995). *Discipline and punish: The birth of the prison.* New York: Vintage Books.

Fox-Keller, E. (1984). *A feeling for the organism: The life and work of Barbara McClintock.* New York: Times Books.

Gibbons, M., Limoges, C., Nowotny, H., Schwartzman, S., Scott, P., & Trow, M. (1994). *The new production of knowledge: The dynamics of science and research in contemporary societies.* New York: Sage.

Jasanoff, S. (2005). *Designs on nature: Science and democracy in Europe and the United States.* Princeton and Oxford: Princeton University Press.

Jonsen, A. (2000). *A short history of medical ethics.* Oxford: Oxford University Press.

Kevles, D., & Hood, L. (1993). *The code of codes: Scientific and social issues in the Human Genome Project.* Cambridge, MA: Harvard University Press.

Koselleck, R. (1979). *Futures past: On the semantics of historical time.* (K. Tribe, Trans.). Cambridge, MA: MIT Press.

Lartigue, C., Glass, J. I., Alperovich, N., Pieper, R., Parmar, P. P., Hutchison, C. A., & Smith, H. O. (2007). Genome transplantation in bacteria: Changing one species to another. *Science, 316,* 632–638.

Luhmann, N. (1997). *Die Gesellschaft der Gesellschaft.* (2 vols., N. Langlitz, Trans.). Frankfurt: Suhrkamp.

MacIntyre, A. (1984). *After virtue: A study in moral theory,* 2nd ed. South Bend, IN: University of Notre Dame Press.

Nowotny, H., Scott, P., & Gibbons, M. (2001). *Re-thinking science: Knowledge and the public in an age of uncertainty.* London: Polity Press.

Pauly, P. (1987). *Controlling life: Jacques Loeb and the engineering ideal in biology.* Oxford and New York: Oxford University Press.

Pence, G. (1998). *Who is afraid of human cloning?* New York: Rowman & Littlefield.

Rabinow, P., & Bennett, G. (2007a). A diagnostic of equipmental platforms. Available online at: http://anthropos-lab.net/documents/wps/ (accessed September 2007).

Rabinow, P., & Bennett, G. (2007b). From bio-ethics to human practices or assembling contemporary equipment. Available online at: http://anthropos-lab.net/documents/wps/ (accessed September 2007).

Rasmussen, S., Bedau, M. A., Chen, L., Deamer, D., Krakauer, D. C., Packard, N. H., & Stadler, P. F. (2008). *Protocells: Bridging nonliving and living matter.* Cambridge, MA: MIT Press.

Shapin, S. (1995). *A social history of truth civility and science in seventeenth-century England.* Chicago: University of Chicago Press.

Stathern, M. (Ed.) (2000). *Audit culture: Anthropological studies in accountability, ethics and the academy*. London and New York: Routledge.

Synthetic Biology Project Web site (2004). Synthetic biology 1.0: First international meeting on synthetic biology. Available online at: http://syntheticbiology.org/Synthetic_Biology_1.0.html (accessed September 2007).

The Synthetic Biology Engineering Research Center Web site. Available online at: www.synberc.org (accessed September 2007).

Weber, M. (1941). Science as a vocation. In H. H. Girth and C. W. Mills (Trans.), *From Max Weber* (pp. 129–156). New York: Oxford University Press.

Weber, M. (1949). Objectivity in the social sciences. *The methodology of the social sciences*. (E. A. Shils & H. A. Finch, Trans.). New York: Free Press.

15

This Is Not a Hammer: On Ethics and Technology

Mickey Gjerris

Do Worry—Be Skeptical

Human minds have a limited capacity for worrying—despite your mother's continuous attempts to prove otherwise. This means that we can handle only a limited number of risks simultaneously. At some point, we just stop caring or begin substituting old concerns with new ones. It is simply too complicated to be simultaneously aware of the risks of food additives, traffic, terrorism, exploding lighters, meteors, loneliness, overweight, untied shoelaces, strangers with candy, crime, smoking, stupidity, bird flu, genetically modified organisms, flying, bathing, walking, breathing, and living. As I write these words, I am in a plane ready for takeoff from Nairobi to Amsterdam; I have just been informed that the plane will be sprayed with some kind of insecticide. Is this a risk? I do not know, and I am too occupied with the risk of the plane falling out of the sky to care. I focus on one risk and resign myself to the other.

We divide the many risks that fill our minds in the beginning of this century into categories to be able to live with them without going insane. The most widespread strategy seems to be denial, but in this chapter I just focus on the division between familiar and unfamiliar risks. A familiar risk is one that we know exists and has existed for some time. It might not be described as a friend, but to some extent it can be comforting to know these well-established risks are around. These are risks like driving or smoking. Nobody, not even the people engaging in the risky activities, would claim that these activities could not go wrong. But they still seemingly happily engage in the activity. The risk is familiar and they feel they can control it, however irrational that belief might be. Humans also have an astounding capacity for believing that bad things will happen only to other people.

But there are other, less familiar risks, which we[1] tend to be less relaxed about. I remember from my childhood how my grandparents were sure that I would lose my eyesight, if I watched too much television. My mother is worried that computers will eventually lead the human race to a painful and catastrophic end, and I cannot help worrying that my children will end up fundamentally uneducated due to the popularity of SMS technology. I did not go blind, the human race is still on, and SMS-technology will probably not be the end of education. These examples show that familiar risks are not necessarily less serious than unfamiliar risks, but that we tend to perceive them as less serious. In general, one can say that the more familiar a risk is to us and the more we feel that we have control over it, the more likely we are to take the risk (Hansen et al., 2003).

One should not, however, jump to the conclusion that all perceptions of risks related to unfamiliar phenomena, such as new technologies, are grounded only in a fear of the unknown, as some advocates of, for example, nanotechnology are prone to say (SmallTimes, 2004). It would be fairer to say that we meet new technologies with a kind of basic skepticism. And it would seem that the more promises that are made on behalf of a technology, the more skeptical we tend to be. I address some of the possible reasons for this skepticism in the next section. But there seems to be a certain kind of wisdom in weighing the promised benefits of a technology against the number and severity of the potential problems that the technology might cause. There is always another side of the coin—and the larger the coin, the larger the other side. Two of the more curious features about modern society—the seeming arbitrariness about what risks we take seriously and the number of risks we ascribe to new technologies—can perhaps be at least partly explained by our limited capacity to worry and the fact that new technologies are often hyped as revolutionary omni-problem-solvers.

The notion of risk that runs so deeply in Western culture today and our obsession with new technologies and their potential benefits and harms might not be as irrational as is often claimed (Bergler, 1995). The skepticism so prevalent toward certain kinds of technologies might just prove to be a very rational and sound way to interpret our past experiences—especially with science and technology (Lee, 1981).

A Changing Perception of Science

There has been a shift in our perception of science and technology during the past fifty years. This shift can be seen both in our general attitude toward science and technology and in the way these things are depicted in popular media, commercials, movies, and books—especially in the science fiction literature. My description of this shift will rely on my own experience with these sources. I should warn that the picture painted below is rather simplified, in order to draw out the central points needed for the rest of this chapter. Much more remains to be said about this subject.

The change in our perception can best be described as a shift from the view that science and technology were important in and of themselves; they were an unproblematic part of the general materialistic progress that Western societies experienced from 1850 to 1950. Doing science and introducing technology meant making progress, and making progress meant increasing opportunities and prosperity. Science moved the world and the only way was up, forward, and toward new opportunities. Some fears, however, were also associated with science and technology. It would be strange for such powerful factors not to induce some concerns about their power. But the concerns were not about science and technology as such; they were linked to the use of technology by individual scientists, such as the figure of Dr. Frankenstein, or the misuse of technology by unscrupulous criminals, the classic enemies of many an American superhero.

The reasons for this uncritical perception of science were manifold. In general, we used to have a certain trust in authorities, including science and scientists. Whether this trust was habitual or grounded in experience will not concern us here; neither will the question of the ethics of power relations. What is important is that, when technology was introduced, it was introduced by people in white lab coats bringing the miracles of science into the living room. Thus, television, Teflon, and refrigerators became part of everyday life and brought progress in a very tangible way. So not only did technology come from people whose education gave them an immediate authority, but the products that entered everyday life were seen as useful. People usually are willing to run greater risks and ask fewer questions about a new technology that they perceive as useful. As long as science was seen as producing useful technologies, we generally did not question it.

A contemporary example of this last point is mobile phone technology. Although some concerns have been raised about the risk of mobile phones frying your brain cells, very few people actually stopped using mobile phones for this reason. And I do not think that reassurances from scientists that cell phones present no risks are the reason for the lack of attention to their dangers. Rather, the technology is so convenient and useful that actually taking the concerns seriously would be too much of an obstacle to our everyday lives.

Nobody is doing Mobile Phone Ethics and nobody is engaging the public in a dialogue about the societal consequences of mobile phones. Considering the changes that mobile phones have brought to our lives, both socially and individually, this is actually very surprising, when one looks at how such technologies as nanotechnology and biotechnology are embedded in such discussions. Part of the explanation for this, I think, is that how mobile phones might be useful to us is immediately clear to us, on a very practical level, whereas the benefits of bio- and nanotech are perceived as being more speculative and distant.

A final reason for the earlier uncritical reception of science and technology that should be mentioned has to do with the general notion that science and technology were necessarily linked to positive progression. The human race had been developing since god-knows-when and now, with the help of science and technology, things were finally beginning to happen. *Homo sapiens* was becoming *Homo technicus*. One heard criticism of the fundamentalist belief in the future as a place filled with science, technology, and eternal bliss only from a few romantics and sentimentalists, mirroring the thoughts of Henry David Thoreau, who, in his famous *Walden* (1854), wrote: "Most of the luxuries and many of the so-called comforts of life are not only not indispensable, but positive hindrances to the elevation of mankind" (1997, p. 15).

Since these golden days of science and technology, belief in authority has generally been lost. The idea that the differences among humans in knowledge and expertise should be reflected in the kind of authority they have in matters concerning their knowledge or expertise is constantly under attack. This is not to be lamented, since the process has pulled many false and self-established authorities off their pedestals. However, it must be admitted that there has been a tendency to reject not only false authorities, but the whole notion of authority, leading to severe problems within the educational system and medicine, for example (O'Neill, 2001). Important for this chapter, though, is the general loss of trust in science and technology that this sinking belief in authorities has caused. Claiming that a piece of knowledge is scientific simply does not carry the weight it once did.

At the same time, the usefulness of science and technology has become easier to question. The practical results of the research into high-tech medicine, biotechnology, nanotechnology, and nuclear power are not always very easy to see. Science, especially bio- and nanoscience, has had a tendency to hype its research and promise far too much too soon. No doubt this constant overselling of scientific results leads to a certain weariness and skepticism when yet another science comes along promising to solve the problems of mankind (Meyer, 2005). Undoubtedly science and technology are still producing many useful products, and undoubtedly we are ignorant of many of them. But at the same time it must be admitted that, for instance, the miracles of biotechnology have been in the pipeline for decades longer than initially promised.

It is now clear that the belief that science and technology always bring us forward to new prosperity and richness is wrong. Like any other human endeavor, they can be misused, and considering their economic potential, they are perhaps more prone to misuse than many other human activities. At the same time, numerous large-scale catastrophes with technology have occurred. Bhopal, Chernobyl, Seveso, and Three Mile Island are all places locked in our collective memory as examples of the not-so-benign face of science and technology. Further experiences—with industrialized farming; chemical pollution of air, soil, water, and living beings; asbestos; air pollution—have likewise hurt the reputation of science and technology.

It would be wrong to say that science and technology are generally contested as important contributors to human life. Both continue to ease our lives and help alleviate suffering and pain, and both scientists and developers of new technologies are usually seen as respectable members of our societies (O'Neill, 2001). But we no longer look at them with uncritical awe. We have become skeptical consumers of the results of science and technology. We are ever more concerned about their consequences and implications with regard to human health, the environment, more abstract socioeconomic matters and the divide between North and South, and our increasing estrangement from nature.

Proto-What?

Into this context of skepticism toward science and technology comes a research program in which the goal is to design and produce living cells that can be made to perform specific tasks. These are called *protocells*, and this research program will undoubtedly give rise to a lot of discussion, if it lives up to the claims the scientists make (Pohorille & Deamer, 2002; European Center for Living Technology, 2004).[2] As the history of biotechnology has shown, many people have rather strong values, opinions, and feelings about life. So it is foreseeable that ethical concerns about protocell technology will arise.

This is not the place to explain the technologies involved in producing[3] protocells. Nor will I go into a lengthy discussion of what life might be in a biological or ontological sense—although such a discussion could clarify the claims made by scientists working in the field. What I will do here is briefly review some aspects of protocell technology and discuss how the concepts discussed earlier (usefulness, risk, familiarity) apply to them.

First of all, it is worth noticing that the technology, if one dares use such a mature word, is only in its infancy. So far, the possible applications border on pure speculation and wishful thinking. So, if one is to discuss the ethical aspects of producing protocells, it will be necessary to engage in a discussion about fictional opportunities. This is not wrong in itself, but it creates the risk of two extremes: either imagining exaggerated worst-case scenarios, or imagining that the technology can be used only for benign purposes and entails no risks to anybody whatsoever. Thus, both the actual risks and the actual usefulness of protocell technology are very hard to discern today; one must take that into consideration when setting up any kind of collective reflection (discussion, dialogue, etc.) on the technology (Grinbaum & Dupuy, 2004).

Another important aspect to take into consideration is the biological nature of the technology. As mentioned previously, the history of biotechnology has taught us, if nothing else, that life is especially important to humans. Thus, a technology that proposes either to redesign living cells to perform specific tasks or actually to build artificial constructs that, in important aspects, resemble living cells will be (and should be) considered controversial, just because it presumes that life is something to be redesigned or designed from scratch. It might emerge in the discussion that the aspirations of the technology somehow justify technologification of the concept of life. But at the outset, any technology that proposes to use life as a mere means to reach some end irrelevant to the life used will undoubtedly be ethically provocative to many people.

The sheer novelty of protocell technology and its alleged huge potentials will ensure that, if it ever leaves the labs and turns up in the marketplace, it will attract a lot of public attention. As with such controversial and unfamiliar technologies as animal cloning, it will be subject to public scrutiny, discussion, rumors, hopes, fears, and skepticism (Lassen, Gjerris, & Sandøe, 2005). It is therefore not without reason that scientists involved in protocell research, such as the organizers of the EU's Programmable Artificial Cell Evolution project (see note 3) and the editors of this volume, have already begun looking into the possible ethical considerations. It is only prudent that the stakeholders behind the technology (researchers, industry, and financial interests) try to anticipate the reactions toward the technology. Calling for ethical reflection and dialogue seems the only wise thing to do—but as I argue in the next section, it is painfully necessary to qualify this call, if it is to be more than mere PR strategy.

The Call for Ethics and Dialogue

I have noted that new technologies such as biotechnology and nanotechnology are often hyped (Berube, 2006). But so are the practices of ethics and dialogue. Once it was a hallmark of any self-respecting grassroots organization with a critical stance toward technology (so-called nongovernmental organizations or NGOs) to demand that the industries and scientists involved with a given technology reflect on the ethical implications of their endeavor and engage in a public dialogue about them. Today, almost everybody agrees that we should do this. Scientists and industry, governments, advisory bodies and research funding bodies, stakeholders and citizens—all join in calling for ethical reflection and dialogue (Roco & Bainbridge,

2002; The Royal Society, 2004; European Commission, 2005). They all agree on the importance of being and doing good, but what exactly this entails is very controversial. Not surprisingly, the same can be said about dialogue. Everybody agrees that dialogue is good, but there is no agreement about such basic questions as who is to participate, what the subject should be, when in the development of new technologies it should happen, and what the goal should be.

An obvious example of all this is the current rhetoric surrounding nanotechnology. Important industrial stakeholders, consulting agencies, public advisory bodies, government-appointed committees, transnational political entities, and investors write myriad reports on the possibilities in nanotechnology. A typical report states that "Nanotechnologies will be a major technological force for change in shaping . . . business environment across all industrial sectors in the foreseeable future and are likely to deliver substantial growth opportunities" (Allianz, 2005, p. 3). In addition, one can almost always find statements of this nature:

The Government agrees that properly targeted and sufficiently resourced public dialogue will be crucial in securing a future for nanotechnologies. The Government's aim for public dialogue around nanotechnologies is to elicit and understand people's aspirations and concerns around the development of these technologies. Through the dialogue process, scientists and the public can jointly explore existing and potential opportunities, and policy-makers will want to hear about, and then respond to, public concerns related to ethical, social, health, safety and environmental issues. (HM Government UK, 2005, p. 21)

Interestingly enough, the reasons stated for establishing a dialogue are invariably the same. Sandler and Kay (2006) have read through a lot of reports on nanotechnology and conclude that one of the most mentioned fears is that nanotech will end up in the same catastrophic situation as genetically modified (GM) crops did in the 1990s, with consumer rejection and public controversies. The way to avoid this is invariably understood to be connected with the idea of "doing" ethics and "having" a dialogue. However doubtful it may be that a lack of ethical dialogue caused the situation for GM organisms, or GMOs (especially plants), for human consumption, and however dubious it is to call this a disaster or catastrophe,[4] it should be clear that there is a widespread call for ethical dialogue about new technologies. This call is so widespread that even the opponents of nanotechnology hold that the way to proceed is through ethical reflection and dialogue (ETC Group, 2003).

The reason for this apparent unity is not, regrettably, that the different stakeholders from scientists to NGOs agree on some kind of participatory ideal of democracy. The apparent unity is possible only because it is unclear what is actually meant by ethics and dialogue. In the next section I try to unfold these concepts and show how differently they may be interpreted.

An *Ethical* Dialogue?

There are many ways of understanding ethics and dialogue. In this section, I focus on two very different understandings of ethics that arise in the discussion about new technologies like biotechnology and nanotechnology. In the next section I describe three understandings of dialogue.

Ethics can be understood in two basic ways. The first is as a problem solver—a hammer, so to speak, that can be used to solve the ethical problems that might arise when we develop new technologies. Thus, research projects often have a small amount of funds allocated to the so-called ELSI (ethical, legal, and social issues), the idea being that this will somehow "solve" ethical problems arising form the research or prevent the research from raising such problems. The second way of understanding ethics is as a flashlight, that is, a means of reflection that can clarify things, but probably also complicate them, and very seldom "solve" anything. Ethical concerns seldom disappear when one reflects on them. To understand this, it can be helpful to look at the history of modern applied ethics.

The discipline of medical ethics arose in the aftermath of the atrocities of so-called doctors in the concentration camps of World War II. This was the first time, at least in the modern era, that ethics examined distinct ethical questions about a specific area of human actions. Ethics went overnight from academic oblivion into governmental committees, editorials, and public meetings, to such a degree that the British philosopher Stephen Toulmin, in the mid-1970s, could write an article on "How Medicine Saved the Life of Ethics" (Toulmin, 1982).

Next came *environmental ethics,* following the discovery of the destructive consequences of the industrialization of the Western world, and then *bioethics* as an answer to the technological developments within medicine and agriculture. By now we have also witnessed the advent of nursing ethics, business ethics, legal ethics, and so on. For a variety of reasons, *ethics* has become popular within most disciplines. One of these is that ethics is now viewed as something good to have on your side, whether you are a private person, a politician, or a company. It simply pays off to be "ethical," which in this sense means to do (or pretend you do) as the majority of your society thinks one ought to. And at the same time, ethics has been specialized to the extreme that almost all sectors of human behavior are seen to demand their very own kind of ethics.

But even though ethics has been specialized to a sometimes almost comical degree and is being funded as never before, there are no signs that this has reduced the amount of ethical disagreement. This is illustrated by the fact that, although the ethics of GMOs has been discussed for the past 25 years in the Western world, no agreement whatsoever has been reached. The participants in the debate have just dug the trenches deeper (Gjerris, 2006). Obviously, the ethical concerns about a technology do not disappear just because they are discussed. What seems to happen is that participants become more and more convinced about the strength of their own arguments.

The underlying problem is that we all live our lives governed by values we cannot defend rationally. There simply is no "scientific" way of deciding whether socialism or liberalism, utilitarianism or deontology, or any other basic ethical framework is "best." It is easy to agree that one ought to be "good," but almost impossible to agree on what "good" actually means (Engelhardt, 1996). To some people good means using all available technology to improve human welfare, whereas to others it means limiting the use of technology. Ethics cannot tell us which of these views is right.

However, we can reasonably expect ethics to help us clarify our own values and better understand the values of others. This does not necessarily lead to agreement, but it might, with a bit of patience, lead to a greater respect for the opinions of others. The Danish philosopher Ulli Zeitler has phrased it this way: "Essentially, the task and reasonable expectation of philosophical activity is not to solve problems, although we may advance considerably by clearing up the central concepts, but opening our eyes to previously unconsidered problems. The last function is crucial for giving new directions to future inquiries" (Zeitler, 1997, p. 39).

Ethics is thus not a hammer but a flashlight that we point at reality to discover what our disagreements are really about and perhaps illuminate problems that were hidden.

Another problem that the development within ethics raises is the specialization into different areas of human behavior. It is obviously proper to investigate whether any specific human practice entails specific ethical challenges that must be thought through and discussed. The problem is the tendency to disregard the ethical challenges that are common to different kinds of behavior. To some degree, ethics today is a discipline in which only specific problems are judged worth discussing. General problems are recognized, but it is somehow never the right time to discuss them. Because they belong to several contexts, they end up belonging to none. An example of this is the claim often made in the ethical discussion of animal biotechnology that, since the animal welfare problems encountered with cloned and transgenetic animals are similar to those encountered within other kinds of animal reproduction technology, there is no need to discuss these problems in the specific context of animal GM and cloning (Gjerris, Olsson, & Sandøe, 2006). But then, when should the question be discussed? In the case of nanotechnology, the repeated claims that we should discuss only actual applications such as specific nanoparticles, and not the visions fueling the huge investments in nanotechnology, is a similar reaction.

The visions are said not to be specifically tied to concrete applications of nanotechnology and should therefore not be discussed within the context of nanotechnology.[5] When and where we should discuss the far-reaching visions of both an Eric Drexler[6] (Drexler, 1996) and a Mihail Roco[7] (Roco & Bainbridge, 2002) is left unsaid and, more often than not, undiscussed. The same problem will in all likelihood arise concerning protocells.

Is there, then, any definitely right way to do ethics? I doubt it! Ethics will always be something that is done from a specific perspective. There is no "view from nowhere" (Nagel, 1986) from which we can gain an objective insight into the ethical aspects of new technologies. All we can do is be honest about our fundamental values and think through what courses of action they recommend to us. And then we can try to listen to those with other fundamental values, not necessarily agreeing with them, but at least taking their thoughts into consideration. Again, this might not solve any problems, but nobody ever promised us that philosophy could do that.

An Ethical *Dialogue*

It is widely accepted today that the public should participate in forming the policies for emerging technologies such as biotechnology and nanotechnology. But although the concepts of public participation and dialogue are widespread in the Western world, the meanings of these notions are so far from being self-evident that they border on the obscure (Nielsen, Lassen, & Sandøe, 2004).

The meaning of "public participation" can generally be said to depend on the reasons for supporting the idea in the first place. Some support public participation as a way of ensuring that the public has some sort of democratic or semidemocratic influence on how research and applications of new technologies are supported. Others see public participation as a way of legitimizing the technologies in the eyes of the public. In the first case, the goal is to live up to certain democratic ideals without influencing the result of the participation, whereas in the latter case the whole point is to get the technologies accepted.

In the real world, the motivation for seeking public participation is seldom clear-cut. But as a rule of thumb, it would be fair to say that the closer one is to being motivated by democratic ideals rather than by interest in having the technologies accepted, the more meaningful the concept of dialogue becomes.

This can be seen most clearly in the notion of the knowledge deficit model. According to this model, the reason for public skepticism toward new technologies is the lack of knowledge that characterizes the public's relationship to, for example, scientists. The remedy to "cure" the skepticism is therefore to provide information;

acceptance will follow. Although the model has been refuted regarding both biotechnology and (although with less empirical material) nanotechnology (Cobb & Macoubrie, 2004; The Danish Board on Technology, 2004; The Royal Society, 2004), some still believe that dialogue can be transformed into a monological information stream that will result in wider acceptance of a given technology.

At the same time, another extreme form of dialogue seems to be forming; one that is just as monological but in which the public rather than the scientists do all the talking. Some stakeholders (especially policy makers and industry) believe that the way to have a dialogue is to ask the other person what he wants, and then give it to him. So when facing technological developments that will have an uncertain public reception, one could ask the public (typically through quantitative polling) what kind of development they want and then try to bring that development about.

And what, you may ask, is the problem with that? I have just lamented that one form of "dialogue" concentrates merely on generating public acceptance. Why not welcome this attention to the public? The reason is basically that it turns dialogue into marketing research. One can always ask how deeply the public should be involved in decision making—whether they should be polled just about their top-of-the-head opinion or be offered some chance to deliberate about the issues. But in the end, it is still just marketing research, done to a greater or lesser degree of eloquence and sophistication. One thing is for certain: As long as the answers one obtains from such research are not thematized, analyzed, and evaluated in some kind of critical discussion, but are just used as weathercocks—ways to figure what to do and especially what not to do—the word *dialogue* remains empty.

Dialogue is a complicated concept with a long and entangled history within the disciplines of philosophy and ethics. I do not want to end up in a Habermasian discourse ethics, where the description of the ideal dialogue somehow ends up answering the ethical questions that the dialogue originally sought to resolve. However, I believe it is very important to clarify what kind of dialogue should be used to involve the public in decisions about new technologies. As always, attempts to answer the "how" questions always raise the "why" questions!

Dialogue is basically a means of ethically balancing two very important considerations. One consideration is to respect other people, and the second is to take responsibility for one's own views of the world and attempt to do what one sincerely believes to be in the best interest of other people (whoever they may be). A classical example of this conflict can be found in the relationship between the physician and the patient. It is a mistake to think the physician's job is just to provide neutral information and respect at all costs the patient's right to self-determinacy. Physicians also have an ethical duty to take responsibility for their patients' well-being. The challenge for physicians is to balance taking responsibility and being respectful.

The concept of dialogue implies that there are two or more opinions about something and that the people holding these opinions are willing to discuss them, holding a small window open in the back of their minds to the possibility that they may be wrong. A dialogue in which one party decides from the outset that only one party (usually the other one) could end up changing his or her minds is no dialogue, but could better be described as a caricature of energetic and zealous religious proselytizing.

I readily admit that these remarks give no concrete suggestions as to how the concept of dialogue should be operationalized, if the aim is to have a "true" dialogue. But when engaging in philosophy and ethics, it is sometimes necessary to visualize the ultimate goals that humans could choose to seek—to describe the ideal state of things. Afterwards, we can always begin cutting this ideal down to size with all our practical reservations. So, although the ideal dialogue might not be possible, it is nevertheless important as something to aim for. Dialogue does not mean simply an exchange of views between two parties, eyeball to eyeball in the conference chamber. In its fullest sense, it is what the early Romantics called *sym-philosophy*, a term I loosely translate as "lovingly seeking wisdom together" (Patterson, 1998, p. 148).

The conclusion to these remarks is thus rather simple. Before entering into dialogue with the public, it is paramount (from an ethical point of view) that we—and who this "we" are, is indeed a huge question in itself—ask the "why" question. Our answers to the "why" question will, to a large degree, answer the "how" question, too. We should perhaps begin to do this. The possibility that overhyped concepts such as dialogue and public participation could elicit only a yawn increases the more these concepts are seen as rhetorical devices used by politicians and scientists whom "the public" has perhaps never trusted as little as they do these days.

Notes

1. In this chapter I use the term "we" to refer to humans in general, whom I believe share many basic experiences. But, of course, people's experiences differ, and they arise in different contexts. So disagreements with my claims about humans in general might result from the fact that some readers do not share my philosophical, social, cultural, sexual, or religious context. I can only apologize for this, but maintain that the only way to talk about life is to talk *through* experience *to* experience. This methodology can be labeled *Demonstratio ad ocolus* or "Go and see for yourself, if I'm not right!" (Jensen, 2001).

2. See also the Web site for the EU project on Programmable Artificial Cell Evolution (PACE), at http://www.istpace.org/ (accessed January 2007).

3. Should we call it *creating* or *producing*? This is an important question, since labels influence how we think of things. In this case, arguments can be made both ways. I will stick with *producing*, reserving *creating* for the act of bringing to life more open-ended entities than mere cellular factories.

4. From a market economic perspective, it was a great success. Somebody produced something and sold it on the market. The buyers did not want it and the producers had to retract the product and find something else to sell. That is successful capitalism in its purest form. Strangely, most people who lament the outcome of the GMO debate are usually strong supporters of market economy.

5. See, for instance, this quote from *Scientific American*: "If the nano concept holds together, it could, in fact, lay the groundwork for a new industrial revolution. But to succeed, it will need to discard not only fluff about nanorobots that bring cadavers back from a deep freeze, but also the overheated rhetoric that can derail any big new funding effort. Distinguishing between what's real and what's not in nano throughout this period of extended exploration will remain no small task" (Stix, 2001, here quoted from Selin, 2006, p. 48). The most interesting thing to note here is how the author finds that there is some objective way of settling what is real and what is not—when it is exactly that question that is contested.

6. Dr. Drexler is one of the founding fathers of nanotechnology and, exactly because of his far-reaching visions, one of the most contested persons within the nanotechnological environment.

7. Dr. Roco chairs the National Science and Technology Council's subcommittee on Nanoscale Science, Engineering and Technology, and is Senior Advisor for Nanotechnology at the National Science Foundation. He is perhaps the most influential figure in the nano-community in the United States.

References

Allianz (2005). *Small sizes that matter: Opportunities and risks of nanotechnologies. Report in co-operation with OECD International Futures Programme*. Allianz. Available online at: www.oecd.org/dataoecd/4/38/35081968.pdf (accessed September 2007).

Bergler, R. (1995). Irrationality and risk—a problem analysis. *Zentrallblatt Für Hygiene und Umweltmedizin, 197* (1–3), 260–275.

Berube, D. M. (2006). *Nano-hype: The truth behind the nanotechnology buzz*. New York: Prometheus Books.

Cobb, M. D., & Macoubrie, J. (2004). Public perceptions about nanotechnology. *Journal of Nanoparticle Research, 6,* 395–405.

The Danish Board on Technology (2004). *Public views on nanotechnology*. Copenhagen: The Danish Board on Technology.

Drexler, E. (1996). *Engines of creation*. London: Fourth Estate.

Engelhardt, H. T. (1996). *The Foundations of bioethics*, 2nd ed. Oxford: Oxford University Press.

ETC Group (2003). *The big down. Atomtech—technologies converging at the nanoscale.* Available online at: http://www.etcgroup.org/en/materials/publications.html?id=171 (accessed January 2007).

European Center for Living Technology (ECLT) (2004). ECLT—Press release opening. Available online at: http://bruckner.biomip.rub.de/bmcmyp/Data/ECLT/Public/documents/ECLT-PressRelease.pdf (accessed January 2007).

European Commission (2005). Nanosciences and nanotechnologies: An action plan for Europe 2005–2009. Communication from the Commission to the Council, the European Parliament and the Economic and Social Committee. European Commission. Available online at: http://europa.eu.int/comm/research/industrial_technologies/pdf/nano_action_plan_en.pdf (accessed January 2007).

Gjerris, M. (2006). *Ethics and farm animal cloning: Risks, values and conflicts.* Copenhagen: The Danish Centre for Bioethics and Risk Assessment.

Gjerris, M., Olsson, A., & Sandøe, P. (2006). Animal biotechnology and animal welfare. In *Animal protection and welfare*, Ethical Eye series. Strasbourg: Council of Europe Publishing.

Grinbaum, A., & Dupuy, J. P. (2004). Living with uncertainty. *Techné: Research in Philosophy and Technology, 8* (2), 4–25.

Hansen, J., Holm, L., Frewer, L., Robinson, P., & Sandøe, P. (2003). Beyond the knowledge deficit: Recent research into lay and expert attitudes to food risks. *Appetite, 41,* 111–121.

HM Government UK (2005). *Response to The Royal Society and Royal Academy of Engineering report: "Nanoscience and nanotechnologies: Opportunities and uncertainties."* Available online at: http://www.dti.gov.uk/files/file14873.pdf (accessed January 2007).

Jensen, O. (2001). At hente rummet ind igen. Teologiske betragtninger over vort naturforhold. In L. D. Madsen & M. Gjerris (Eds.), *Naturens sande betydning—om natursyn, etik og teologi* (pp. 78–105). Copenhagen: Multivers.

Lassen, J., Gjerris, M., & Sandøe, P. (2005). After Dolly: Ethical limits to the use of biotechnology on farm animals. *Theriogenology, 65,* 992–1004.

Lee, T. R. (1981). The public's perception of risk and the question of irrationality. *Proceedings of the Royal Society of London. Series A, Mathematical and Physical Sciences, 376* (1764), 5–16.

Meyer, G. (2005). *Ethical challenges in bio-scientific projects: Extracts from a study of ethical and societal aspects of animal disease genomics.* Copenhagen. Danish Centre for Bioethics and Risk Assessment.

Nagel, T. (1986). *The view from nowhere.* Oxford: Oxford University Press.

Nielsen, A. P., Lassen, J., & Sandøe, P. (2004). Involving the public—participatory methods and democratic ideals. *Global Bioethics, 17,* 191–201.

O'Neill, O. (2001). *Autonomy and trust in bioethics.* Cambridge: Cambridge University Press.

Patterson, G. (1998). *The end of theology—and the task of thinking about God*. London: SCM Press LTD.

Pohorille, A., & Deamer, D. (2002). Protocells: Prospects for biotechnology. *Trends in Biotechnology, 20*, 123–128.

Roco, M. C., & Bainbridge, W. S. (2002). *Converging technologies for improving human performance. Nanotechnology, biotechnology, information technology and cognitive science*. The National Science Foundation. Available online at: www.wtec.org/ConvergingTechnologies/Report/NBIC_report.pdf (accessed January 2007).

The Royal Society (2004). *Nanotechnology: Views of the public. Quantitative and qualitative research carried out as part of the nanotechnology study*. London: The Royal Society.

Sandler, R., & Kay, W. D. (2006). The GMO – nanotech (dis)analogy? *Bulletin of Science, Technology & Society, 26* (1), 57–62.

Selin, C. (2006). *Volatile visions: Transactions in anticipatory knowledge*. PhD Thesis, Copenhagen Business School, series 08.2006.

SmallTimes (2004). *As nanotech grows, leaders grapple with public fear and misperception*. May 20, 2004. Available online at: http://www.smalltimes.com/Articles/Article_Display.cfm?ARTICLE_ID=269457&p=109 (accessed January 2007).

Thoreau, H. D. (1997). *Walden; or, life in the woods*. Edited with an introduction and notes by Stephen Fender. Oxford: Oxford University Press.

Toulmin, S. (1982). How medicine saved the life of ethics. *Perspectives in Biology and Medicine, 25*, 736–750.

Zeitler, U. (1997). *Transport ethics. An ethical analysis of the impact of passenger transport on human and non-human nature*. Århus, Jutland: CeSaM/University of Århus.

16

Toward a Critical Evaluation of Protocell Research

Christine Hauskeller

Protocell science aims to engineer cellular prototypes, in other words, to manufacture simple living units. Two strategies are pursued to achieve this end: building protocells from nonliving materials (bottom-up approach) and reducing complexity in natural cells (top-down approach) (Rasmussen et al., 2004, p. 963). The long-term goals are to create novel living products for medical and technological uses in, for instance, gene therapy, computing, and industry, and gain knowledge of the basic bio-physico-chemical processes underlying life.

Any attempt to create minimal cells or entities that are described as living requires a definition of *living*. Protocell research employs certain definitions of life that are common in current biological science but appear reductionist in the sense of being functionalist, technical, and lacking qualitative criteria such as those involving concern for health or well-being. Thus, while all powerful technologies are liable to change living conditions, protocell research could become so powerful that, in addition to exerting a strong influence on future social reality, it changes our notions of life. On both counts, the development of protocell research calls for ethical evaluation.

I begin this chapter by reviewing some product-directed aims of protocell research and discussing which ethical issues are likely to be the most salient to the governance of protocell science. I argue that the most important issues are the risk of destroying natural environments and threats to human lives from medical and weaponry use, familiar problems of bioscience that are already on the agenda of ethical and regulatory discourses. Then I discuss the epistemic aims of the field and the socially controversial questions to which protocell research is expected to deliver answers. There are cognitive dissonances between the understandings of life in this field and those predominant in the general moral and political culture of Western democracies.

The starting points for this analysis are our predominant moral values, which imply recognition of general human rights and equality and which are based on the regulative ideal of human dignity. In contrast to this social setting of values, knowledge societies marginalize those humanities and social science fields that have techniques for discussing such themes. Instead, sciences and technologies that conceptualize life as a complex biochemical mechanism are generated and promoted. The opposition between the value-based thought style in modern societal practice and reductionist scientific understandings of life becomes problematic when solutions to problems in the first are expected from developments in the second. In my view, the boundary between these levels is the site of the most interesting philosophical problems associated with a bioscientific research area that aims to engineer new forms of life.

Assessment of Protocell Science

Emphasized among the many knowledge and engineering goals of protocell science are the potential benefits for clinical applications and high-technology products (Pohorille & Deamer, 2002, p. 127; Noireaux & Libchaber, 2004, p. 17674). Justifying science with reference to therapeutic utility is very common these days, presumably because health matters to people, and also because its improvement is regarded as a measure of political success in current democracies. Public funding is allocated primarily to science that may enhance public well-being. Inherent in this promise of therapy is a biomedical self-understanding of protocell research, which in turn brings with it public accountability of the research.

Protocell research is not a precisely defined field of science or technology with outcomes that can be readily assessed. Rather, it is an assembly of multidisciplinary projects with diverse aims and methods, but currently without outcomes. This means that the visions and ethical issues that might arise are opaque and uncertain, which presents problems for an ethical evaluation. The speculative character of any prospective assessment of social change resulting from the invention or application of new knowledge and technologies is its most pressing credibility problem. In biomedicine, social attitudes often change simultaneously with technological change. Social and legal changes accompany the emergence of the science independent of its success in health product delivery—successes that could, if achieved, become weighed against the ethical and economic investments. Consider, for example, the policy changes and emerging regulatory rules around embryo research, and the management of female human and animal gametes in the context of advancing stem cell science in Britain, which are directed toward accommodating cell nuclear replacement technologies and the creation of human-animal chimeras. Public sociopolitical struggles have occurred concerning the deregulation of global trade restrictions in human egg cells and embryos or the creation of human embryos. In Britain, these struggles are usually in the form of public consultation exercises (HFEA, 2004, 2006) or statements by advisory bodies (Scottish Council on Human Bioethics, 2006). Wider public and media debates follow along the lines set by the former ethical governance institutions. In those struggles, the prospective benefits from the research are evaluated against existing regulatory practice, which is considered to reflect the dominant moral intuitions and understandings.

From its early days in pathology, scientific medicine has been posing challenges to social understandings of human bodies and altering established concepts of human health, life, and death. Any assessment of potential is uncertain; this applies both to the supposed prospects and achievements of a scientific field and to the social changes that might arise from it. The important advantage of the pro-science attitude of (often naive) optimism and progressiveness is that it promises a better future, while the caution and ambivalent scenarios presented by philosophers, social scientists, and ethicists evoke little excitement or confidence. In comparison to genetics or stem cell research, public awareness of protocell research is still minimal. This may change quickly, should inflated promises of miracle cures and colorful rhetoric in presenting the science attract wider public attention and press coverage.

This chapter presents a classification of the aims of current protocell research and assesses the likelihood of the science to achieve them in a beneficial way. Tentative evaluation of the relevance to protocell research of important ethical issues that arose in the context of other bioscientific technologies is followed by discussion of the epistemic goals and the appropriateness of this science to answer epistemological questions. The conflict between understandings of life that currently exist, parallel but separated, in different social thought communities is at the center of this discussion.

The Problem-Oriented Approach to Science and Technology Evaluation

Ethical assessment of science has to begin early. If a science or technology is found to be problematic after implementation, it is extremely difficult to take it back, given the infrastructure created during its development, and the fact that its products often benefit some. Complicated negotiation over the avoidance of negative impacts, then, becomes the main activity of ethics, and regulation becomes the outcome of a series of usually flawed compromises rather than clear policies on the development and public funding of science. Byproducts derived in the process of trying to model and modify nature in new ways can have long-lasting effects on medicine, the development of biological weapons, social morality, or the just distribution of access to therapies and wealth. The starting points for evaluation are the particular societal problems that a specific research or technology promises to address and the side effects accompanying its development. Ideally, science policy compares alternative approaches to solving the general problem and their respective social and economic costs.

Problem-oriented assessment means, for example, that when we expect improved healthcare and treatment options for common diseases from stem cell, genetics, or protocell research, we must ask whether those approaches to lessening suffering from, for example, diabetes, liver cirrhosis, myocardial infarction, or Alzheimer's disease are among the most promising both medically and in terms of social policy. Other solutions and policy options need to be considered, which might range from stricter food, drink, and drug policies or general education programs to various biomedical approaches or more liberal euthanasia regulations in certain cases. Isolated ethical assessment of any one approach to better health and life conditions tends to be insufficient, because the many alternative ways in which the social goals could be achieved never come into sight. Often, less fashionable approaches to improved health, that might be less costly in monetary or ethical terms, become marginalized.

An obstacle to systematic attempts at problem-oriented science evaluation is the inevitable tension between the short-term focus of political decision making in Western democracies, the actual long-term effects of science and technology, and the speculative character of any evaluation regarding the likely successes and sociopolitical effects of the latter. Despite these obstacles, the problem-oriented approach described by Wolfgang Bender (1988, pp. 174–185) is the only acceptable evaluation method. To understand the reasons why life sciences, particularly the manipulation and creation of life, evoke ambivalence and adverse reactions in many people, a critical philosophical perspective is helpful. Critical theory and anthropology provide arguments that capture the particular situation in which protocell research is placed in the wider cultural environment, and what its aim to create life means with respect to both specific cultural traditions and the self-understanding of humanity.

What Is Life?

Any announcement that life has been made from a nonliving substance presupposes a set threshold according to which it is decided when something is alive. A transition from nonliving to living has to have happened somewhere between the raw materials used and the supposedly living product. The three criteria for life employed in protocell research are metabolism, self-replication, and evolution: "[T]here is general agreement that a localized molecular assemblage should be considered alive if it continually regenerates itself, replicates itself and is capable of evolving" (Rasmussen et al., 2004, p. 963). Other descriptions speak of "metabolism, self-assembly, and self-replicating proto genes" as the basic features of an evolving chemical system that can be called alive (Nilsson Jacobi, 2006).

Metabolism is another word for regeneration in the sense of sustaining life by exchanging materials and energy with the environment. *Capability to evolve* means that a system maintains a recognizable identity while adapting to changing environments. Self-assembly means the ability to integrate its elements together to form a living entity without direct external help (e.g., electrical or chemical manipulation). Self-replication is the ability to reproduce. These criteria for life demand that a system survive over time in a defined environment, sustain a certain degree of stability of characteristic features while still adapting itself to changing conditions, and produce more of its kind. These criteria are compatible with certain common medical measures of human health, and thus are as much a cultural product as any other human activity. However, just as physicochemical deficits in function that have effects on health may be described in molecular terms, the effects are measured not on the molecular level but rather on the normative level of human well-being or suffering. Medical states rarely correspond exactly to molecular states, as we know from genetic testing or other medical tests. Genetic findings, or the presence of certain microbes or viruses, for instance, do not provide reliable prognoses for individuals; they deliver explanations only when molecular symptoms coincide with being unwell. The definitions of life accepted in molecular biology and in medicine may partly overlap, but they are distinguished by the subjective component in the latter. In the medical context, molecular markers are useful only in relation to other surrogate markers, which again only indicate quality of life and mortality.

Affective Dissonance in Attitudes Toward the Life Sciences

Many people observe the developments in the life sciences with ambivalence, and many respond with strong emotions such as desire, hope, fear, and angst. Protocell science is affected by this ambivalence. Several philosophical theories can be used to explain such a reaction to science and technology, and to the biosciences in particular. Philosophy aims at achieving better understandings of the nature of humanity in relation to itself and the world around it, and tries to explain the human ability to theorize and reconfigure both the external world and itself. Critical theory and philosophical anthropology are two approaches that differ with respect to their emphasis on and evaluation of human power over human and external nature.

The *Dialectic of Enlightenment,* by Max Horkheimer and Theodor W. Adorno (1933), describes the ambivalence between the human desire to rule over nature and the human dependence on nature as natural beings in the context of the breakdown of German high culture into barbarism during the Third Reich. Science is understood in this text as power over nature in the terms of instrumental rationality. Instrumental rationality aims to overcome and abandon the forces of nature and transcend human naturalness. However, this overcoming of nature seems unavoidably accompanied by barbaric actions that destroy the bond to nature and to the dimension of the self that is part of nature. The desire for morality and civilized community contradicts the denial inherent in the exercise of power over nature. The text thus evaluates triumph of instrumental rationality over communal and natural integrity as dangerous and negative.

A different and less negative interpretation of the human ability to change its own nature and that of its surroundings can be found in the philosophical anthropology of Helmut Plessner. He describes a similarly twofold understanding of human existence as both centric and ex-centric (Plessner, 1975). However, Plessner thinks that human cognitive ability is a natural but not denaturalizing element of human existence (Lindemann, 2007). Therefore, the human creation of artificial life and the reshaping of human bodily existence into artificial forms are not in themselves problematic or in conflict with the centric positioning.

More recent literature, in particular by feminist scholars, has offered accounts of the tension between modern scientific rationality and the sources of compassion and a good life (e.g., Merchant, 1982; Fox Keller, 1985). The feminist literature includes contradictory interpretations concerning the liberating and oppressive potentials of biotechnology, particularly reproductive technologies, for more social equality and self-identity of women (Sherwin, 1992; Purdy, 1996).

Those diverse philosophical understandings help explain the ambivalent public responses to the life sciences. They show that science needs to be understood as a cultural product that reflects and interferes with how the relationship between humanity and nature is conceptualized. Science promises more independence from and control of nature (e.g., with regard to alleviating suffering from disease). Its way of addressing life in its attempts to achieve these goals is generally reductionist and technical, and this can clearly be said for the criteria of life underlying protocell research in comparison to what it means for humans to be alive and to lead a good life.

I personally do not consider it very likely that the manipulation of genomes, the insertion of chimeric or artificial elements into cells and bodies, or forms of artificial life as such pose a threat to human life in civilized societies. However, I do agree that serious ethical problems accompany bioscientific invention. In my view, ethicists should be less concerned with abstract human naturalness and identity, and devote more attention to the commoditization and trade in human tissues and the exploitation of women and poor people that comes with them, or the voluntary and involuntary killing of animals and humans in laboratory experiments and experimental medicine. The life sciences have often been criticized for their hubris, and for a number of good reasons, particularly with respect to the sociopolitical conditions within which they develop and that shape them in specifically problematic ways. In public debates, however, fears are often expressed that seem less rational and substantial than worries about exploitation in global organ trade. These fears influence the cultural state of the biosciences, and need to be addressed. Sensible ethical assessment of protocell research has to take into account the conflicts between existing concepts of life and the public fears that affect scientific development, and public support for and understanding of the latter.

The aim to synthesize or create life from nonliving matter has cultural predecessors, for example, in the image of the Homunculus, the cabbalistic myth of the Golem, the creator image raised in science by splitting embryos in the nineteenth and early twentieth centuries, and in cloning experiments sixty years later. Life creation is neither unforeseen nor culturally neutral. Widespread religious and ethical views forbid it, because creation of life is considered God's work alone. Others argue that human engagement in the making of life undermines the basic existential equality among humans. They identify the given biological equipment of humanity as a natural residue of the intersubjective recognition of equality, the loss of which might irretrievably destroy the roots of democracy in this fundamental equality (e.g., Habermas, 2003).

Like every biological science, protocell research is grounded in a mixture of traditional and recent molecular definitions of what life is, and its presentations often conflate various understandings of life. Biology and medicine sometimes purport to change our understanding of life and what it is to be human. See, for example, the announcement of the Human Genome Project: "We have powerful new ways to see what it is that makes each individual unique—but perhaps more importantly we have new tools to see what it is that makes us all the same. Our common humanity is set out in the wonderful spiral staircase that is our DNA, and at last we can read its letters" (HGP, 2000). The current convergence of biology and medicine into biomedicine, and the manipulations of bodily structure that genomics and stem cell research intend, use a molecular understanding of life supposedly to improve the quality of life (Hauskeller, 2005). However, viewing life as a physicochemical composition that can be manipulated purposefully does not automatically enable one to improve the quality of life as subjectively experienced. The conditions for achieving better quality of life are complex; recent attempts in biology (and protocell science, in particular) to define life may foster discussion about relevant notions of life and their social meaning.

Aims and Directions of Protocell Research

Protocell research combines knowledge from different strands of biology, material science, chemistry, and physics with biochemical materials and components such as reconstructed genomes, autocatalytic RNA networks, self-replicating lipid aggregates, and enzymes. The direct aim is to assemble these elements into a minimal living system.

The bottom-up approach in protocell science seems the more radical. All the elements needed for life are human-made, and their interactions seem more controlled than in the top-down approach. The stepwise creation of synthetic components to a minimal cell supposedly delivers in-depth knowledge of what is involved in the process of life, at least the (no doubt questionable) assumption that if we can build it, we know what it is. In comparison, the top-down approach operates with, for example, the knockout of certain gene complexes or the removal of a cellular component from a living cell that then remains alive or not in the senses of life previously defined. This does not imply that the function of the remaining elements is well understood. In theory, the computer modeling of cellular life and its interactions, which is in practice important for both approaches, allows simulation of a multitude of changes in parameters that influence cell behavior. But simulation is only as reliable as the knowledge that has gone into constructing the parameters in the first place. In any case, simulation is never enough; wet (laboratory) testing is always required.

In its current early stages, the research is mainly geared toward successful engineering of special functions of cell elements. Examples are the engineering of lipid vesicles formed through catalysis of encapsulated clay particles with RNA on their surface (Rasmussen et al., 2004), or of vesicle bioreactors (Noireaux & Libchaber, 2004). The genomes and infrastructure of minimal standardized cells could, in a second step, be supplemented and enlarged to produce specific metabolic effects in a sick organism, for example.

The following rough classification of alternative pathways, knowledge aims, and desired technologies is intended to analyze their credibility as problem solvers. I suggest two major categories: product-oriented aims, and epistemological aims for better knowledge of life. Both may be found in statements about the goals of a research project, probably also for pragmatic reasons such as justification and funding purposes. Since World War II, many countries, especially Western-style democracies, have come to see science funding as a major responsibility of the state. This leads to public investment in science, and subsequently a strong request for the public accountability of science. With this development, public morality becomes a major element in science development. The U.S. government argued in 2000 that it refused federal funding for embryonic stem cell research because of the wide public rejection of the involved destruction of early human embryos (Best & Khushf, 2005). Apart from such scruples, the limited funds create a moral obligation for public funding agencies to select wisely and with the interest of the public in mind. Priority research areas have to be identified from a number of political viewpoints, and the practical result is an increasing tendency to demand measurable utility from scientific inquiries in all disciplines.

In effect, this trend undermines the conceptual boundary between basic and applied research. The former is less likely to receive funding, and the temptation has grown to overstate the potential utility of any science project. Despite the common amalgamation of product and knowledge aims, their analytic separation is helpful for ethical evaluation, because these goals go hand in hand with different technological needs and hence face different potential risks and related moral objections.

Protocell science strives for new forms of life and is therefore likely to affect our understandings of both life and the actual forms of life present. Even if protocells never become a reality, the byproducts of protocell research may change society. For example, a method of combining strands of artificial DNA strings into larger units and thereby creating minimal functional genomes, which are necessary for certain lines of protocell research, is currently being developed (e.g., Noireaux & Libchaber, 2004, p. 17673). Once perfected, such a method could alter the existing technologies for genetic modification of plants and animals, as well as gene therapy.

The distinction between epistemic and product goals is not simply a restatement of the difference between basic and applied science. Instead, it distinguishes different kinds of social utility: knowledge and ideology versus marketable goods and products. This difference matters because protocell research does not raise the usual ethical problems concerning the risks of its products, but it does create significant confusion in ideological battles, as I hope to demonstrate here. Nevertheless, there are moral issues concerning the products of protocell research, for example, because of the gap between the molecular view of life and the subjective experience of the quality of life, because protocell products might cause pollution, and because protocells might be used as weapons. The difference between the top-down and bottom-up approaches is also significant for ethical and risk evaluation. The former seems more likely to produce products sooner, such as cells that are manipulated to fulfill specific medical or technical functions. In such cases, smooth integration in the bodily tissue is of supreme importance; of less importance is a full understanding of why a certain viral or bacterial container is effective for gene transfection, as long as harm can be excluded. Functional protocells, by contrast, may be better understood, but their integration into the natural environment or a complex organism may be more risky.

Product- and Technology-Oriented Goals

The potential products of protocell research can be grouped in three categories. First, protocell research could create products that make it easier to control the insertion of a gene in a gene-therapy patient or in genetically manipulated animals and plants. They might enhance present tissue engineering through better control of the differentiation pathways of pluripotent stem cells and their responses to the environment in the laboratory or in the recipient. Generally, this research and engineering activity might change bioenvironments in multiple ways. Such changes are currently envisaged on the molecular scale, but this might change should any form of industrial production and application of protocells emerge.

A second category is the development of biophysical instruments for the medical sector and its device industry, and for control or active management of natural bioenvironments. Examples are new viral or bacterial agents that could alter agriculture or the metabolism of plants to solve problems of air or water pollution, or implants that release therapeutic elements into the bloodstream, similar to insulin pumps. Gene sequences or other functional units replacing missing or defunct elements (e.g., mitochondria) could potentially be contained in entities absorbable by a specific tissue structure. Also, prototypes of minimal cells with certain defined properties might become useful as standardized research tools for industries exploring metabolism in the environment or in organisms.

A third product category could be nanoscale self-advancing or self-responding products for nonmedical applications. Industries such as fabric and material production, computing and communication, warfare and space industry, and the government agencies that fund them, are potential investors in such research and technology and its potential users.

These categories are neither very specific nor exhaustive; they cannot be, because this research along different pathways is in its early stages. Product- and technology-oriented goals might or might not be achieved, and the actual use of an innovation is always difficult to predict. While no living product has yet been built according to a strict bottom-up approach, the top-down approach is closer to delivery. Knockout technologies and genetic enhancement of cells and viruses have been studied for a while, and the boundaries of what counts as top-down protocell research are blurry. Manipulated cells are already in production. The envisaged products of protocell science could be used anywhere where the small size of the living product, its potential adaptability and susceptibility to changes in its environment, and its highly predictable autocatalytic responses are sought.

The Ethical Assessment of Product-Oriented Protocell Research

One can learn something about the ethical assessment of protocell research by examining the range of criticisms against other recent innovations in physical, chemical, engineering, and biosciences and the ways these become implemented in society. The following ethical assessment of protocell research is incomplete, and the issues raised are not all equally important. Their applicability and relevance depend on the future course of protocell research. In general, the most important problem of bioscience is the risk that its knowledge and products might harm humans instead of improving the quality of their lives.

The Risk of Novel Kinds of Bioweapons

A general problem with research into manipulating and reconstructing cell function is what has come to be called "dual use." Concerns about dual use first arose in the context of nuclear energy, one byproduct of which is nuclear waste that can be readily engineered into an ingredient of atomic bombs. The term "dual use," however, has now been applied to all sorts of unwanted technology applications, and this has diluted the specific reference to use for weapons of mass destruction. I use the phrase here with its original reference in mind. The dual use of microbiology has been studied intensely, particularly with regard to biological research into vaccination and immune response. Protocells that respond in a controlled manner to changes in their external conditions could be used for both managing blood sugar levels, for instance, and producing specialized bioweapons. Curative and life-threatening drugs take the same routes through the body. To decrease the risk for evil use, research should be conducted in a fashion that minimizes the potential for dual use (Bender & Hauskeller, 2003; Nixdorff, 2005).

The previous analytic distinction between epistemic and product-oriented goals is helpful in distinguishing between project designs in protocell research that are more or less suitable for ready transfer of expertise and mass production of cells. Research with mainly epistemic goals does not require industrial production, and scientific knowledge can be, but need not be, accompanied by technical innovation. The confirmation or rejection of scientific hypotheses can be achieved with a few independently conducted experiments with similar findings. A few hundred protocells would suffice to establish, in principle, the function of cellular mechanisms or the self-assembly potential of elements under certain laboratory conditions. In contrast, the effective delivery of a marketable consumer product requires not only a suitable engineering concept, but industrial dimensions. For assessment, the difference between functional product and explanatory value means that knowledge-oriented protocell science is inherently less likely to produce—and can be directed to prevent—easy opportunities for misuse, whereas the option for assemblyline production modes is a basic criterion for research that aims at marketable products.

The potential utility of protocells as bioweapons is an ethical challenge demanding institutional oversight. This, however, is true for many other biomedical research fields. The moral and political problem here is the lack of effective institutions of global control and governance resulting from nontransparency of research protocols and designs, particularly in national defense.

Aspects of a Broad Assessment of Potential Protocell Products

Studies on technological inventions have shown that new products or tools are always multiuse and can always cause unpredicted effects in society. The aspects discussed in this section are familiar from other research areas in the life sciences, and it appears at present that protocell research does not open up radically new dimensions for ethical evaluation. Given the diversity of current research projects and the early stages they are in, it is difficult at this point to identify which of the following themes will become most important. Generally, the issues of moral concern evoked by transplantation medicine, genetics, stem cell science, and nano and brain research recur in a similar gestalt. Only in-depth analysis of specific project designs and goals could deliver more precise insights regarding the specific relevance of any aspects. Here, I discuss seven areas of potential concern: ethical problems regarding trade and commoditization of research materials, health and environmental risks, competition and its effects on product safety, the long route to a generally available or affordable asset, patenting and its impact on availability, hubris and hype, and language and self-representation.

Ethical problems regarding trade and commoditization of research materials (especially human tissues) relate to the exploitation potential inherent in the perception of human or animal body parts and cells as trade objects, and in a potential habitualization of assigning them monetary values. Many current biosciences evoke this concern. Regarding protocell research, however, it seems that bottom-up approaches are at no risk of becoming problematic in this respect, and top-down approaches that use living materials will be at risk only if animal and human cells should ever become their target for manipulation—which is not currently the case. Apart from human or other morally relevant material donors to the research, other kinds of material sources can become morally significant, too, namely, the bio-physico-chemical raw materials used for cell engineering, particularly in cases of mass production. Such materials tend to become morally relevant when their use bears ecological costs or worsens the global economic situation toward greater injustice. This latter aspect seems worthy of further examination, although not in this chapter.

Health and environmental risks and reversibility are a second set of issues of moral concern. Protocells could be used for a multitude of environmental applications, including the remediation of environmental damage. However, the prospect of release outside a security-grade laboratory raises concern because of the potential for unforeseen interactions between protocells and the natural environment (and its inhabitants). In the case of human health, the risks of the experimental phases for medical and other applications are judged as particularly important—trial deaths are morally unacceptable, and other health risks need to be considered and prevented.

The long-term effects of the release of a new evolving life form are unpredictable, and the high degree of integration of a cellular implant or an artificial bacterium into an environment makes its removal difficult in the case of adverse effects. The option to engineer the possibility to reverse a certain innovation to a previous benign state has often been called for, but such engineering is impossible with most modern technologies, given the kinds of changes they introduce in nature and society. In this respect, protocell research strategies might require mechanisms of control.

A third aspect of ethical evaluation is competition and its effects on product safety. Global competition in product development is high, particularly in the health industry. This affects all areas of the life sciences. People and governments believe that both national recognition and a huge potential for wealth depend on holding the best marketable patents. The subsequent pressure for scientific and engineering success has had repercussions, for example, in gene therapy, when improper and rushed clinical trials killed patients (Thompson, 2000), and with misconduct in cancer research (Bostanci & Vogel, 2002) or fraudulent stem cell science (e.g., Couzin, 2006). Such instances affect the credibility of a field and raise public ambivalence toward the biosciences. In order to maintain its current social role as a high-ranking problem solver, science has to avoid fraud, misconduct, and deaths. The current request for ethical assessment from within the field can be seen as addressing this problem, and mechanisms of public control have been established and are under regular review, especially after an incident of irresponsible conduct. Applications of protocell science will be subject to these controls, and special attention will have to be paid to safety and reversibility issues.

Fourth, the route to a generally available cure or affordable asset is often long. The principle of the equal value of all human lives provides good reasons to be critical both of exclusive, privately owned products from publicly funded research and of directions of inquiry in the life sciences that favor cures for the diseases of the rich rather than increased chances of a good life for all. Public spending on science competes with that on education, security, or social work. The strong focus on the life sciences, which many countries currently adopt, is ethically problematic because much of the population might benefit more from other investment policies. But it is difficult to identify the most beneficial spending routes in the long run; spending policies have to take into account many conflicting interests. The orientation of science policies toward funding research that promises to solve problems of high concern to the public can suffice only as a guiding principle in evaluating protocell research.

The many potential products that might arise make protocell science appear a better investment in some respects than stem cell science, for example, which so far mainly addresses typical diseases of civilization. Protocells for curing common human or environmental maladies, even if the trial phases are passed, would probably still be far from generally available to patients. All modern treatment options share this socioethical problem. Even though a cellular therapy might be equally useful for patients globally, during the initial phases of its marketability, mainly well-off patients will have the opportunity to benefit. Only mass production and reliable distribution routes and services could enable general availability. Venture capitalists examine a research field regarding factors such as suitability to grand-scale production, established transport routes, and the transferability of the technology to different places and institutional structures. Consequently, research practices become adapted during the process of development such as to satisfy the requirements for industrial mass production and global distribution. When the method of scientific inquiry does not follow those criteria, investors tend to hold back, leaving a research field to public funding. This was the case in stem cell science between 2000 and 2003, when only individualized medical applications seemed achievable in the short term (Vogel, 2001).

From an ethical point of view, one could say that both private and public funding ought to be interested in the same features of a research strategy, namely, one oriented toward efficiency and mass production. However, the ideal of equal access to good therapies stands against the safety concerns and the openness to reversibility of research designs.

A fifth area of concern is patenting and its impact on availability and competition. Patents secure the inventor of a new item the right to exploit its commercial use by charging fees for it. Patent-related price increases of products limit the reach of innovations. Protocells, if achieved, seem ideal products for patenting; they are well-defined artificial living machines with describable effects and functions. Given that patents for nonhuman living systems are granted readily at present, the patentability of protocells of the currently envisaged minimal types would probably be easy, and subsequently property rights sought and prices increased. Other ethically important effects of patenting are that, particularly with respect to publicly funded science, it compromises general utility and stifles competition, which can put lives at risk.

Hubris and hype is another area of moral concern: Many current problems of the life sciences arise from inflated claims and dubious interactions between science and industry. Consider the common uses of genetic testing kits for various purposes, most of which are only vaguely related to what the tests can actually show. Protocell science evokes fears about the hubris of humankind, fears that are deeply engraved in popular culture and certain thought styles dominant in Western societies. Many worry that the dream of science to uncover, open up, and take apart the elements of nature and life might produce monsters, and that scientists who imitate the creation of life transgress the healthy boundaries set to human action by morality and respect for nature. As described earlier, hubris and hype can have problematic consequences, especially in a climate of competition. Among these problematic effects are the temptation to fraud and attitudes that can lead to irresponsibly early releases of cells or living products into the environment or to premature and rushed animal or patient trials. In this climate, moderation is required to maintain the potential for ethically acceptable products.

A seventh area of moral concern is language and self-representation. Protocell science is often described as aimed at creating life from nonliving matter, or "from scratch." This is partly an overstatement, insofar as the materials artificially constructed and assembled are developed by imitating natural cells. In this respect, nature delivers the master drawing and the structural design for protocells. In other contexts, such work would not count as creative, but rather as dilettante copying that leaves the marks of the constructor's intentions still on the copy, while nature-like functionality is not fully achieved. The use of highly loaded terms goes hand in hand with the evocation of fears of hubris. Current protocell science relies more on assembly skills and physical technology than on new concepts or understandings of life processes. Viewed from this perspective, protocell research reinforces the scientific paradigm of segmentation into elements and sections of integrated units and organisms, and contradicts current claims that society and good science policy need clearer and more meaningful discussion about what it is to be alive. The molecular view of life employed by protocell research implicitly includes many entities in the realm of the living that are not commonly regarded as morally relevant life forms, as such. A more adequate presentation of protocell research that takes its epistemological assumptions about life into account might reduce the fear that this science is overambitious in creating new life.

Epistemic Goals

In debates about scientific risks and potential, scientific freedom is usually a strong argument, and for good reasons. However, knowledge and specific thought styles can be perceived as risky in themselves, insofar as they call into question traditional understandings. Often, scientific knowledge is sought with the aim of clarifying ideological disputes in society. Biological understandings of life, elements of evolutionary theory and the conflict between creationism and evolution are candidates for epistemological aims of protocell research, which can be broken down into the following three types of knowledge goals:

1. Improving the current state of biological knowledge: Top-down and bottom-up approaches can both be used to confirm the completeness or deficiency of current biological understandings of the essential components that contribute to the performance of life in a certain cellular composition.

2. Understanding self-organization and epigenetics: Knowledge about the elements of life and the complex interactions among them could provide better understandings of self-organization and of the cell (genome)–environment interface. This might subsequently lead to improved biomedical and other marketable products.

3. Understanding or modeling the origin of life—an ideological conflict: The laboratory imitation of various possible environmental scenarios at the beginning of life on Earth could challenge or support theories on the origin of life. Precise knowledge of the conditions under which well-characterized chemical or biological elements engage in symbiotic interaction, or join together to form complex integrated systems and stable units, can produce better understandings of the way in which environmental changes impact the formation of primitive life forms.

Assessment of the Potential Knowledge Gains and Their Social Meaning

I will discuss each of these potential knowledge gains, starting with the first: improving the current state of biological knowledge. Biological knowledge has many actual and potential therapeutic uses in genomics, transplantation medicine, tissue engineering, and so forth. It is therefore important to assess whether this knowledge relies on adequate assumptions about cellular and organism function. In-depth understanding of cellular processes is one of the most obvious means for protecting patients and society against inadequate uses of biotechnology, as happened in premature clinical trials with gene therapy. Better knowledge of genome-cell interaction could improve current therapeutic attempts, such as gene therapy cures for children with a rare X-chromosome–related immunodeficit syndrome. Depending on the kind of vector used, the likelihood increases that the curative gene could insert into a genome region whose interruption can lead to leukemia (Hacein-Bey-Abina et al., 2003).

From within the molecular approach to life and medicine, such better knowledge appears as the obvious route to progress. This implies that any knowledge gain aimed at this type of understanding for improved manipulation is inherently complicit with the molecular approach to life. From a broader perspective, however, it can be argued that any investment in life-saving science development must be evaluated with respect to the dependency of human life on biomedical science and industry that emerges with this knowledge and its application. The increase in the domination of and detachment from nature can be seen as, in turn, producing novel dependencies on global industries and human-made facilities, the consequences and desirability of which need to be taken into account.

A different aspect concerning the utility of protocell science for improved biological knowledge is that positive confirmation of biological theories about the function of particular cell elements cannot be achieved through this route for methodological reasons. For example, it is impossible to prove that the knockout or silencing of certain genome sections had only the functional effects observed. This unavoidable incompleteness of knowledge might turn into a problem in practical applications of engineered cells in vivo, when innumerable external parameters influence genome and cell action and interaction. Similarly, there is no question that the bottom-up approach can teach a lot about basic chemistry and functions in cells. However, not even the successful self-assembly of a cell from nonliving materials would prove that anything much has been understood about the natural processes that organize cellular life, because natural cells do not usually self-assemble from nonliving matter. The relevant similarity of the synthetic and natural cell cannot be positively confirmed. Only approximation in various characteristics can be measured. This implies that protocell science might well improve our knowledge of how to construct living systems and thus contribute to better biotechnology, and yet not provide evidence for how nature constructs living systems.

I turn now to the second kind of epistemic goal: improving knowledge about self-organization. It seems possible that in vivo bottom-up science could build itself up from protocells to prokaryotes to eukaryotes to increasingly complex organisms. The basic principles of self-organization in different circumstances and stages of development could, in principle, be tested on such artificial life forms through millions of variations in different environments. This does not apply in the same way to top-down approaches, because supply and safety issues arise more prominently—which is certainly one of the reasons why in silico modeling has become such a prominent support tool for artificial life, protocell science, and stem cell research. The epistemological difference in the shift from confirming current prevailing hypotheses to the production of new knowledge entails the development of novel technologies and a stronger move toward molecular concepts of life.

This development can be interpreted in two ways: first, as an even stronger (problematic) shift of the dominant human concepts of life into a purely molecular vision of life, or second, as many philosophers of biology hope, as a step toward a better understanding of the complexities of the molecular processes of self-organization in nature, which might lead to an integration of human and molecular accounts of life (e.g., developmental systems theory; see Oyama, 2000, ch. 11; Moss, 2007). I doubt whether such a synthesis of concepts starting from the biological viewpoint can be achieved. The ideal is to bridge the gap between reductionist, causal, and functionalist accounts of life, on the one hand, and the subjective understanding of life and biology in the tradition of philosophical anthropology (e.g., Plessner, 1975; Lindemann, 2007) or critical theory, on the other. Self-identities of living entities (Rehmann-Sutter, 2005, pp. 56–59) are not reducible to purely biological processes of metabolism, self-assembly, evolution, and reproduction; without further qualification, the purely biological conception of life entails no normative demands and expectations. It is questionable whether the many ways of looking at life can or should be integrated, as I explain in the next section. In any case, protocell research is not likely to contribute much to any such integration, because its attempt to create novel forms of life delivers little solid insight into the processes of natural life, and its perspective on life is purely reductionist.

The third epistemic goal of protocell research is understanding and explaining how life can originate from nonliving elements. The competing worldviews of evolutionary theory and creationism or intelligent design both hope for support from such inquiries. It is unsurprising that biologists and creationists each hope that protocell research will strengthen their stance, because both are ambiguous about humans' place in nature (as life forms and as dominators of nature). Each relies on metaphysical assumptions about life, as I explain later, using the idea of metaorganisms.

Protocell science seeks to find out about the conditions under which life might have originated, by asking the following kind of question: How did elements integrate into larger units of various kinds, which then entered into symbiotic or parasitic relationships with other such units and synthesized into early forms of life, such as prokaryotes and other primitive unicellular organisms, that later evolved into more complex organisms? In principle, this bio-physico-chemical explanation of the origin and development of life is coherent; however, it includes a metaphysical presupposition from the point of view of creationism. The scientific explanation contradicts creationism, according to which life has an intentional origin from outside the material world. I will go into some detail regarding whether protocell research might affect the public and social credibility of the sides in this ideological conflict.

To clarify the utility of protocell research in answering metaphysical questions, it is necessary to analyze how its understandings of life relate to those in the theories of evolution, intelligent design, and creationism. As mentioned earlier, the definitions of life in protocell research are not molecular as such, but they are tested with the technologies of molecular biology. The metabolism of a cell and its ability to reproduce and evolve are determined through the application of the specific techniques of molecular biology. There is no other way to determine such details about entities on the micro scale; observation with the naked eye is obviously insufficient. However, the three criteria are not definitive in deciding which kinds of things are alive. The criteria are vague translations of everyday measures into seemingly scientific parameters; they are vague about what they refer to and where they draw boundaries. Moreover, they establish a space of micro living entities that is very inclusive, very counterintuitive, and ethically not very relevant.

The microbiological understandings of metabolism and the ability to reproduce and evolve do not necessarily imply that only certain kinds of entities can be counted as living. The three criteria include entities not commonly considered living, and erase traditionally significant differences between kinds of living systems. The units of ethical concern do not match the units that protocell science establishes as living systems. Life, as it is construed by protocell science, disagrees with understandings of the identity of living organisms and with the organocentric understandings from which common measures for health and well-being have been derived. This matters, when connections are drawn from one to the other, the possibility of which is a major basis for legitimizing protocell science.

In recent years, genomics has expanded the boundaries of definitions of life based on indicators similar to those previously mentioned. Examples are recent ideas of metagenomes (e.g., Craig Venter's Project on the Sargasso Sea; see Venter, 2004; Venter et al., 2004) or metaorganisms. Such biological metaentities currently emerge in molecular and microbial genomics and philosophical theories of their conceptual meaning. Such accounts can be interpreted as propositions challenging predominant cultural concepts of what objects can be considered living entities. John Dupré describes metaorganisms as follows:

The clearest context in which to present the idea of microbial metaorganisms is with the phenomenon of biofilms. Biofilms are closely integrated communities of microbes, usually involving a number of distinct species, which adhere to almost any wet surface. Biofilms are ubiquitous, from the slimy rocks and stones found under water and the chemically hostile acid drainage of mines, to the internal surfaces of drinking fountains and catheters; indeed biofilms are where most microbes generally like to be. In addition to . . . genetic exchange . . . the constituents of biofilms exhibit cooperation and communication. These are most clearly exemplified by the phenomena of quorum sensing, in which microbes are able to determine the numbers of cells in their communities and adjust their behavior—including reproduction—in appropriate ways. (2006, pp. 8–9)

To consider dental plaque, for instance, as metabolizing, self-assembling, evolving, and maintaining itself, and in this sense as a living entity, is clearly beyond the common understanding of life. Although many would agree that my dental plaque belongs to my body, it is not usually seen as a living entity in its own right—and it can certainly not be described as autonomous. Analytically, there is no reason why multicellular organisms such as biofilms should not be counted as living entities, according to the criteria for life employed in protocell science. This, however, takes away the markers that identify plants, and, in particular, animals and humans as the first-order living entities. The special traits they have as living phenomena lose any significance and hence their normative implications disappear.

Different conclusions can be drawn from this argument concerning the constructions of the realm of the living. Two epistemological problems stand against the relevance of protocell research to theories of the origin of life. One is that the pathway from primitive cells to higher organisms is not established by evidence that primitive cells can coincidentally self-assemble alone. The formation of a cell is indicative for the historical assembly of higher organisms only if cells are seen as the interacting building blocks of those organisms. But higher organisms, and more specifically humans, are at the center of creationist concerns, not prokaryotes or primitive cells. The definitions of life in protocell research draw a line between living and nonliving entities that is radically different from the one underlying creationist understandings of species boundaries and the origin of life.

The ideological conflict between creationism and scientific theories on the origin of life is partly caused by thought styles that speak past each other. The construction of a primitive cell to test the assumption that life can grow from nonliving matter can be both an irrefutable argument against the need for a creator and no evidence at all, if these primitive cells are not credited with sufficient significance to be counted as life. The definition according to which entities such as biofilms are living entities is not compatible with what creationists are concerned with when they fear human interference with the creation of nature. The gap between these two attitudes toward life prevents validity of the conclusions of one approach for the other. Although from within each paradigm of ideology the other may be refuted by the success or failure of protocell construction, the epistemological differences between them make it unlikely that those who are not already convinced will alter their view through protocell evidence.

What appears as a fundamental ideological conflict, however, does not lead individuals to consistently take either one or the other side. Empirically, however, different thought styles can coexist quite happily, even within individuals. Examples include Catholic scientists working with human embryos in the laboratory. In *Genesis and Development of a Scientific Fact,* Ludwik Fleck (1981) provides a solid sociological explanation for this phenomenon. He describes the ways in which facts in a certain thought community are created as dependent on their underlying thought system. A scientific community, just like a religious or other social thought community, forms with and through the development and maintenance of its thought system. Thought communities derive and create truths and insight through the use of their basic set of assumptions. According to Fleck, individuals belong to several thought communities and act as "vehicles for the inter-collective communication of thought." His sociological explanation is that "[t]he stylized uniformity of his [the individual's] thinking as a social phenomenon is far more powerful than the logical construction of his thinking. Logically contradictory elements of individual thought do not even reach the stage of psychological contradiction, because they are separated from each other. Certain connections, for instance, are considered matters of faith and others of knowledge. Neither influences the other, although logically not even such a separation can be justified" (1981, p. 110).

The social disputes over explanations for the origin of life are, according to Fleck, power struggles for the predominance of thought styles and the related social status of the thought collectives that represent them. In a situation in which science funding depends on social recognition and the management of potential ethical challenges, the credibility of the thought style that carries projects such as protocell creation needs to be fostered and, hence, confrontational dispute with opposing thought styles is unavoidable. According to Fleck, the collision between two styles is greater the greater their differences: "The principles of an alien collective are, if noticed at all, felt to be arbitrary and their possible legitimacy as begging the question. The alien way of thought seems like mysticism. The questions rejected will often be regarded as the most important ones, its explanations as providing nothing or as missing the point" (1981, p. 109).

How protocell science defines life can be seen as just as transcendental and undefined as commonsense and creationist understandings. However, these styles occasionally coexist in individuals, and the question they address is an important problem of human understanding of the self and world: What is life—or, more precisely, what forms of life count? For assessing the social impact of protocell science, this dimension of the compatibility of its concept of life with the societal one is relevant, because much of social morality is grounded on normative understandings of (certain forms of) life. The clash of thought styles and the questionable possibility of translation between them deserves attention, especially with the lack of explanation for the widespread ambivalence toward the life sciences.

Without the metaphysical premises of creationism, the general understanding that human life is not described adequately, when meaningful assumptions about quality of life, equality, dignity of living beings, and so on are excluded, still makes sense. Is anything that shows signs of metabolism, evolution, and regeneration considered a living system, and if we agree that this is a sufficient definition, on what basis are different normative values established for living systems with additional properties such as the ability to experience pain, fear, or death? One might argue that the distinctiveness of life that matters gets lost in this perspective.

Protocell scientists, as almost all modern biophysical scientists, may claim that their criticized reductionist understanding of life is not a problem with respect to the sphere of traditional life. On the contrary, its reduction to the simplest possible definitions is a technique employed to achieve new living systems to improve quality of life in the fuller sense. Its aim is not just to create life for its own sake. The translation process between both understandings is anything but clear, however. The route from the person who presents herself as a potential customer or patient to the repair technologies under development in laboratories worldwide, and then back again with a remedy to help this customer, has many hurdles of miscommunication and misrepresentation. They usually take the form of both reduction of health symptoms to biochemical processes and the generalization of certain biochemical explanations to overall cures for a complex set of disease causes and experiences in patients. The well-being (or sickness) of persons presents itself in ways that are not reducible to the factors and measures employed in science that operates on the molecular level. The retranslation has also shown considerable failure rates in most cases of testing and disease.

According to the protocell definition of life, inserting living entities with defined properties into any organism alters the molecular identity and composition of the receiving unit, which may be seen as one or many mutually parasitic living systems rather than a patient. This organism may reject the alien elements, it may not manage to survive the intrusion, or it may adapt to this change and ideally do so in the desired form, benefitting from the process. This implies that evolution, even human evolution, through alien artificial living systems, is part of the kind of cure biotechnological medicine offers with protocells. But to perceive any benefit of the treatment for a patient, the conventional understanding of living entities has to be employed. The ultimate test for well-being is that the person concerned states herself that she is well or better—even though, according to some other thought systems, she might no longer be quite the same in terms of the living systems that coexist in her body.

Concluding Remarks

This chapter has considered many aspects of ethical concern and epistemological utility that affect protocell research in different ways. They are relevant because protocell science is a cultural activity, and as such is both intentionally engaged in solving specific cognitive and social problems and fully embedded in the social, economic, and ideological conditions of life in Western societies. I have argued that protocell science hardly promises the novel epistemological breakthroughs that some hope for, although it could contribute to the cognitive paradigm to which it belongs. Furthermore, I found that the ethical dimensions of the research, particularly its product-oriented goals, are familiar from other, more established biotechnological and biomedical fields. Some dimensions appear more relevant than others, and most significant are risk of weaponry use and risks to human health and the environment. Bioethics is seen by many as the appropriate instrument to reflect on and control these risks, and little concerning potential strategies of good governance is specific for protocells.

One ethical issue, however, stands out more prominently in protocell science than in related fields: the wide gap between the understandings of life held by society in general and by current biosciences. This gap prevents the smooth passing from new understanding at the molecular level to the meaningful creation or improvement of life. Although this gap is familiar from other discussions in the philosophy of science, it has not yet been addressed sufficiently in ethical evaluations of science.

The criteria for life that molecular biology has long employed are distinctly reductionist compared to concepts in society in general. The different meanings of "life" in different contexts cause confusion. Evelyn Fox Keller described a similar slippage and its effects on the understanding of basic scientific concepts with regard to the term *gene*. She speaks of the powerful effects that blurry and multiple-meaning terms have on the development of the life sciences and the ways in which people respond to them (Fox Keller, 1995). Lenny Moss has developed similar arguments criticizing the conflation of radically different understandings of genes as preformationist units or as elements in development (Moss, 2003). The conflation occurs in both scientific presentation of aims and expressions of desired social benefits in areas such as medicine. It is not sufficient merely to mark clear distinctions between different ways of thinking about life, for each viewpoint ought to become influential in and for the other.

The challenge of protocell research is the force with which it invites addressing the disparity between concepts of life. Genomics represents itself mainly as passive in this respect—its enhancement section remains segregated, and projects such as the Human Genome Project, the Human Genome Diversity Project, and the blooming business of biobanks are about understanding life, not creating or manipulating it. Stem cell science is usually criticized for destroying (not producing) life, and began with the image of using the natural healing powers of the organism to develop new cures. It operates only within the space of living entities and, at most, reassembles parts to produce particular properties such as stem cell clones. Protocell research, on the other hand, admits upfront that it aims to create novel life on the molecular scale. This openness to defining life should be used to enter into a wider discussion of understandings of life and what they mean for individuals, society, medicine, and science, including whether integration or orders among them are necessary or even possible.

I assume that the concern and ambivalence that many people direct toward the biosciences are related to this underaddressed reality of radically distinct life concepts. Protocell science opens the door for discussion among science, philosophy, and the general public, and I hope this invitation will be accepted; if not for intellectual reasons, then for the sake of more in-depth and prudent evaluations of what the life sciences mean for society and cultural development, particularly with respect to human self-understanding.

Acknowledgments

I wish to thank Mark Bedau and Emily Parke for inviting me to engage with protocell research in the first place, which I find very interesting. Moreover, I wish to thank Mark and Emily for their helpful critical comments on an earlier version of this chapter. I am grateful also to Gesa Lindemann who responded with inspiring criticism and comments, to Barry Barnes for discussions on the Sargasso Sea and dual use, and to John Dupré for advice on biofilms. This research was part of my work at Egenis, the ESRC Centre for Genomics in Society at the University of Exeter, funded by the UK Economic and Social Research Council.

References

Bender, W. (1988). *Ethische Urteilsbildung.* Stuttgart: Kohlhammer.

Bender, W., & Hauskeller, C. (2003). Der Stammzelldiskurs: Kriterien der Technikbewertung. In K. Mensch & J. C. Schmidt (Eds.), *Technik und Demokratie* (pp. 179–196). Opladen, Germany: Leske und Budrich.

Best, R., & Khushf, G. (2005). The stem cell controversy in the United States: Scientific, philosophical, and theological aspects. In W. Bender, C. Hauskeller, & A. Manzei (Eds.), *Crossing borders* (pp. 241–262). Munster: Agenda Verlag.

Bostanci, A., & Vogel, G. (2002). German inquiry finds flaws, not fraud. *Science, 298* (5598), 1531–1533.

Couzin, J. (2006). . . . And how the problem eluded peer reviewers and editors. *Science, 311* (5757), 23–24.

Dupré, J. (2006). *The constituents of life: The Spinoza lectures, University of Amsterdam.* Amsterdam: Van Gorcum.

Fleck, L. (1981). *Genesis and development of a scientific fact.* Chicago: University of Chicago Press.

Fox Keller, E. (1985). *Reflections on science and gender.* New Haven, CT: Yale University Press.

Fox Keller, E. (1995). *Refiguring life: Metaphors of twentieth-century biology.* New York: Columbia University Press.

Habermas, J. (2003). *The future of human nature.* Cambridge: Polity Press.

Hacein Bey-Abina, S., Von Kalle, C., Schmidt, M., McCormack, M. P., Wulffraat, N., Leboulch, P., et al. (2003). LMO2-Associated clonal T-cell proliferation in two patients after gene therapy for SCID-X1. *Science, 302* (5644), 415–419.

Hauskeller, C. (2005). Science in touch: Functions of biomedical terminology. *Biology and Philosophy, 20,* 815–835.

Horkheimer, M., & Adorno, T. W. (1933; 2002). *Dialectic of enlightenment.* G. Schmid Noerr (Ed.). Palo Alto: Stanford University Press.

Human Fertilisation and Embryology Authority (HFEA) (2004). Public consultation exercise. *The regulation of donor assisted conception.* Available online at: http://www.hfea.gov .uk/cps/rde/xbcr/SID-3F57D79B-3C08502E/hfea/SeedConsult.pdf (accessed June 2007).

Human Fertilisation and Embryology Authority (HFEA) (2006). Public consultation exercise. *Donating eggs for research.* Available online at: http://www.hfea.gov.uk/cps/rde/xbcr/ SID-3F57D79B-3C08502E/hfea/donating_eggs_for_research_safeguarding_donors_consul- tation_FINAL.pdf (accessed June 2007).

Human Genome Project (HGP) (2000). *The first draft of the book of humankind has been read.* Sanger Centre Main press release from June 26, 2000. Available online at: http://www .sanger.ac.uk/Info/Press/000626.shtml (accessed June 2007).

Lindemann, G. (2007). Medicine as practice and culture: The analysis of border regimes and the necessity of a hermeneutics of physical bodies. In R. Burri & J. Dummit (Eds.), *Biomedi- cine as culture.* London: Routledge.

Merchant, C. (1982). *The death of nature: Women, ecology and the scientific revolution.* London: Wildwood House.

Moss, L. (2003). *What genes can't do.* Cambridge, MA: MIT Press.

Moss, L. (2007). Textures of life: Detachment theory and the microbial universe. In *Studies in history and philosophy of biology and biomedical sciences,* unpublished manuscript.

Nilsson Jacobi, M. (2006). Home page. Available online at: http://frt.fy.chalmers.se/cs/people/ jacobi (accessed June 2007).

Nixdorff, K. (2005). The difficulties in applying ethics to BW-relevant life sciences research. *Epidemiology, 16* (5), S36.

Noireaux, V., & Libchaber, A. (2004). A vesicle bioreactor as a step toward an artificial cell assembly. *Proceedings of the National Academy of Sciences of the United States of America, 1001* (51), 17669–17674.

Oyama, S. (2000). *Evolution's eye: A systems view of the biology-culture divide.* Durham, NC: Duke University Press.

Plessner, H. (1975). *Die Stufen des Orgnaischen und der Mensch.* Berlin, New York: de Gruyter.

Pohorille, A., & Deamer, D. (2002). Artificial cells: Prospects for biotechnology. *Trends in Biotechnology, 23,* 123–128.

Purdy, L. M. (1996). *Reproducing persons: Issues in feminist bioethics.* Ithaca, NY: Cornell University Press.

Rasmussen, S., Chen, L., Deamer, D., Krakauer, D. C., Packard, N. H., Stadler, P. F., & Bedau, M. A. (2004). Transitions from nonliving to living matter. *Science, 303,* 963–965.

Rehmann-Sutter, C. (2005). *Zwischen den Molekülen: Beiträge zur Philosophie der Genetik.* Tübingen, Germany: Francke Verlag.

Scottish Council on Human Bioethics (2006). Consultation response to the Human Fertilization and Embryology Authority. December 8, 2006. Available online at: http://www.schb .org.uk/index.htm (accessed June 2007).

Sherwin, S. (1992). *No longer patient: Feminist ethics and health care.* Philadelphia: Temple University Press.

Thompson, L. (2000). Human gene therapy: Harsh lessons, high hopes. *US Food and Drug Administration Consumer Magazine,* September–October 2000. Available online at: http:// www.fda.gov/fdac/features/2000/500_gene.html (accessed June 2007).

Venter, J. C. (2004). J. Craig Venter Institute Web site. Available online at: http://www.jcvi .org (accessed June 2007).

Venter, J. C., Remington, K., Heidelberg, J. F., Halpern, A. L., Rusch, D., Eisen, J. A., et al. (2004). Environmental genome shotgun sequencing of the Sargasso Sea. *Science, 304,* 66–74.

Vogel, G. (2001). Stem cells lose market luster. *Science, 299,* 1380–1381.

17

Methodological Considerations about the Ethical and Social Implications of Protocells

Giovanni Boniolo

Many people are currently working on a scientific and technological program aimed at realizing protocells, through either top-down or bottom-up approaches (see the introduction to this volume). But this has not yet been achieved. Therefore, we are in a rather strange situation: We have begun discussing the social and ethical implications of something that does not exist, though it seems that it will exist soon (exactly how soon is unclear). In the history of science and technology, it is unusual to discuss the implications of something that does not yet exist. As far as I know, all the discussions of new scientific and technological results have occurred after the results were known. It should be noted, however, that many of these discussions were driven primarily, sometimes negatively, by pressure to respond to public reaction to some frightening aspect of these results.

In the case of protocells, we have the chance for a freer discussion, even if it is necessarily abstract. We currently cannot know exactly what features protocells will have, and, therefore, exactly what their implications will be. Nevertheless, we have the opportunity to arrange the proper framework for the debate that will inevitably arise once protocells do exist. In particular, we can give proper attention to methodological preliminaries, without the need to produce an immediate answer to a pressing problem.

In what follows, I address these methodological preliminaries, with the goal of helping to influence the logically and rhetorically correct argumentative framework. In particular, after a short introduction, I examine the strategies along which the debate could move. I then turn to the major arguments that can be used. Finally, I emphasize the roles and responsibilities of the scientific community, proposing a sort of international agency that should monitor and prepare analyses of what is occurring.

What Are We Debating?

To begin with, I wish to recall some points with which we are well acquainted but that allow me to set forth my view correctly. When we talk of the social and ethical implications of protocells, of course, we are not talking about protocells in and of themselves, since in a strict sense neither the protocells nor any other scientific and technological entity has social or ethical implications. It could have physical, chemical, or biological effects, but not social or ethical implications. Rather, people's actions concerning the creation or use of protocells have social implications. We ethically judge intentional actions, that is, the creation or use of protocells, and not the protocells in and of themselves. Therefore, we should clearly differentiate between an entity, its scientific and technological description, the actions involving that entity, and the moral judgments about these actions. Entities, and scientific or technological descriptions of entities, are socially and ethically neutral. However, the actions concerning those entities are not neutral; nor, of course, are our moral judgments about those actions.

One could claim that protocells might have positive effects, since they might be used, for example, as drug delivery systems or other pharmacological devices, as diagnostic or therapeutic tools, or as microsystems capable of cleaning dangerous chemicals from the environment. Yet, one could also claim that they might have negative effects, since, having the capacity to self-reproduce and evolve, they could outcompete "natural" species in their niche; moreover they could be used by bioterrorists. But, if we want to set the questions correctly and precisely, from a methodological and ethical point of view, this way of speaking is misleading.

Protocells are intentionally created by humans with the purpose of performing a certain function. Second, humans intentionally place them into given environments in order to realize their desired functions. Thus, we have two different actions: creating protocells and using protocells. These two actions could have different social and ethical implications, depending on how and in what environment they are performed. Someone could create protocells with a certain function while working for a state agency or for a private company. They could sell protocells to those wealthy enough to pay for them, or they could make them available free of charge to anyone who wants them. Protocells that were created to achieve one function might also achieve a second, different function. Achieving this second function would alter their social impact. But the difference is not in the protocells themselves; the difference is in the contextualized actions concerning those protocells.

The same can be said for the negative aspects. Consider bioterrorism. Suppose someone creates protocells with a certain function that he knows to be beneficial in one environment, E, but extremely lethal in another environment, E'. If they are used in environment E', the social and ethical difference is not in the protocells but in whoever intentionally and purposefully uses them in E', instead of in E. It is the action that has social and ethical impact, not the protocells. The same holds in the case of someone who intentionally and purposefully creates protocells with a lethal function: These actions have social impact!

Following this line of thought, the ethical implications of protocells actually concern the moral valuations of their creation and use, and not the protocells themselves, or their scientific descriptions. None of these latter things are good or bad in and of themselves. Rather, it is the set of our ethical beliefs that allows us to ethically evaluate the actions on protocells, that is, actions that are directed to create or use them. If we do not want to mix the levels of analysis, especially if we want to avoid the naturalistic fallacy, we are obliged to conclude that the attribution of a moral sense is grounded in our way of judging, and not in what we judge (see Boniolo, 2006; Boniolo & De Anna, 2006).

The Three Strategies

How can we engage in the moral valuations of creating and using protocells when protocells do not yet exist? It seems that there are three debating strategies that could be adopted: the by-analogy strategy, based on analogy between protocells and other technological and scientific entities; the ontological strategy, concerning the possibility of formulating social and ethical considerations starting from an analysis of what a protocell is; and the type-token strategy, according to which we consider protocells as examples (tokens) of scientific and technological innovation, and therefore to be discussed as any other example of innovation (types).

Let us examine these three strategies, to test their validity.

The By-Analogy Strategy

Analogy is an inductive way of reasoning, since it increases knowledge by passing from what is known to what is unknown. It is based on the following structure: Given A (the starting analog), with the properties $p_1, \ldots, p_n, q_1, \ldots, q_l, r_1, \ldots, r_m$; if B (the arrival analog) shares with A the properties $p_1, \ldots, p_n$ (even if not the properties $q_1, \ldots, q_l$), then B probably has the properties $r_1, \ldots, r_m$. The conclusion of the argument is only probable, and its robustness depends on four parameters:

1. The number n of the analogous properties $p_1, \ldots, p_n$: The more they are, the stronger the conclusion is.

2. The number m of inferred properties $r_1, \ldots, r_m$: The fewer they are, the stronger the conclusion is.

3. The relevance (relative to the context in which the analogical reasoning is performed) of the analogous properties $p_1, \ldots, p_n$: The more relevant they are, the stronger the conclusion is.

4. The relevance (relative to the context in which the analogical reasoning is performed) of the disanalogous properties $q_1, \ldots, q_l$: The more relevant they are, the weaker the conclusion is.

In our case, the arrival analog should be protocells, whereas the starting analog might be modified genes, genetically modified organisms, nanoentities, and so on. This means that, to infer the probable existence of unknown properties of protocells, our analogical reasoning should commence with some known similarities between given material or functional properties of protocells and given material or functional properties of the chosen starting analog (modified genes, nanoentities, etc.). Unfortunately, this line of thought would not be as strong. First of all, since protocells do not exist yet, we could not know any of their material and functional properties exactly. Therefore, at maximum, we could commence loosely and vaguely, claiming that some material and functional analogies exist. Moreover, with reference to what we are discussing, interest would not be in conclusions about probable unknown material or functional properties of protocells, to which we should arrive by means of the analogical reasoning. Instead, we would be interested in conclusions about the actions of creating and using protocells, and the moral value of these actions. That is, we would have an odd analogical argument running as follows: Protocells would have (even if loosely and vaguely) some material and functional properties analogous to those possessed by the starting analog (modified genes, nanoentities, etc.), so:

• Since creating and using the starting analog has a certain social impact, creating and using protocells would probably have the same social impact;
• Since creating and using the starting analog should be ethically valued in a certain way, creating and using protocells should be ethically valued in the same way.

It is quite evident that, even if the starting analogs were extremely strong and relevant, as indicated by parameters 1 and 3, the conclusion of the analogical reasoning would be problematic. For we should begin (1) from the supposed material and functional analogical properties $(p_1, \ldots, p_n)$ possessed by both the protocells and the starting analog (modified genes, nanoentities, etc.), and (2) from the social implications $(r_1, \ldots, r_m)$ of the actions of creating and using the latter. Then we should infer probable analogous social implications of creating and using protocells. But we do not say that the same social implications necessarily exist for creating and using two different things that share some supposed common properties.

The matter is also problematic when we turn to the ethical implications. In this case, we should begin (1) from the supposed material and functional analogical properties, and (2) from the ethical valuation of the actions concerning the creation and use of the starting analog. Then we should infer the analogous ethical valuation of the actions concerning the creation and use of protocells. Also in this case, we do not say that the ethical valuations of creating or using two different things that share some supposed common properties have to be similar.

There is a problematic aspect to note here. Since both the social implications of the creating and using actions and the moral valuation of these actions strongly depend on the cultural context, and since we do not know what such a context will be when protocells really exist, the preceding argument must be judged even weaker.

There is at least another questionable point in this way of arguing. The reasoning starts from an analogy between some material or functional properties of protocells, and those of the starting analog. Yet, as parameters 3 and 4 indicate, its robustness is based not only on the relevance of the analogies (parameter 3), but also on the irrelevance of the disanalogies (parameter 4). Unfortunately, in the case of protocells, the disanalogies are extremely relevant, for a protocell is an entity characterized by self-organization, evolution, self-assembly, and capacity to harvest what it needs from its environment. These are extremely important properties, and furthermore they are properties possessed specifically by protocells. Note, they are not only extremely important for individuating what a protocell is (see the conditional definition of *protocell* given in the next section), but also for discussing the social and ethical implications of the creation and use of protocells. The worries we can have about their creation and use concern exactly these particular properties (Could we control their capacity to self-organize, self-assembly, evolve, compete with existing forms of life, and harvest what they need from the environment?). Therefore, because of the peculiarity of protocells, the robustness of any analogical argument dealing with them is very doubtful from the beginning. This is a case in which the material and functional disanalogies are more relevant than the material and functional analogies.

The Ontological Strategy

The ontological strategy, which starts from what a protocell really is, might be a good one for facing the issue of the social and ethical implications of protocells. Of course, this strategy directly involves the definition of *protocell*. Immediately, problems arise because there are many types of definitions, and which one we choose strongly influences the *definiens* (i.e., the terms of the definition). Let us stay for a while on this point.

There are at least two approaches to definitions: connotative and denotative. According to the former, a definition indicates the properties characterizing the *definiendum* (what is being defined, in this case *protocell*), as in the Aristotelian definition *per genus proximum et differentiam specificam*. That is, we individuate, first, the genus whose subclass is the species denoted by the *definiendum*, and then the specific feature differentiating that particular species from the others of the same genus. An example of the latter is given by the conditional definition. We can define X by indicating the individually necessary but jointly sufficient conditions $C_1, \ldots C_n$ that have to be satisfied to consider something an X. This seems exactly the kind of definition we need in case of protocells: "X is a protocell if it is microscopic, self-organizing, evolving, assembled from simple organic and inorganic materials, and capable of harvesting raw materials and energy from its environment and converting these into forms it can use."

Beyond the possible different approaches, a definition can have different epistemological statuses, depending on our epistemological commitment to the *definiens*. It can be essentialistic, if we think that the *definiens* essentially indicates the characteristics of the *definiendum*. This is Aristotle's way of defining the term. However, an essentialistic definition seems very problematic within empirical sciences and technology. In the empirical sciences, definitions are supposed to be stipulative. In other words, we simply stipulate that the *definiens* is good for that *definiendum*, without having any particularly strong metaphysical commitment.

Unless we do not pursue the strange idea of attributing an essentialistic status to the definition of *protocell*, since we share the position that dealing with protocells is a matter of science and technology and not metaphysics, it seems wise to consider the preceding conditional definition as stipulative. Note that there should be no discussion on a stipulative definition; if anything, the theoretical consequences of that definition should be discussed. Nevertheless, if we want to arrive at a methodologically correct valuation of the actions concerning protocells, a possible shared stipulative definition is not enough. Certainly we must know what *protocell* means, but the value judgment is also grounded on other information: our social and ethical beliefs, our scientific and technological knowledge, the influence we have from the cultural context in which we formulate the valuation, and so on.

The debate over the ontological nature of protocells can also affect questions about what nature is, what an artifact is, what a function is, the differences between proper function and secondary function, what being alive means, and so on. But, are these interesting problems really connected to the discussion of the social and ethical implications of protocells? I am not so sure. That a protocell is created with a certain proper function, but used with a secondary function, that we can precisely identify the difference between an artifact and something natural, that we are able to indicate unambiguously what nature is, that a protocell must be categorized as "living" or "nonliving" technology—surely these are relevant ontological questions, but they seem to affect only tangentially the debate on the social and ethical implications of creating and using protocells. In other words, the discussion of the social and ethical implications can proceed without any ontological presupposition. In this sense, we really could do sociology and ethics without ontology. Note, on the other hand, that if we wanted to ground our social and ethical considerations on stated ontologically "good" or "bad" properties of protocells, we would fall into the naturalistic fallacy. Again, the goodness and the badness do not lie in the world, but in our ethical way of valuing the world.

If what has just been proposed is plausible, ontological considerations may be set aside when we address the real problem at hand. Sometimes those who propose this kind of ontological analysis are falling victim to the straw man fallacy and the misplaced precision fallacy. The straw man fallacy concerns the possibility of rhetorically diverging from the real problem by discussing another similar or closely linked problem. The misplaced precision fallacy concerns holding a highly accurate discussion of something that has nothing to do with the real point to be discussed, even if it is rhetorically made to appear extremely relevant. If we discussed ontological aspects instead of limiting ourselves to the social and ethical questions concerning protocells, we would be targeting a "straw man" instead of the "real man." We could use the best analytical tools available, but our precision would be misplaced, since the point does not concern the proper function of a protocell, or whether we should apply the label *artifact*, or whether we are dealing with "living technology." The real point regards the social implications of our actions of creating and using protocells, and how these are ethically judged.

The Type-Token Strategy

Let us turn to the third strategy. According to this strategy, we should not base our discussion of the social and ethical implications of protocells on their supposed analogies with other scientific or technological entities, or on what a protocell is in a strong ontological sense. Rather, we should consider a protocell as an instance or "token" of a new type of scientific and technological entity.

As with any other innovation created and used by humans during the approximately 6000-year history of our culture, protocells (rather, our actions involving protocells) have social and ethical implications. Of course, protocells have particular features, as does any other new entity. The Greek *gnomon* and the Roman *hypocaust* (a sort of central heating) were groundbreaking technological innovations in ancient times, as were the compass, gunpowder, the telescope, the microscope, and the computer chip. Each of these new technological entities (rather, the actions performed on them) had strong social and ethical implications (of course, relative to contextualized culture and society).

From this point of view, protocells are just another example in a long and continuous process of scientific and technological advancement. Why should they be discussed in a different way? *Mutatis mutandis*, they can be discussed exactly how the hypocaust, the compass, gunpowder, the telescope, the microscope, or the chip were and are discussed. The cultural context has changed, but the general problem is the same: How should we face a new type of scientific or technological entity, of course taking into account its specific features?

This is the real problem. If we focus the discussion on protocells in and of themselves, we run the risk of forgetting that humans have faced situations like the one we are facing throughout the 6,000 years of our cultural history. We have always had to take into consideration, to varying degrees, the possibility of creating something dangerous, or something that could be used in a dangerous way. Independent of the particular features of the protocells, it seems wise to discuss their social and ethical implications from this point of view, that is, by considering their being a new instantiation—a token—of an old type of story.

Three Arguments Against Protocells

We can plausibly anticipate three common objections to the actions of creating and using protocells: The first (the argument about transcendence) is *a priori*, that is, independent of any empirical reference; the other two (the slippery slope argument and the precautionary principle argument) are *a posteriori*, that is, based on observations or experimental data.

The Transcendence Argument

It is sometimes said that only God should create life forms from scratch.[1] Infringing this injunction would be to act with hubris against what religion considers natural, sacred, or permitted only to God. Objecting to creating protocells on these grounds involves resorting to an argument containing implicit or explicit premises such as "God does not want this," "God is offended by your actions of creation, since you are doing what only he is supposed to do," or "Do not play God!" But do these arguments work?

As is evident, these arguments assume the existence of God. But this is a matter of faith, not of reason. It follows that any argument about transcendence is, first of all, rationally irrelevant. If we want to conduct a rational discussion on the social and ethical implications of the actions of creating and using protocells, we must rule out any premise that is rationally irrelevant. Thus, since arguments about transcendence involve faith and hence are rationally irrelevant, we should abandon them.

This kind of argument has another problem. There is a difference between the *argumentum ad auctoritatem* (appeal to qualified authority) and the *argumentum ad verecundiam* (appeal to unqualified authority). Both forms of argument propose that a certain position must be accepted on the basis that some prominent authority (be it an individual or a community) accepts it. There is, however, a clear distinction between the two, since the former is a valid argument and the latter is not. An *argumentum ad auctoritatem* claims that something is to be accepted because a real *auctoritas*, or authority in the relevant field, accepts it. For example, when we say that DNA is a double helix based on Watson and Crick's statement, we introduce a correct *argumentum ad auctoritatem* because Watson and Crick are recognized as *auctoritates* in the field of molecular biology. On the other hand, if we claim that abortion must be banned because Watson and Crick said so, we are introducing a fallacious *argumentum ad verecundiam* because Watson and Crick are not *auctoritates* in this field. Now, those who believe that God exists would see the appeal to God in the transcendence argument as a valid *argumentum ad auctoritatem*, at least in certain cases. But this holds only for those who believe in God. Those who do not believe that God exists will view this argument as a fallacious *argumentum ad verecundiam*.

There is something else to note. Which God are we invoking? The Catholic God? The Protestant God? The Hebrew God? The Islamic God? One of the gods of Olympus? One of the gods of Walhalla? It is evident, indeed, that the invocation of God is never abstract and general. It is always an invocation of the God of a particular religion. It follows, therefore, that only for the believers in that religion is the invocation of that God a valid *argumentum ad auctoritatem*; for the believers in a different religion, it is a fallacious *argumentum ad verecundiam*. Why should we accept the *auctoritas* of a God other than the one in which we believe?

To conclude this matter, I will discuss a particular linguistic aspect that can lead to both equivocations and nominalistic—and therefore void and useless—discussions. Virtually all religions have a unique tale of the genesis of the nonliving and living worlds, and a unique doctrine of God's power, in particular a unique doctrine concerning what *creating* means. This is an interesting side topic concerning protocells, since (at least according to the bottom-up approach) they are "created from scratch."

Let us consider, for example, the Catholic religion. Here, the term *creation* means the work of God who, by means of only an intentional action, brings to existence what did not exist before. It is creation *ex nihilo*, that is, "from nothing": an action that only God can perform. Those who are not God cannot do it; only God can violate the law according to which *ex nihilo nihil fit*, that is, from nothing nothing can be. The doctrine of creation is accompanied by the separate doctrine of generation. Although only God can create anything, both God and humans can generate things (e.g., God generated Christ, humans generate other humans). And, of course, God is neither created nor generated.

Having said this, let us come back to the locution "protocells are created from scratch." Here, the term *creation* is used differently from its use in the Catholic religion. Now it means something like "construction," or "making by a human." It follows that all of those who object to protocells from the point of view that only God can create are committing the fallacy of equivocation, since they mix together two different meanings of the same term. Furthermore, many interpreters of the Catholic tale of genesis say that God created the nonliving and the living world "from scratch." Here, the locution from scratch means *ex nihilo*, that is, from nothing. Therefore, it has a totally different meaning from its meaning when we speak of protocells. Now it means from nonliving materials, but not literally from nothing. Note that we could use something analogous to this second meaning of *from scratch* in any other case that involves putting things together to obtain something with emergent properties. In this sense, if we assembled two molecules of hydrogen and one of oxygen, we would "create water from scratch," that is, from materials that were not in any way water. It follows that those who object to the protocells because only God can create "from scratch" are committing the fallacy of equivocation.

The Slippery Slope Argument

The slippery slope argument and the precautionary principle argument (to be discussed in the following section) are both *a posteriori*, because they consider the possible empirical consequences of certain actions. There is a good deal of literature discussing their pros and cons, and I do not propose to enter this huge debate here. I simply will show how they can be applied or misapplied when we are discussing the actions of creating and using protocells.

Let us begin with the slippery slope (Walton, 1992; Volokh, 2003). In its stronger version, it claims that if an action A occurs, then by a gradual series of small steps, through the actions B, C, . . . , eventually Z will occur. But since Z is valued negatively and should not be caused to occur, A should not be done. Because of this "slippery slope" from A to Z, we should not allow the action of creating protocells, since that would allow a series of successive actions leading to, for example, a form of artificial life that could successfully and dangerously compete with already existing forms of life.

If this argument is conceived in such a way that the concatenation of successive actions is necessary, it must be considered a fallacy. The previous action does not necessarily imply a link in the chain of successive actions. Some concede this, and soften the argument by introducing probability: The creation of protocells will probably open a concatenation of actions, which will probably lead to, for example, a new form of life that will drive existing forms of life to extinction; so, to prevent such a possibility we should ban the creation of protocells. This probabilistic version of the slippery slope argument avoids the fallacy of the earlier version (if we use a radical interpretation of the precautionary principle, as we shall see in the next section). Unfortunately, however, the argument is now too weak. It would be strongly persuasive, if we were capable of assigning a high probability to the passage from A to Z. But we cannot do this, since we are discussing actions and events situated in possible future contexts that we do not know at all.

It should be noted that the slippery slope argument would still be very weak even if we were able to establish the high probability of any steps. For, if we accepted the argument, we should also admit a symmetrical argument, which might be called the *desirable rack-and-pinion argument*. This is in accordance with the rhetorical rule that if you use an argumentative structure, you must concede the same argumentative structure to your opponent. The desirable rack-and-pinion argument is structured symmetrically with the slippery slope argument, since it uses a chain of possible actions and events. But, in contrast to the latter, which goes from an initial action to its possible undesirable consequences, the former goes from an initial action to its possible desirable consequences. In particular, the initial action concerns the creation of protocells, and it would be the starting point of a chain of probable positive actions and events ending with a better way of living.

In summary, the slippery slope is a fallacy if the succession of steps is considered necessary. But if it is considered only probable, this argument is extremely weak, since there is no possibility of determining *a priori* the probability of any step. Moreover, even if we accepted this weakened form, we would be obliged to accept the symmetric argument of the desirable rack-and-pinion, too.

The Precautionary Principle Argument

According to the precautionary principle argument, since protocells are a new form of technology, we do not have the scientific certainty that they will not harm the environment and whatever lives in it, and therefore the actions of creating and using protocells must be cautiously constrained until it is shown they are harmless. Moreover, the burden of proof for showing that protocells are harmless must fall on those who advocate creating and using them.

There are many objections to this argument, but there are many counterobjections, as well (see O'Riordan & Cameron, 1996; Deville & Harding, 1997; Morris, 2000; Goklany, 2001; Harremoës et al., 2002). Nevertheless, it should be clear that this argument is quite vague. It admits a lot of interpretations, ranging from the most radical (we must not perform any actions concerning new technologies until we are fully certain that they are harmless), to the weakest (a commonsense argument, something like "Be cautious when you create or use something when you are unsure what its consequences will be"). Note, however, that it is usually applied according to the interpretation offered by the regulators in charge at that moment and in that context. And, of course, if we have very ideological regulators, we would have very ideological interpretations of the precautionary principle.

Certainly the precautionary principle stresses that the environment and whatever lives in it must be preserved. It also underlines the finiteness (we cannot know all the infinite consequences of our actions) and fallibility (we can make mistakes) of human understanding, and therefore our inability to always know what new scientific or technological products will lead to. Of course, it is common sense to bear all of this in mind, as we should bear in mind many other wise everyday suggestions (be honest, help others, do not kill animals without necessity, love children, respect the law, do not sit on freshly painted chairs, etc.). We might also agree on the grounds of one of the main principles of applied ethics—that is, the principle of nonmaleficence, which asserts the obligation not to inflict evil or harm on others— that the burden of proof for showing that a new technology is safe should fall on those who propose to create or use the new technology, and not on those who propose caution or, even worse, on those who live where the new technology will be released. Nevertheless, I believe that we must take the precautionary principle with a grain of salt, that is, wisely. There might be a risk in creating or using new technologies, but we must not deny that banning those actions could also have a cost. Moreover, nobody knows all the consequences of any action with certainty.

From the long history of Eastern and Western cultures, we know that a really wise man is not always wise; accordingly we should consider the precautionary principle argument with precaution (that is, wisely). Sometimes we should take a risk, if the risk is under control; sometimes we should not, if it is not under control. But sometimes we should not take a risk, even if the risk is under control; and sometimes we should take a risk, even if the risk is not under control. It depends on the particular context. It would be unwise to ignore this context dependence and seek a single formulation or application of the precautionary principle that is valid for all times, all cultures, and all new scientific and technological innovations.

Responsibility

Responsibility derives from the Latin *responsus*, which means "to give an answer." Thus, without being guilty of the fallacy of the etymon (i.e., abusing etymology), we can say that any time we claim scientific and technological researchers must be responsible for what they do, we ask that they answer to the rest of us for their behavior. For being socially and ethically responsible for an action means taking the burden of its possible social and ethical implications on oneself. Therefore, others are entitled to ask for a "social and ethical answer" for researchers' actions and their implications. In particular, whenever an individual researcher is supposed to have taken responsibility for his or her actions concerning the creation and use of protocells, that researcher is also supposed to "give us a social and ethical answer" about what he or she is doing; but this is not always compulsory. For the researcher could be interested in taking on only the burden of the economic responsibility, that is, "giving an answer" only to those who pay for his or her creation or use of protocells. Do not forget that this is a responsibility, too. Or the researcher could be not at all interested in "giving any answer," that is, in taking the burden of any responsibility.

A plea for individual social and ethical responsibility is, of course, something good, but we should be well aware that it might be totally ineffective. We may hope that all individuals are good, honest, socially cognizant, ethically coherent, and responsible for what they do; but we know that this is not the case.

Therefore, what to do? Interpret the precautionary principle radically and ban any kind of research on protocells? Maybe there is another way out. But before considering it, let me summarize the path covered until now.

First of all, we should realize that speaking about the social and ethical implications of protocells is really speaking about the social implications of the actions of creating and using protocells and the moral judgments on such actions. Then, though both the by-analogy and the ontological strategies for evaluating these actions seem very weak or fallacious, the type-token strategy appears to be more practicable and correct. Moreover, I have argued that the transcendence argument against the actions of creating and using protocells is a fallacy. I have also shown that the slippery slope argument is either fallacious or very weak, and therefore useless against protocells. The same goes for the precautionary principle argument, as this critical principle is too vague and ambiguous. Unless the principle is given an extremely restrictive interpretation, it is just a commonsense but ineffective suggestion to be cautious. Finally, I have emphasized that social and ethical responsibility cannot be left to the goodwill of the individual researchers: They are human and, as such, are subjected to different external stimuli. On the other hand, it would be inappropriate to ban the research merely because not all people are good and honest.

There is another possibility: appealing to the scientific community's responsibility. Instead of leaving the responsibility to the individual scientist's goodwill, or to politicians, ethicists, environmentalists, and others who are totally external to the community working on protocells, it should be this community that internally and autonomously decides to take on the burden of social and ethical responsibility. This could be done by creating an internal and autonomous agency that periodically (yearly?) monitors what has been and will be done, and prepares social and ethical debate on the actions concerning these results, intended as tokens to instantiate the long story of scientific and technological research. The establishment of such an agency should also be positively valued in light of the fact that protocells do not yet exist. This fact enables us to reflect constructively on what is happening step by step along the route to protocells, before they are ever created. As far as I know, there are no other cases in which a scientific community, researching something to come, has decided to constitute an internal and autonomous agency devoted to monitoring what is occurring and preparing debate on its social and ethical implications. This means that there is no institutional model to follow; but, on the other hand, those who work in the field of protocells are less constrained in creating it. There might be an objection: If you create an internal agency, a conflict of interest could arise, since the controllers of the research and the researchers would be the same people. This could be true, but the agency should also be autonomous, that is, not only exempt from external influences and impositions, but also independent of the researchers. That is, it should be composed of both researchers chosen in the community according to certain criteria of "wisdom" to be decided, and external members chosen on the basis of their competence in sociology or ethics, independence, moral rigor, and other criteria. All of this, however, should be carefully discussed.

Why not try to pursue this path, which seems to be wise and rational? Why not decide that only the community working on protocells has the necessary and sufficient competencies to take on the onus of giving a social and moral answer to the implications of creating and using protocells?

Acknowledgments

I would like to express my sincere thanks to Mark Bedau and Emily Parke for their extremely useful comments on a first version of this chapter.

Note

1. Consider this English translation of the Bible passage concerning Paul addressing the Areopagus: "The God who made the world and everything in it, this Master of sky and land, doesn't live in custom-made shrines or need the human race to run errands for him, as if he couldn't take care of himself. He makes the creatures; the creatures don't make him. *Starting from scratch*, he made the entire human race and made the earth hospitable, with plenty of

time and space for living so we could seek after God, and not just grope around in the dark but actually find him. . . . One of your poets said it well: 'We're the God-created.' Well, if we are the God-created, it doesn't make a lot of sense to think we could hire a sculptor to chisel a god out of stone for us, does it?" (Peterson, 2002; Acts 17:24; italics mine).

References

Boniolo, G. (2006). The descent of instinct and the ascent of ethics. In G. Boniolo & G. De Anna (Eds.), *Evolutionary ethics and contemporary biology* (pp. 27–40). Cambridge: Cambridge University Press.

Boniolo, G., & De Anna, G. (2006). The four faces of omission: Ontology, terminology, epistemology, and ethics. *Philosophical Explorations*, 9, 276–293.

Deville, A., & Harding, R. (1997). *Applying the precautionary principle*. Annandale, Australia: Federation Press.

Goklany, I. M. (2001). *The precautionary principle: A critical appraisal*. Washington, DC: Cato Institute.

Harremoës, P., Gee, D., MacGarvin, M., Stirling, A., Keys, J., Wynne, B., & Guedes Vas, S. (Eds.) (2002). *The precautionary principle in the 20th century: Late lessons from early warnings*. London: Earthscan Publishers.

Morris, J. (2000). *Rethinking risk and the precautionary principle*. Amsterdam: Elsevier.

O'Riordan, T., & Cameron, J. (1996). *Interpreting the precautionary principle*. London: Earthscan Publishers.

Peterson, E. (Ed.) (2002). *The message*. Colorado Springs: NavPress Publishing.

Volokh, E. (2003). The mechanisms of the slippery slope. *Harvard Law Review*, 116, 1026–1134.

Walton, D. (1992). *Slippery slope arguments*. Oxford: Oxford University Press.

About the Authors

Mark A. Bedau is professor of philosophy and humanities at Reed College in Portland, Oregon, and Chief Operating Officer of ProtoLife Srl. He has a PhD in philosophy from University of California, Berkeley, and is actively researching various topics in artificial life and philosophy of biology, with many publications in each. He is editor-in-chief of the *Artificial Life* journal, and recently coedited the volumes *Protocells: Bridging nonliving and living matter* and *Emergence: Contemporary readings in philosophy and science*, both published by MIT Press.

Gaymon Bennett is director of ethics at the Synthetic Biology Engineering Research Center (SynBERC), University of California, Berkeley. He has a master's degree from the Graduate Theological Union in Berkeley. His theology thesis is a reinterpretation of the place of dignity in the contemporary world. He is currently getting a doctorate in anthropology, focusing on the ethics of emerging scientific practice.

Giovanni Boniolo, chair of philosophy of science at the FIRC Institute of Molecular Oncology Foundation (Milan, Italy), coordinates the PhD program in Foundations of Life Science and Their Ethical Consequences at the European School of Molecular Medicine. He coedited *Evolutionary ethics and contemporary biology* (Cambridge University Press, 2006), and is currently publishing *On scientific representations* (Palgrave, 2007).

Carl Cranor is professor of philosophy at the University of California, Riverside, and has written widely on philosophic issues concerning risk, science, and the law. He has published *Regulating toxic substances: A philosophy of science and the law* (Oxford, 1993) and *Toxic torts: Science, law and the possibility of justice* (Cambridge, 2006), as well as coauthoring *The identification and regulation of carcinogens* (U.S. Congress, 1987). He served on California's Proposition 65 Science Advisory Panel as well as its Electric and Magnetic Fields Science Advisory Panel. In 1998, he was elected as a fellow of the American Association of the Advancement of Sciences and, in 2003, was elected a fellow of the Collegium Ramazzini.

Bill Durodié is coordinator of the Homeland Defence Research Programme at the Centre of Excellence for National Security in Singapore. He was previously senior lecturer in Risk and Corporate Security at Cranfield University, director of the International Centre for Security Analysis, and senior research fellow in the International Policy Institute, within the Fifth Research Assessment Exercise–rated War Studies Group of King's College London. His main research interest is the causes and consequences of our contemporary consciousness of risk. He is also interested in examining the erosion of expertise, the demoralization of élites, the limitations of risk management, and the growing demand to engage the public in dialogue and decision making in relation to science.

Mickey Gjerris is assistant professor at the Danish Centre for Bioethics and Risk Assessment. His work centers on ethical issues arising from the interface between living beings and technology, especially biotechnology and nanotechnology. He holds a PhD in bioethics from the University in Copenhagen and an MA in theology.

Brigitte Hantsche-Tangen, MA, studied literature, sociology, and political sciences. She worked as a social scientist at various university and research institutes (Göttingen SOFI, Department of Medical Sociology, Hannover (IFG), Osnabrück Uni), focusing on the topics of labor, gender, family, and health. She is coauthor of the *German Federal Report on Women's Health*, 2001.

Christine Hauskeller is senior research fellow at the ESRC Centre for Genomics in Society and senior lecturer in philosophy at the University of Exeter in the UK. She has an MA in philosophy, sociology, and psychoanalysis (Frankfurt, 1992) and a DPhil in philosophy (1999). Her recent research focuses on the political philosophy of current bioscience, in particular on the shaping of stem cell science through its cultural environment and the analysis of the role of genomics in society.

Andrew Hessel is a consulting biologist who works at the interface of industry and academia to facilitate genomic initiatives. Since 2003, he has focused on synthetic biology and open-source biology, ideas he believes could significantly reshape the biotechnology industry. He is based in Toronto, Canada.

Brian Johnson was, until recently, senior advisor on biotechnology to the British statutory nature conservation agencies, and was head of the Agricultural Technologies Group at English Nature (now Natural England), one of the UK government's advisors on nature conservation. He has been closely involved in the debate on potential effects of GMOs on biodiversity and other aspects of the environment. After pursuing academic research in population genetics and ecology, he has spent the last twenty years in nature conservation. He has been a member of several advisory committees concerned with biological research, regulating the release of GMOs into the environment, and the development of more sustainable farming methods. He is currently a lead author of the International Assessment of Agricultural Science and Technology for Development, an initiative launched by the World Bank/UNEP.

George Khushf, PhD, is the humanities director at the Center for Bioethics, associate professor in the Department of Philosophy, and a member of the Nanocenter at the University of South Carolina. He conducts research on the philosophy and ethics of emerging technologies, and currently has an NSF-funded project to assist Nanoscale Science and Engineering Centers (NSECs) in the development of upstream ethics initiatives.

Emily C. Parke is currently business manager at ProtoLife Srl. She has worked as editorial manager of *Protocells: Bridging nonliving and living matter* (MIT Press, 2008), and editorial assistant for the *Artificial Life* journal. She has a BA in philosophy from Reed College, and wrote a thesis in bioethics on the precautionary principle.

Alain Pottage is reader in law at the London School of Economics. He has been a visiting professor at the Ecole des Hautes Etudes in Paris, the University of Sydney, and Cornell Law School. His work focuses on various aspects of patent law, with particular reference to the biological sciences, and he is currently completing (with Brad Sherman) a history of the concept of invention.

Paul Rabinow is a professor of anthropology at the University of California, Berkeley, and a principle investigator at SynBERC. He is editor of the series *In/Formation* at Princeton University Press. His most recent book is *Marking time: On the anthropology of the contemporary* (Princeton University Press, 2007).

Per Sandin received his PhD from the Royal Institute of Technology, Stockholm, Sweden, in 2005. The topic of his dissertation was the precautionary principle. His current research interests include the philosophy of risk, military ethics, and the ethics of crisis management and disaster preparedness.

Joachim Schummer is Heisenberg Research Fellow at the Technical University of Darmstadt. He graduated both in chemistry and philosophy and received his PhD (1994) and Habilitation (2002) in philosophy from the University of Karlsruhe. He was visiting professor at the University of South Carolina (2003–2004) and visiting fellow at the Australian National University (2006–2007). His research interests focus on the history, philosophy, sociology, and ethics of science and technology, with emphasis on chemistry and, since 2002, nanotechnology. He is the founding editor of *Hyle: International Journal for Philosophy of Chemistry* (since 1995) and serves on various international committees, including the UNESCO expert group on Nanotechnology and Ethics.

Mark Triant has a BA in philosophy from Reed College. His contribution to this volume was written when he was an undergraduate at Reed working with Professor Mark Bedau and supported by a Ruby Fellowship.

Laurie Zoloth is professor of bioethics and humanities, and of religion, at Northwestern University, the Feinberg School of Medicine, and the Weinberg College of Arts and Sciences. She is director of the Center for Bioethics, Science and Society and of the Brady Program in Ethics and Civic Life. From 1995 to 2003, she was professor of ethics and director of the Program in Jewish Studies at San Francisco State University. Her current research interests include emerging issues in medical and research genetics, ethical issues in stem cell research, and distributive justice in health care.

Index

Basic Bioethics
Glenn McGee and Arthur Caplan, editors

David H. Brendel, *Healing Psychiatry: Bridging the Science/Humanism Divide*

Jonathan Baron, *Against Bioethics*

Michael L. Gross, *Bioethics and Armed Conflict: Moral Dilemmas of Medicine and War*

Karen F. Greif and Jon F. Merz, *Current Controversies in the Biological Sciences: Case Studies of Policy Challenges from New Technologies*

Deborah Blizzard, *Looking Within: A Sociocultural Examination of Fetoscopy*

Ronald Cole-Turner, ed., *Design and Destiny: Jewish and Christian Perspectives on Human Germline Modification*

Holly Fernandez Lynch, *Conflicts of Conscience in Health Care: An Institutional Compromise*

Mark A. Bedau and Emily C. Parke, eds., *The Ethics of Protocells: Moral and Social Implications of Creating Life in the Laboratory*

The Ethics of Protocells

Moral and Social Implications of Creating Life in the Laboratory
edited by Mark A. Bedau and Emily C. Parke

bioethics/philosophy

Teams of scientists around the world are racing to create protocells—microscopic, self-organizing entities that spontaneously assemble from simple organic and inorganic materials. The creation of fully autonomous protocells—a technology that can, for all intents and purposes, be considered literally alive—is only a matter of time. This book examines the pressing social and ethical issues raised by the creation of life in the laboratory. Protocells might offer great medical and social benefits and vast new economic opportunities, but they also pose potential risks and threaten cultural and moral norms against tampering with nature and "playing God." *The Ethics of Protocells* offers a variety of perspectives on these concerns.

After a brief survey of current protocell research (including the much-publicized "top-down" strategy of J. Craig Venter and Hamilton Smith, for which they have received multimillion dollar financing from the U.S. Department of Energy), the book treats risk, uncertainty, and precaution; lessons from recent history and related technologies; and ethics in a future society with protocells. The discussions range from new considerations of the precautionary principle and the role of professional ethicists to explorations of what can be learned from society's experience with other biotechnologies and the open-source software movement.

Mark A. Bedau is Professor of Humanities at Reed College in Portland, Oregon. He is the coeditor of *Emergence: Contemporary Readings in Science and Philosophy* and *Protocells: Bridging Nonliving and Living Matter*, both published by the MIT Press in 2008. Emily C. Parke is Business Manager at ProtoLife Srl.

Basic Bioethics series

"Bedau and Parke's *The Ethics of Protocells* is a seminal work on the ethical and social implications of creating synthetic life. It should be required reading for anyone entering this field."
—Linda MacDonald Glenn, Department of Medical Education, Alden March Bioethics Institute, Albany Medical Center

"*The Ethics of Protocells* maps out the territory of the scientific and cultural implications of the emerging field of synthetic life. This book will be a key reference point for those working in fields such as science and technology studies and bioethics as well as policy makers and the interested public."
—Ruth Chadwick, Department of Philosophy, Cardiff University, Wales

The MIT Press
Massachusetts Institute of Technology
Cambridge, Massachusetts 02142
http://mitpress.mit.edu
978-0-262-51269-5

9 780262 512695

90000